The
Coming Oil Crisis

by

C.J. Campbell

Multi-Science Publishing Company & Petroconsultants S.A.

Note that preface is dated 1997
& notes in most ch. date
as late as 1996.
Perhaps this is a typo & 5/6 1998?
No. Look closely & it becomes clear: 1988 is the year (or by which of the "Copyright, Design & Patent Act" under which authority) the actual © is established. As noted about, the actual © could not have been obtained in 1988.

CONTENTS

LIST OF ILLUSTRATIONS

For
Bobbins
Jack and Julia
Anne and Patch
Simon and Oddny
Emma and Clara

PREFACE

It turns out that writing a book isn't easy. You set forth and try to nurse the thing along, not knowing quite what you are going to say until you say it. No sooner have you said it than you regret what you have said and wonder if you have got it right. Why did I try?

Chance has given me certain experiences and insights which provide compelling evidence that oil production is about to peak. As it becomes scarcer and controlled by fewer hands, it will inevitably become much more expensive. That in turn may herald a major discontinuity in the way the world lives.

This is not a new subject. It was much debated in the 1970s after the oil shocks of that period, but the fears then raised, while valid enough in principle, were not realized because the world was still a long way from the midpoint of the depletion of its oil which more or less corresponds with peak production. Twenty years later, we are now close to midpoint: so, this time the shock will be for real.

It is an immensely important subject which deserves to be more widely understood. That is why I have written this book. I have tried to make it non-technical and readable. I don't want to sound didactic in expressing the views I have but rather to invite debate and discussion. I am therefore especially grateful to those who have contributed the interviews which are contained in the book. They express a different slant on the same subjects based on different experiences. There is an autobiographic streak in the book. Its purpose is not to indulge in an ego trip but rather to explain the experiences that led to the conclusions. I desire to be objective in the analysis and to state the premises.

Absolutely nothing is straightforward as you will soon discover if you read on. For example, I have only recently grasped the role that double-taxation treaties had on the price of oil. There are probably many more strange relationships that I have not spotted.

In tables, graphs and appendices, I have provided my current assessment of what the world's endowment of *Conventional* oil is. The numbers are quoted as computed which gives the impression of greater than justified accuracy. There is one thing that can surely be said about these estimates: they are wrong. The issue is how wrong? In a sense, they challenge those with counter views to come out and state them with details and evidence. Silence amounts to tacit acceptance. There may be surprises in individual countries, but it would be a mistake to imagine that all surprises will be positive. The numbers tell the story and deserve to be considered closely. Another related and very important issue is to understand the difference between *Conventional* and *Non-conventional* oil: I will leave you to read on to do that.

The World is not running out of oil. Or rather, not for a long time. What it is running out of is cheap oil, and soon. It will still be cheap to produce, but it will be expensive to buy because it will be increasingly scarce and controlled by a few Middle East countries. This introduces an enormous political factor to the issue. I wonder if the US government would fire missiles at Iraq if it properly understood the resource constraints.

Some may think that the end of cheap oil-based energy, which has fuelled our consumer age, is a Doomsday message. The *Coming Oil Crisis* will be just that because the transition will not be easy, but I sometimes think that the World needs a change in direction in any case. From the ashes of the oil crisis may arise a better and more sustainable planet. It must at least become more sustainable as Mankind lives out his allotted life-span in the fossil record. Whether or not it is better depends on how well we manage the transition. We don't have long to prepare.

I hope that you will find what follows useful in planning your future. More than that, I hope that it will encourage you to plan: the crisis is imminent. Those who anticipate can do well from the economic and political discontinuity; those who react can survive; but those who continue to live in the past will suffer. I hope, too, that governments will somehow get the message and not leave everything to the open market that is ill-equipped to deal with the depletion of a finite resource on which we have all come to depend. Even if they cannot bring themselves to prepare, perhaps this will at least help them know what hits them when the crisis strikes, as surely it must.

Milhac, France
8th February, 1997

ACKNOWLEDGMENTS

This book is the outcome of a cooperative effort to which many people have contributed. I would like particularly to acknowledge the following:

Jean Laherrère has been a great help throughout, providing insights, many graphs, encouragement and friendship. He has been patient with someone almost devoid of mathematical skills.

Tom Jamison, Robert Harris, George Leckie Ken Chew, Jean-Michel Frautschi and Bogdan Popescu of Petroconsultants in Geneva have stimulated and encouraged the preparation of this book, being generous in releasing their precious data and insights, as well as contributing to its publication. It has been a pleasure to cooperate with them.

Ron. Swenson has followed the study with keen interest from the standpoint of alternative renewable energy; Dr R.C.Duncan has provided stimulating reports with his insights; and Buzz Ivanhoe has continued to correspond and provide valuable material, as he has done for many years. His creation of the Hubbert Center for Petroleum Supply Studies promises to keep the flame of resource study alight.

Walter Ziegler, Alain Perrodon, Jean Laherrère, Bill Pace, Richard Hardman, Leik Woie, Buzz Ivanhoe, Clive Needham, Ron. Swenson, and Richard Duncan contributed greatly with their interviews.

Roger Bentley of Reading University has taken an active interest and provided many useful comments.

Brian Fleay and Charlie Richardson have raised the flag in Australia and Jennifer Consadine is doing her best in Canada.

Ann Slettebø used her artistic talents to illustrate the book and to catch its meaning.

Liz Atkinson Hardman was a great help.

Bill George helped greatly with proof reading and commentary.

Bobbins put up with a lot of doomsday chat, making an invaluable contribution with editing and proof-reading. Oddny was forgiving when we *"snakkered om olje"*.

DEDICATION

Fig. 0.1. Harry Wassall.

HARRY WASSALL
Founder of Petroconsultants

This book is concerned with how much oil there remains to produce, and its depletion pattern. No one would be able to answer these questions but for the work of Harry Wassall who died on November 25th 1995. I met him not long before he died, and discussed the issue of oil depletion with him. It was a subject of great concern to him, and it is therefore a pleasure to dedicate this work to him.

He was an American citizen born in 1921, the son of a New York commodities trader. He graduated in geology at the University of Tulsa in 1942, and joined the Navy to serve out the war in the Pacific on board a Landing Ship Transport.

In 1946, he joined Shell as a field geologist before moving to Gulf Oil, accepting a foreign assignment to Cuba, where he worked as a field geologist from 1950 to 1956. There he married the beautiful Gladys, a well-born Cuban lady. He enjoyed the life of the island, and when Gulf wanted to transfer him, he decided to start his own consultancy, Harry Wassall & Associates. In addition to normal consultancy work, he started the Cuban Scout Letter, a newsletter covering industry developments in the country. He was by nature an entrepreneur and enthusiast, and soon added to his endeavours a drilling as well as a chemical company.

The Scouting Newsletter had expanded through much of Latin America, and a nucleus staff had been recruited by the time Fidel Castro's revolution brought enterprise in Cuba to a standstill in 1959. Harry resolved to extend the newsletter's coverage to the Eastern Hemisphere, and opened a small office in Geneva to build the business. The Wassalls themselves moved to Madrid in 1962, where he

developed various consultancies, not all related to oil.

He then dedicated himself to building up the Geneva office, which grew from strength to strength. It was time for a new name: Petroconsultants. The scope of operations expanded rapidly, with a whole range of products beyond the newsletter, which was now global in scope.

I well remember, when working on international new ventures for Amoco, that the first port of call in evaluating a new proposal was always the Petroconsultants material to find out what wells had been drilled; to look at the concession maps, and to find out the background.

Petroconsultants established for itself an enviable reputation, performing an essential service for the industry, with which it maintained first class connections.

Harry saw the growing potential of the computer, and his company was one of the first to computerize its database. It secured contracts to manage the major companies' data on a confidential basis.

Over the years, the Company has expanded its network of contacts around the world, including the former Soviet Union, and it maintains links with specialists on a wide range of subjects. So it is very well placed to provide authoritative studies on world resources.

The Company now has offices in Geneva, London, Houston, Singapore and Sydney, employing a multinational staff of 250.

This is a great achievement that owes almost everything to its founder, Harry Wassall. He was an entrepreneur in the best sense of the word; a man of courage, insight and imagination. But above all, he is remembered as a man of great kindness, humour and intense loyalty to his friends and staff.

As the world approaches the peak of its oil and gas production, knowledge of the resource base becomes an increasingly essential issue for planning the future, with colossal political and economic implications.

It is a pleasure therefore to dedicate this book to a man who contributed so much to building the foundation for all realistic estimates of future oil and gas supply.

Harry Wassall's death necessitated a restructuring of the Company's shareholding. This has now been achieved with the IHS Group of Colorado, a leading publisher of technical databases, which has taken up a controlling position. It augurs well for future growth, and means that the foundations will be well remembered.

Chapter 1
INITIATION

This book is about the world's endowment of oil. It is a very important subject, considering that cheap oil-based energy has been the lifeblood of the world's economy over the best part of this century. Its influence has been enormous: it has driven the way people live; what they expect; how they order their lives. It is an intensely political subject, and it is a romantic story: "black gold" has turned many rags to riches, for individuals and for nations. It has spawned jealousies and arrogance, and has been a key factor in several wars: the Gulf War being the most recent. The great American economic miracle, with its mindless consumerism, owes much to its endowment of oil: the country would have been much less of a world power without it. Yet, American production peaked in 1971[1], and is now in terminal decline.

Yes, it peaked in 1971. No one much noticed at the time: there was still plenty of gasoline at the filling station. But it was a critical and fundamental inflexion: a discontinuity. Before you can produce oil, you have to find it, and the 1971 peak in production in fact reflected an earlier peak in discovery that also went un-noticed. We can now look back and see the importance of these events: they marked the beginning of the end. Oil will not last forever. It was formed only rarely in the long geological history of the Earth, and only in a few places under most exceptional conditions: it is decidedly a finite resource, being consumed at a rate now increasing above two percent a year. In 1996, the world consumed about 24 billion barrels. A barrel holds 42 US gallons, several tankfuls for an average car. We use a lot of oil, which is vital for transport and agriculture, which means food.

The 1971 peak was a local discontinuity in the United States, the world's most mature oil country, but it did not much affect that wealthy nation which was able to import increasing amounts of oil from other less depleted countries. Again, with no one particularly noticing, imports have been rising ever since: such that a country that once supplied the world now imports more than half of its needs. But what about the other less depleted countries? Can they go on supplying the world forever, or will they too soon face the same inflection to falling production as the United States has already experienced? That is one of the questions that I will endeavour to answer in this book. The short answer is no, but it is not as simple as that. Another related question is how much will it cost.

You would think that it was a fairly straightforward issue to resolve; surely economists have access to data banks that hold the answers; surely the great international oil companies know what the situation is; surely governments are now planning what to do – unfortunately not. The subject is clouded in mystery. As you begin to dig into it, you find a maze of conflicting information and disinformation, as well as imprecise and confusing definitions. There are few hard and fast facts. Judgment and experience are therefore called for to choose a path through the minefields. That is an apt analogy, for there are colossal vested interests with motives to distort and confuse: oil is money, and money is power. There is a great deal at stake.

But judgment itself is part analytical and part intuitive, and in the latter regard is subject to bias. Many people's intuitive knowledge of oil is based on the weekly trip to the filling station. It has been there for as long as they can remember; and their intuitive judgment tells them that it is likely to continue to be there for the foreseeable future. My intuitive knowledge of oil, by contrast, has been based on looking for the stuff. In the course of my life, I have evaluated hundreds of prospects around the world, and my intuitive judgment, built on that experience, tells me that only very few prospects will fully meet the geological criteria to succeed. I stress *fully* because a prospect depends on many factors. One weak link in the chain means an expensive dry hole. These factors are themselves not easily determined as they depend on geological circumstances far beneath the Earth's surface, whose precise nature can be interpreted only with difficulty. The unravelling of geology relies on observing what can be observed, and then using geological principles to make a logical extrapolation from the known to the unknown. I will follow this soundly based scientific principle in this study.

It makes sense therefore to evaluate the subject through the eyes of an explorer, and recount how his knowledge and appreciation of the situation evolved. Perhaps one should say rather how this explorer came to grasp the nettle. So, I will adopt an autobiographical framework around which to weave the story. Along the way, I will interview others who can contribute their knowledge; and I will digress frequently into political and technical matters that lie outside my own professional expertise. I want to try to make it a readable sort of detective story. Explorers are taught to look for clues.

TRINIDAD

The *Regent Springbok* tanker tied up at the loading jetty at Pointe-à-Pierre, Trinidad on 9th April, 1958. It had arrived in ballast from England to load petroleum products from Texaco's giant refinery with which to supply the company's marketing chain in Europe. In addition to its normal crew, it carried a supernumerary crew of about twelve passengers, who were Texaco staff coming out as new recruits, or returning from the home leave to which they were entitled after three years' service. I was one of the new recruits coming out to his first job, fresh from Oxford University, where I had just completed my D.Phil. thesis in geology.

Texaco is a major American oil company that had recently bought the British-owned Trinidad Leaseholds Ltd., which continued to be British-staffed. Trinidad itself had, not long before, become independent after many years of British colonial rule, and seemed to be a happy multi-racial society of Negroes, Indians (originally brought in as indentured labour to work the sugar estates after the abolition of slavery), Chinese, Lebanese and others of French, Portuguese, Spanish, British and mixed extraction. There were splendid old white Trinidadian families with names like *de Verteuil, Rostow, Fernandez,* who made up a colourful and dynamic element, although vaguely conscious that they were an anachronism whose days were numbered. The average age of Trinidadians was fourteen, promising to make it a very crowded island.

To be more accurate, it was not quite my first job, as I had spent a year in Borneo as a member of an Oxford University Expedition, mapping the remote Usun Apau Plateau in the interior of Sarawak. Rocks, I knew, but precious little about oil.

Once the personnel formalities were over, I was taken to the Geological Office, a long low green building with mosquito screens over the windows. It stood beneath a spreading samaan tree on a slight elevation above the refinery and tank farm. I was introduced to Ken Barr, the Chief Geologist, a silver-haired and gentle-voiced man who welcomed me to his department.

Fig. 1-1 Dr K.W.Barr, Chief Geologist of Texaco Trinidad Inc.

Before long, he said that he wanted me to meet Dr Kugler, the legendary father of Trinidad oil, who had had a key role in developing the island's oil industry since his arrival in 1913. We set off down the corridor, and soon I

began to hear an animated conversation in German: it was Dr Kugler, discussing some aspect of Trinidad geology with Karl Rohr, a fellow Swiss field geologist. I was ushered into the great man's presence. He was slight of build, and had a sort of rolling gait as he got up to greet me. His silver hair was swept back above a chiselled face. He was dressed in khakis and a white shirt as was normal in the tropics in those days. He turned his one good eye on me, and began to quiz me about my experiences in Borneo, immediately revealing a great breadth of knowledge, not only about its geology, but its people and natural history. It did not take long to realize that here was a remarkable man. He was to have a lasting influence on my career, and indeed life.

Fig. 1-2. Dr Hans Kugler, father of Trinidad geology, who greatly influenced my career.

So began my career in the oil business. I spent only two years under his tutelage, but it was enough to instill in me a scientific interest in petroleum geology. At Oxford, I had had sybaritic tendencies and, after the Borneo experience, had not really wanted to return to the arduous life of a field geologist in the tropics. But under Hans Kugler's influence, I found my feet and became dedicated to my new profession.

Several weeks later, I was assigned to my first mapping project in the Rock River area of the Trinity Hills. It was thick tropical rain forest, virtually devoid of natural outcrops. So, the work involved controlling teams of Indians, drilling augur holes from which samples of bedrock were recovered. I would sit in the forest under a huge beach umbrella draped in mosquito netting while I described the samples; labelled them; numbered them; and plotted the location of the augur hole on a map. Typical entries in the brown field notebook read something like:

*Cb238. Clay, greenish-grey, micaceous, silty,
slightly calcareous, with carbonaceous fragments.*

*Cb239. Clay, light grey, micaceous, with silty
calcareous streaks.*

The work continued for several months, and the samples were studied in the Geological Laboratory under John Saunders[2] a micro-palaeontologist. His job was to extract micro-fossils called Foraminifera, which allowed him to date the sample by reference to a palaeontological zonation. Trinidad's geology is extremely complex and difficult, being made up of highly folded and faulted sequences of rather monotonous clays. It is only by means of their fossil content that the different formations can be identified. Mapping the strata so recognized then made it possible to unravel the structure, and thereby identify possible reservoirs and traps for oil.

Fig. 1-3. A mobile land rig for repairing wells.

**Fig. 1-4. My good friend, John Saunders,
a micro-palaeontologist with whom I
worked closely.**

Once the determinations were made, I set about the interpretation, and little by little, the pieces of the jigsaw fell into place. The juxtaposition of two normally separated formations, identified by their micro-fossil fauna, could be traced from one augur line to the next, pointing to the existence a thrust fault[3], which could in turn be correlated with a dip discontinuity[4] in a well to the north. In the well, the thrust was overlain by some reservoir sands which were water-bearing, but such sands were missing at the surface. It meant that they had died out somewhere between the well and where I had been mapping. Perhaps oil had migrated up-dip to accumulate in the truncation. But I had also identified a cross-fault that offset the thrust. Exactly where the truncation was located was hard to predict, but my best guess was that it lay just outside Texaco's lease in an adjoining block operated by BP. In the United States, such information would have been a closely guarded secret, but in Trinidad in those days, the companies maintained a friendly cooperation. When Kugler had studied my maps, he sent me down to Palo Seco to show them to Sam Wilson, the Chief Geologist of BP's Trinidad operation. In due course, he tested the idea with a well, and found oil. My first step into the oil industry had been a success, but finding oil for another company did not exactly earn medals from the management.

Trinidad days passed happily, not only during working hours but in the friendly society of an expatriate oil camp,

**Fig. 1-5. Dr Hans Kugler, as I remember him best:
observing nature in the rain forests.**

which somewhat resembled an expensive country club, save for the sulphurous smell of the refinery. Weekends were often spent water-skiing or on the beaches of the beautiful Northern Range beyond Port-of-Spain. Rum-punch parties and Carib beer were there in abundance for the benefit of a young bachelor. At one such party, he met the beautiful Bobbins Ludford, a secretary who had stepped ashore from one of the subsequent tankers that brought staff from London. We were soon to be married.

Then, one day in September 1959, Kugler called me into his office and asked if I would like to be transferred to Texaco's operation in Colombia. He had been talking to the Manager for Western Hemisphere Operations in New York who was impressed by Trinidad's scientific approach and wanted to apply it throughout the Company. Kugler had also greatly impressed the Chairman of the Board, Augustus C. Long[5], when on a visit to Trinidad. So, I was entrusted with the missionary work to bring palaeontology to the Texans running the Colombian operation. Thus, opened one of the most colourful and interesting chapters of my life.

But before we leave Trinidad, let us look back from today's vantage point, and see what its oil resource situation was, and has become. At the time, none of us had such a perspective, being preoccupied with detailed day-to-day projects and studies. There was no idea in the back of our minds that the resources were finite or had a particular distribution. These subjects, the theme of the book, will be developed in later pages. Trinidad had had a long oil history. Columbus had caulked his ships with tar from the famous Pitch Lake, a natural seepage, and oil operations had already commenced during the last century. Forest Reserve, the largest field was found in 1914, followed by numerous other small onshore fields, most discovered before the war. Efforts then turned offshore, first to be rewarded by the discovery of the Soldado Field in 1954 in the Gulf of Paria, a landlocked embayment that separates Trinidad from Venezuela. They were later followed off the Atlantic Coast, where Amoco found several large oil and gas fields in the late 1960s and 1970s. Trinidad had been in fact the gateway to Venezuela which belongs to the same geological province. Discovery peaked in 1954, twenty-five years before production did so in 1979. It is an example of a country that has had three exploration phases: early onshore; shallow offshore and a late stage deeper water effort in the Atlantic. I will later evaluate the significance of these patterns, but meanwhile will introduce the subject with two graphs, Figures 1-6 and 1-7. Most of Trinidad's oil had been found before 1930. The term *Ultimate* means the total amount that will have been produced when production ends. A wildcat is an exploration borehole that, if successful, results in a new field. The plot of cumulative discovery against wildcats tends to be hyperbolic[6], or a series of hyperbolas, each related to a phase of discovery. Projecting the curve gives an idea of what the *Ultimate* will be, and how many boreholes it will take to get there. The concept of an *Ultimate* recovery is a very important one, because production will one day end. The curves of discovery are flattening almost everywhere.

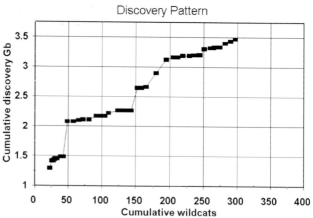

TRINIDAD
Discovery Pattern

Fig. 1-6. Discovery pattern of Trinidad: the larger fields in each exploration phase are normally found first.

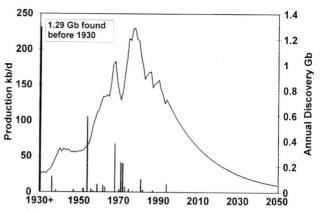

TRINIDAD
Ultimate 3.75 Gb

1.29 Gb found before 1930

Fig. 1-7. Trinidad production profile and annual discovery.

Fig. 1-8. An offshore platform in the Gulf of Paria, which gave Trinidad the second of its three lives.

The Coming Oil Crisis

COLOMBIA

I left Trinidad at 7 am on October 1st 1959 aboard a DC-3 of Lineas Aereas Venezuelanas, and flying at an altitude of a few thousand feet traversed the spectacular Coast Range of the Andes before landing at Maiquetia, the airport of Caracas at the foot of the mountains. From there, a modern highway, full of fast and seemingly huge American cars, led into the Venezuelan capital. It was an exciting place, very different from sleepy Trinidad with its British colonial atmosphere. After a stopover, I continued on a Constellation airliner to Maracaibo and into the interior of South America, catching occasional glimpses of vast stretches of forest cut by meandering rivers. At last, we dropped through the cloud-base to land in a broad upland valley in which lies Bogotá, the capital of Colombia at an altitude of 8300 feet. It was a grey cool day, very different from the tropics I had left behind. An ancient Chrysler taxi with several hundred thousand miles behind it, and driven by a driver in a black trilby hat and a poncho, known in Colombia as a *ruana,* brought me to the Tequendama Hotel. It was like stepping back into Europe, and I was delighted.

Now began a very new and exciting experience: to live in this Andean city, where hung a pervasive smell of eucalyptus smoke, and to work on the magnificent geology of Colombia, cut, as it is, by three ranges of the Andes. The centre of Bogotá dated from the 16th Century when it had been the vice-royalty of a part of the Spanish Empire, known as Gran Colombia covering what is now Venezuela, much of Central America and Ecuador. The inner suburbs of the city were built in red brick, mainly before the War, and were somewhat reminiscent of North Oxford. Farther north, developed the modern tree-lined suburb of El Chico, where the wealthy lived in large Spanish-style houses, surrounded by high walls and wrought-iron fences. Ancient buses of several rival lines raced each other along the roads, passing the occasional trails of donkeys, carrying fire-wood, which were led by a *ruana*-clad Indian. Above the town, rose an impressive range of barren hills, capped by the Virgin of Monsarrat.

Next day, I made my way to Texaco's office on the 14th

Fig. 1-9. Bogotá, a charming Andean town in the days before drugs.

Floor of the Edificio Seguros Bolivar, one of the few skyscrapers in the commercial part of the town, not far from the hotel. There, I met the Chief Geologist, Ken Bishop, a tall laconic American, who reminded me rather of one of those figures you see in American war films, stooping with an intent expression as he beckons his troops to follow him into action. The Exploration Department consisted of about twenty young Americans, who had been mobilized as what was termed a Task Force in one of the Company's periodic new thrusts to explore Colombia. The Company had been established there for many years, operating the Velasquez Field in the Magdalena Valley. I was assigned to a small team evaluating the Sabana de Bogotá area, a high intra-montane depression within the Andes, which was characterised by some huge outcropping anticlinal features. Superficially, they were impressive prospects for oil, provided they contained satisfactory reservoirs and source-rocks, the details of which were not then known. Pat Maher[7] from Minnesota, Ray Robbins and a quiet spoken Texan, Carl Henderson, were my new colleagues on the project. They probably did not know what to make of this *Limey* who had arrived in their midst, but they were open and friendly.

I started work reading reports to become familiar with the geology. I also started understand the Company's style of operation. It was a far cry from the gentle academic hand of Ken Barr in Trinidad, where we had all worked together as dedicated geologists and friends on a more or less equal standing. Now, I experienced something of the hierarchial atmosphere of American business: there were bosses and there were employees, for whom the pay check was the main motivation. It is not such a bad system, as you knew where you stood, which is more than you always did in a European company, as I was later to find out with the Belgians. I did not mind the pay check either: I had joined the then legendary dollar payroll, and was earning what seemed an astronomical amount, far above the Trinidad salary. Much as I enjoyed colourful Colombia, I often thought nostalgically of Trinidad and particularly of Bobbins Ludford to whom I soon sent a telex:

"Letter asking for your hand in mail stop please answer soonest stop love Colin."

She did, and we were married by Adrian Clarence Buxton, the British Consul, on December 19th, 1959. In later years, when I was in management, I often thought of myself as little more than a telex operator feeding the bureaucracy of a large company, and sometimes thought how it was appropriate that I should have also proposed by telex.

Before that happy day however, the Company had what was called a *Wise Box,* at which every project had to be presented with much theatre to two executives from New York, Dashy Bode and Bill Saville[8]. The former had extraordinarily primitive notions of geology, and thought that mapping was simply a matter of measuring the dips of the formations and constructing geometric cross-sections.

Fig. 1-10. Bobbins collecting fossils in Colombia.

He seemed oblivious of the subtleties of stratigraphy: how formations thicken and thin; and change their compositions, reflecting variations in the environments in which the rocks were laid down. He failed to recognize the importance of palaeontology by which to date the formations, and thus correlate them from one area to another. This, I thought, was the moment to don Kugler's missionary hat, and propose that the Company should now apply palaeontology and improve its lamentable lack of stratigraphic knowledge. The message was not at all well received: in the hierarchy, it was not thought fitting for a junior geologist to challenge an executive, least of all a *God-damn Limey*. The message of course carried an implied criticism, which is never appreciated: a subtler approach would have been more successful. It was a lesson I never managed to learn.

Probably as a consequence, I was soon shipped down to do field work in the Magdalena Valley and to map an area of Miocene non-marine strata, conveniently devoid of fossils. My base camp was a bullet-pocked estate house, near the Carare River, which was being used by a seismic crew. In a later chapter, I will explain this exploration technique that involves letting off explosive charges and recording the echoes reflected back from subsurface formations. The bullet holes were from the attentions of bandits who infested the area, and from whom a detachment of Colombian troops was supposed to protect us. Each morning, a helicopter would carry me out to a clearing on one of the seismic lines, where I would be joined by

Hernando Patiño, the surveyor, and my field party of about twelve tough Colombians led by their Capitaz, Abel Alberacín. The days were spent labouriously following overgrown streams, clogged with secondary growth, in the hope of finding an outcrop with a dip to be measured. The men had to hack a path through the undergrowth with machetes. Mosquitoes abounded and led me to sew a flap on the back of my hat to protect my neck from bites, making me look something like a French Legionnaire. One day, the supervisor from the Velasquez Field came out to visit me: he was a splendid Texan by the name of Wayne Marrs, whose opening words as he stepped from the helicopter were

"Stud, where did you get that bonnet?"

The project was a futile exercise, since the structures of interest lay beneath a thrust-plane, such that the dips at the surface had barely any relevance anyway. Later in the office, I discovered that the Company already had a comprehensive report on the area, acquired from Shell years before, when it was still under virgin forest and outcrops were more plentiful. But scientific study was not the style of the Company at the time: reports were neither written nor read under the Task Force mentality.

It took several months to complete this mapping. I then moved south to the Rio Guaguaqui to map the thrust-belt that separated the Magdalena Valley from the Eastern Andes. It promised to be a more interesting project, and soon I was collecting ammonites from the Cretaceous shales. This time, we were on our own without helicopter support, and had to rely on dugout canoes to transport us up the rivers. My notebook is full of sketches of tropical flowers and birds. I felt myself, however humbly, to be in the best tradition of the classical explorers of South America, like Humboldt or Darwin himself. But after several weeks work, I woke one day with a fever and yellow eyes. When resting in camp brought no improvement, I sent some of the men to make their way on foot to the Velasquez Field, some thirty miles away, to get help. At last, a helicopter arrived and flew me to the camp, from whence the Company DC-3 brought me back to Bogotá. I had contracted jaundice and amoebic dysentery, the normal hazards of tropical field work.

The cure was rest at home and Vitamin B injections. It was a welcome break, and it was good to see Bobbins again after my long absence in the jungle.

When, a few weeks later, I had recovered sufficiently to return to the office, I found that Hans Tanner, a Swiss geologist, had come back from vacation to supervise the evaluation of the Sabana de Bogotá and Llanos Foothills, to which I was reassigned. Hinting that he shared my view of the Company's technical weakness, he instructed me to commence a thorough stratigraphic investigation of the eastern flank of the Cordillera Oriental. We needed to know what reservoirs to expect below the surface structures that had been mapped, and how deep they would lie. That could only be worked out from a stratigraphic evaluation.

Fig. 1-11. Field work in the Magdalena Valley.

Fig.1-12. Field camp in the Magdalena Valley.

of ammonites. As we progressed, we came to another massive sandstone scarp of Albian age, which might be a deeper reservoir, although it appeared to be too indurated to have adequate porosity and permeability, at least within the mountains. At the weekends, I returned to Bogotá to organize and photograph the fossil collections, which were identified by Hans Bürgl, an Austrian palaeotologist at the University, with whom I became a close friend. Eventually, the stratigraphy was worked out in detail with the identification of all the classic European stages: *Maastrichtian, Santonian, Campanian, Turonian, Cenomanian, Albian, Aptian, Barremian, Hauterivian, Valanginian, Berriasian and Tithonian* – the names are etched in my memory[9].

Fig. 1-13. Hugo Cely surveying outcrops with the planetable and alidade.

Fig. 1-14. Saddling up the mules.

Now began what was to become the most rewarding chapter of my geological career. The field party was reassembled, and we set forth to map the road section from Bogotá to Villavicencio, a town in the foothills of the Andes at the edge of the Llanos plains that extend far into the heart of the continent. To begin with, we drove out each day from Bogotá, but when the distances became too great, would rent places to stay in colourful villages along the way. I can remember waking to the sound of mules trotting along the cobbled streets: it was a scene from medieval Spain.

Hugo Cely, a Colombian trainee geologist, helped Hernando run the planetable and alidade survey, and the men with stadia rods would be dispatched to mark the outcrops, which were abundant. We started in the Tertiary lake-bed deposits around Bogotá, before coming to the impressive scarp of the Upper Cretaceous Guadalupe Sandstone, now one of the producing reservoirs in the Cusiana Field, of which more later. The sequence was only mildly deformed, and it became possible to map accurately every formation. Furthermore, it was richly fossiliferous. For example, the oyster, *Exogyra squamata*, proved a useful marker for the Cenomanian stage, which could be traced across country. Careful search was rewarded by a rich fauna

This is a good moment to introduce in a few words the geological column and explain the meaning of terms like Cretaceous and Jurassic. The rocks laid down during the last 600 million years of the Earth's history are ascribed to four systems: Palaeozoic (230-600 Ma); Mesozoic (65-230 Ma.); Cenozoic (2-62 Ma.) and Quaternary for the last two million years. Rocks older than 600 million years are attributed to the Precambrian or Proterozoic (Figure 1-16). These systems are further sub-divided, each stage being recognized by its fossil content. The ammonites, for example, died out at the end of the Cretaceous Period, 65 million years ago. The absolute ages in millions of years of most of the intervals have now been determined by studying the decay of radioactive elements.

Fig. 1-15. A Cretaceous ammonite.

Millions Years	ERA	SYSTEM	Characteristic Life
		QUATERNARY	Homo sapiens
2			
		Neogene	
23	CENOZOIC		
		Paleogene	
65			Mammals
		Cretaceous	
141			Dinosaurs
	MESOZOIC	Jurassic	
195			Ammonites
		Triassic	
230			
		Permian	
290			
		Carboniferous	Coal-forming forests
345			
		Devonian	
395	PALAEOZOIC		Fishes
		Silurian	
425			
		Ordovician	Trilobites
500			
		Cambrian	
600			Molluscs
	PRECAMBRIAN		
4600			Formation of the Earth

Fig. 1-16. Geological Column.

Once the first traverse had been completed, I turned to the next road northwards that led through a more remote section of the Andes to Gachalá, famous for its emerald mines. Comparing the two sections, and using the fossils for correlation, began to show the stratigraphic trends: how particular formations were thinning and thickening, and subtly changing in composition. This knowledge would prove essential to predicting conditions in the prospective areas.

From Gachalá, we hired mules and followed an ancient *camino real*, the trails followed by the runners of the Inca Empire, to cross the Farallones de Medina, a massif rising to over 4000 m. I will never forget emerging into the Llanos foothills and riding muleback into Medina. It was a scene from a Wild West movie. As we rode into town, we could sense the hostility and almost feel the rifles being aimed at us from the adobe houses around the square. A gang of rough characters riding into town usually meant only one thing. It was a tense moment before we established our credentials as *petroleros*, not *bandoleros*.

I had become a dedicated geologist, fascinated by Colombia's geology in all its majesty, and I loved this colourful country with its still evident Spanish old world culture. Although I was so often away in the field, Bobbins and I enjoyed our new married life at our small apartment in El Chico. We were wealthy, at least by our standards, and had many new friends amongst the expatriate community[10]. Bogotá was full of splendid European restaurants from the Balalaika, where a barrel-chested White Russian sang; to the Zambra, where its owner, Juanillo, gave renderings of *La Flor de las Clavelles,* before cutting off the tie of one of his guests to add to a collection hanging from the ceiling. Then there was the *Bella Suiza*, a typical Swiss inn at Usaquén, where we would often go for a Sunday lunch of *wienerschnitzel mit roesti*, washed down by locally brewed German beer.

Julia, our daughter was born on January 22nd 1961. It was a splendid life. Professionally, I felt that my missionary work for palaeontology had been vindicated, and I was disappointed to find myself so out of tune with the company I worked for, who did not seem to appreciate, nor give credit for, what was being achieved.

Our tour of duty was now over, and we returned to England for a three-month leave. As the days passed, I increasingly dreaded returning to the baleful environment of the office, which I mistakenly attributed to it being an American company. I decided to approach BP to see if they had an opening, thinking that I would fit better into a European company. To my great delight, Norman Falcon, the legendary Chief Geologist, who with J.V .Harrison, had made the initial surveys of Iran, offered me a job to return to Colombia, where BP was then starting up an operation under a joint venture with Sinclair Oil & Gas. It was too good to be true.

With hindsight, I can see that my reaction to Texaco was an immature, emotional one, most of all to attribute what I perceived to be its weaknesses to its American base. As I was later to learn, all companies can fall into bad hands in

Fig. 1-17. I seem to have had a penetrating gaze.

probably led it to other promising finds across the border in Ecuador.

I would interview the company concerned for information, armed with tracing paper on which I would try to extract the maximum of information from their maps by vaguely surreptitious means, and then rush back to the office to write up an appraisal, fearful of being forestalled by another company. It sharpened my wits and made me more of an oilman than a geologist. It was a very valuable experience. It also gave me a comprehensive knowledge of Colombian geology, distilled from the work of all of these companies, many of whom had been working in the country for many years.

It was a very pleasant office which was staffed by an agreeable group seconded from BP and Sinclair[11].

Fig. 1-18. Freddie Martin visits me in the Cauca Valley.

the ups and downs of their existences: everyone in the corporate jungle faces his own tensions and pressures. They had more to worry about than some young geologist in Colombia with a bee in his bonnet about fossils.

A SECOND BITE OF THE CHERRY

Fossils were not a priority in the new company either, where force of circumstances pushed me into a much more commercial role. As a newcomer, BP had very little acreage in Colombia, and was therefore interested in farming into concessions owned by other companies. That is to say, they would earn an interest in a concession, normally by drilling a well in return for a 50% stake. Many of the world's oilfields have been found by other than the original owners of the concessions in which they lie. The concessional system at the time awarded initial rights as *Applications,* which could later be converted to full *Concession* status carrying drilling obligations. These rights changed hands frequently as companies acquired and released acreage subject to shifting ideas and strategy, often brought about by changes in the tax regime. My job became that of an evaluator of such offers and opportunities. Most I rejected, but occasionally would come across an interesting possibility. Indeed, had BP accepted all the recommendations, it would soon have become the largest company in the country. It had already found the Provincia Field, drilling on Esso acreage, to which could have been added the Rio Zulia and Orito fields which would have

We were to remain in Colombia for another four years until the spring of 1966. Despite the reminiscences, this is not an autobiography but an account of the evolving understandings of a young exploration geologist in relation to world oil resources, the theme of the book. There is no need therefore to recount all the adventures and incidents that made up this happy chapter of our lives, which included the birth of our son, Simon, on October 29th 1962. It is worth mentioning, however, that my final year was spent compiling a regional report on the geology of Colombia, which anticipated in fact aspects of the then evolving geological revolution of Plate Tectonics. It recognized the presence of several major transcurrent faults and proposed a structural evolution in terms of a westerly movement of the continent, overriding accretionary prisms of sediment at its leading edge. The report correctly identified the Llanos as the most promising province in the country, a conclusion that was not to be vindicated until 1983 when Occidental drilled the Cañon Limon giant discovery in the recommended tract. BP did not act upon this recommendation in 1965 when the entire basin was there for the taking, but belatedly succeeded in farming-in to a block in 1988 that yielded the giant Cusiana Field, not far from the town of Medina into which I had ridden on a Texaco mule twenty-five years before. It tends to confirm that there are fewer technical surprises in exploration than is commonly imagined. In addition to assessing the merits of the different basins that make up Colombia, the report also assembled information on the size of fields and made some prognosis as to future discovery. It demonstrated an awakening understanding of resource constraints: the limitations of many of Colombia's basins were already evident. In this regard, it was a useful step in the growing appreciation of this critical subject. Yet, although I could see the local limitations, the world still seemed a very large place, and I had at the time no feeling of any global constraint.

In the same way as we took a preliminary look at Trinidad's production profile, let us also look at Colombia's in Figures 1-19 and 1-20[12]. It too had had an early start when General Virgilio Barco let the early concessions, and when the Tropical Oil Company developed the giant La Cira-Infantas Field in the Middle Magdalena valley in the 1920s. The remote Llanos with its crop of giant fields was not developed until late because of its very remote location on the wrong side of the Andes for the purposes of export. Unlike Trinidad, Colombia has still to reach peak production[13], but once it does, production will decline rapidly.

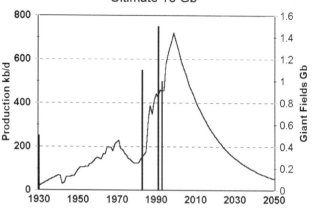

Fig.1-20. Production profile of Colombia, which has yet to reach peak.

notebook. We worked outdoors and in a natural environment, and probably gained an intuitive knowledge of the rocks amongst which we spent our days. As I read professional papers to-day, I am often struck by how artificial they seem, however elegant the intellectual hypotheses expounded. The pioneering explorers also faced many practical daily decisions of where to camp, how many mules they needed, and how to get the best from their labour force. No courses were offered in these subjects in those days: it was more or less a matter of common sense and a growing confidence built on experience.

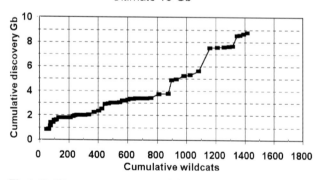

Fig.1-19. Discovery pattern of Colombia: the late stage giant fields in the Llanos were not unanticipated.

Fig. 1-21. Field work in Colombia.

These days in Trinidad and Colombia characterized an epoch in exploration that has virtually ceased to exist. It was onshore, and involved studying the rocks at outcrop: whether in tropical rain forests, the oil lands of the United States or the deserts of the Middle East. The geologists who did this work had a pioneering role, often working far from base and without supervision. It is not surprising that they were individualists. It was a slower pace of working, gradually accumulating knowledge with a hand lens and

Fig. 1-22. On horseback in the Western Andes.

The Coming Oil Crisis

Today, exploration has moved on from the outcrop to rely on seismic surveys and borehole data, displayed on the computer screen, and manipulated by enormous computing power. Geologists now rarely see rocks, and spend their lives glued to their screens in corporate offices under supervisory hierarchies. It must be much less fun, but yet, even now, prospects are still to be found only in the heads of explorers[14]. The days of poking holes into large structures are over, and ever more ingenuity will be required to identify the smaller and more subtle traps, which are all that remain. The need for stratigraphic insights that characterised my days in Trinidad and Colombia has become ever more important. Old Kugler was right. Forgive me, if I sound a bit nostalgic for the "good old days', but perhaps it is useful to explain the foundations upon which this study is based.

Until now, a geologist's mission has been to develop nature's resources for the benefit of Man. He has been only too successful, and as this book will try to explain, he must now adopt a new mission: to explain just how finite these resources are, in the hope that Man will come to understand that he must make better use of them. Geology, unlike engineering, is a descriptive science that imbues a sense of detachment and a long sense of history. We cannot change history nor the resource limitations of the rocks formed millions of years ago. Oil is no more than a liquid rock, whose occurrence is controlled by Nature. However skilled the engineer is at taking the oil out of the reservoir, he is no good at all at putting it into the reservoir. The economist and manager can do nothing without this basic ingredient of economic life: oil has had a crucial role over the past century. When geologists tell us that we are about to face a major change as the resources of available oil dwindle, we had better pay careful attention.

ENERGY

Why was I and others employed to look for oil? Because it was a potent source of energy, which is what makes the world go round, whether we speak of industry, agriculture, transport or life itself. The term comes from the Greek *en* meaning *in,* and *ergon* meaning *work.* It had already been recognized by Aristotle, 300 years before Christ. But it was not until the development of thermodynamic theory in the last Century that the physical principles were understood. The coming of the steam engine and the first railway in England in 1825 brought revolutionary developments in the use of energy, which had a great effect on the way people lived. They moved from the countryside to the towns, often to work in depressing circumstances, and received little of the benefit of the new energy, which effectively introduced capitalism as we know it. Since then, the use of energy in all its forms has expanded greatly. I remember visiting my aunt's farm in Ireland just after the War. They had fifteen farm labourers, horses to pull the carts, and a pony trap in which to do the weekly shopping. A paraffin-driven generator provided electricity for lighting. Now, they are connected to the mains electricity, have tractors and sophisticated harvesters but hardly any labour. They grow Christmas trees for the German market instead of wheat.

The new sources of energy which made this transformation possible involved colossal national investments. It is estimated that something like ten trillion dollars of investment will have been spent to furnish the world's energy needs over the last three decades of the Century. The provision of such energy itself consumes energy on a vast scale, whether we speak of oil production, coal mining or nuclear power, not to mention the electricity generating station itself. It may take up to five years for the plant to produce more energy than its construction consumed. A modern coal-fired power station, yielding one gigawatt of electricity, consumes 10 000 tons of coal a day, but also gives 600 tons of ash and 200 tons of sulphur dioxide, as well as much carbon dioxide released to the atmosphere. We are running out of space for the waste.

Furthermore, all materials and foods require substantial amounts of energy to produce and deliver them to market in usable form. The production and use of energy also carries adverse environmental effects in the form of acid rain, greenhouse gases, nuclear waste, not to mention the blighting of the landscape in many ways. The usage of energy varies greatly from country to country depending on the degree of so-called progress. Pre-industrial countries, such as India, use less than 0.5 tons of petroleum equivalent per capita per year, whereas the inhabitants of the most profligate user, the United States, consume twenty times more. They use twice as much energy as do the Europeans but their per capita income is only fifty percent higher. It seems inconceivable that the growth of usage in the extravagant countries can continue, nor that the less industrialized countries can catch up, even assuming that that was desirable.

Some Theory

For the physicist, energy is a constant in a closed system. It is not consumed but simply transformed from one form to another: heat to movement or movement to heat. A calorie is the amount of energy required to raise one gram of water 1°C. It is equivalent to 4.184 *joules,* another measure of energy, defined under the S.I. system as the kinetic energy of one watt per second. Power is measured in terms of the energy produced or consumed over time in *watts,* being one *joule per second.* There is also the distinction between kinetic energy, which relates to objects in motion, and potential energy, such as the power of a compressed spring.

In fact, heat and mechanical energy are related phenomenon under the principles of modern physics. The laws of thermodynamics govern the efficiency of energy transformation into useful work and heat. A wind-driven generator captures some seventy percent of the energy input, whereas one fired by heat captures only about forty percent. What is not used is released as wasted, and even damaging, heat to the environment.

Food

Food provides both nutrition and energy. The average person requires 2500 kcal per day, or three kWh, to keep him going. That is to say, the equivalent of running a small electric fire for three hours. The continuous power consumed is about one-sixth of a horse-power.

The most primitive tribes of hunters and gatherers expend about one unit of energy to recover sixteen units from the food obtained, but they need plenty of forest to wander through. Modern agriculture reduces the relationship to about one-to-four, through the use of mechanization and fertilizer, which consume energy. It is even down to one-to-two in some very intensive production.

Most food today is processed, packaged, transported, stored, refrigerated, cooked and advertised, which may consume as much as seven times the energy content of the food itself.

Oil provides much of the energy needed. Even where I live in rural France, there is an almost continuous distant throb of tractors at work.

Fertilizers and pesticides are in ever greater use. Some are made from hydrocarbon gas, and energy and oil are consumed in their manufacture, transport and application.

Fisheries are very heavy oil users, but again there is scope for improved efficiency, with drift netting using much less than trawling.

Transport

People are on the move all the time, and goods are shipped all over the globe in ever increasing quantities. Transport uses about twenty percent of the energy consumed in developed countries. The efficiency ranges widely: a bus uses about one-third as much as a car per passenger moved. Furthermore, there are gas-guzzlers and efficient light-weight small cars.

Air travel has proliferated greatly: jet trails are visible somewhere in the sky most of the time almost everywhere. The oceans are traversed by tankers, cargo ships and cruise-liners, all powered by oil. The military too burn it up.

Transport depends heavily on oil, by far the most convenient fuel. About sixty percent of the oil produced goes in transport.

Sources

The energy required for food, heat, mechanical power, and producing materials comes directly or indirectly from the sun and nuclear activity. Much comes from fossil fuels comprising coal, oil, and natural gas, all of which are present on the planet in finite quantities, having been formed long ago in the geological past. They are obviously subject to exhaustion, as this book will consider in respect of oil.

In addition, there are renewable sources of energy, including bio-mass, hydro-electric, wind, wave, tidal, and solar sources[16], which so far have produced little, but must become more important in the future. The scale of the world's dependence on fossil fuels is daunting and dangerous. Oil makes up forty percent of traded fuel. Its production is about to peak and no one seems to realize the implications.

There is a need for some form of energy accounting to properly understand the energy content of things and activities, so as to better see how vulnerable we may be, not only to shortage as such, but to increased prices. In due course, and in the not too distant future, it will be necessary to allocate or ration them, not necessarily only by price. Already Australia sets an example by questioning the wisdom of a new road to the Sydney Airport for the Olympic Games[17]. It would not be difficult to contrive effective fiscal measures aimed at energy efficiency. For example, it is worth knowing that in energy terms importing 1 kg of meat is equivalent to importing 1.3 kg of gasoline. Recycling metals uses much less energy than does primary production: iron, one-third; copper, one-tenth; and aluminium, one-sixteenth. The scrap dealer is a more valuable member of society than is often appreciated.

Electricity is a particularly useful form of energy, but it is not at all an efficient usage. About one-quarter of the energy used in an industrial country is dedicated to producing electricity, but the output is a fraction of that because of the losses in generation and transmission. The terrible waste of pointless street lighting is very evident. Flying over Europe or America by night shows glittering lights from horizon to horizon: all for no real purpose and burning up a non-renewable resource inefficiently.

SOURCE	%	OUTPUT	USE
Renewable		Electricity	Commercial, industrial,
Bio-mass	n.a.	generation	household heat, light &
Wind	--		power.
Wave	--	Direct	
Solar	--	fuel	Transport
Hydro	3		
Non-Renewable			Agriculture & fisheries
Geothermal	--		
Nuclear	7		
Gas	23	Waste heat and wasteful usage	
		(about 70%)	
Conventional Oil	40		
Non-conventional Oil	--		
Coal	27		

Fig. 1-23. Energy input and output.

Energy is a critical factor in the world's future, and oil is the most vulnerable to early exhaustion. It is time to think about the consequences.

Energy Concentration

In considering energy, it is important to realize that what really matters is energy concentration. All of it, apart from nuclear energy, comes ultimately from the Sun. Looking at a local mediaeval walnut mill, I wondered how the thing could work on the little trickle of a stream, which was all that flowed through the valley. The answer is that the stream filled a reservoir which provided a sufficient head of water to drive the mill for a day or two before having to be refilled. It was a form of energy concentration. Look at a log burning on a fire. The tree from which it came had been concentrating sunlight for twenty years while it grew. We are able to concentrate that into heat for an hour or two while the log burns. Fossil fuels are even more concentrated forms of sunlight that fell upon the Earth for long periods of time millions of years ago.

In early years, Man used human and animal muscle power, itself derived from food grown in sunlight. Imagine for example the work of the shipwrights building the Santa Maria in which Columbus sailed to America. They had to cut down the trees by hand, transport them by horse, and saw them into planks by hand: it took a long time. Now, the same work is achieved rapidly by the use of power saws and drills, using highly concentrated energy.

As we use up the concentrated energy of the past, as provided by fossil fuels, things will have to again move more slowly and sustainably.

NOTES
(For references, see Bibliography)

1. The USA peaked in 1971 with 9.65 million barrels of oil a day, according to Oil & Gas Journal data for oil only. The inclusion of Condensate and/or NGL gives a slightly different date.

2. I can't list all my other associates, but should nevertheless record my indebtedness to them all for their happy comraderie and cooperation in those early days: they include George Higgins, my supervisor; Max Cater, Vernon Hunter, Jimmy Frost, and Howard Bennett, not to forget Polly, the Secretary, the team of draughtsmen, and Watty the fossil collector.

3. See Appendix 1 for a short glossary of geological and oilfield terms.

5. See Texaco p. 37.

6. Those of you who are mathematical will have no difficulty in knowing what hyperbolas and parabolas are. Simply stated, they are plots in which the degree of curvature gradually diminishes. Several examples are illustrated herein.

7. He went on to form the Lone Wolf Oil Company in Denver and become a millionaire, before the oil glut of 1985 wreaked havoc with his endeavour.

8. I don't know exactly what their roles were. Bill Saville had been a field geologist in Colombia and was an impressive character: a tall, gaunt frame with a moustache, who could have played the part of a sheriff in a Western movie. He had greater sympathy for what I proposed, but in the corporate politics was probably unable to act. I, of course, had no insight into corporate politics at the time. Another figure in the office was Don McGirk, a pleasant, distinguished-looking man in a tweed suit, who was the Exploration Manager, but seemingly not part of the Bishop-Bode axis.

9. I later published a paper with Hans Bürgl on this key section, which is one of the best exposed in South America (see Campbell, 1965).

10. Our particular friends included the Barry's whose paths we crossed many times later; the Ross's also from Texaco Trinidad, and the Tanner's.

11. Moose Bernard, a larger than life American, was in charge and Freddie Martin was Exploration Manager: a quintessential upper class Englishman with a nervous cough and a love of night life. Others were Bill George, the Deputy Manager, who later helped found Nordic-American; Richard Hardman (now a director of Amerada and President of the Geological Society); and David Walker (later to become Chairman of the British National Oil Company). Jim Spence and John Harrison were also there when I first joined. The draughtsmen, López, Obando and Prieto, and the secretary Magda Luz Gutiérrez, were especially close, helping me produce my reports, which in the days before photocopying and word processing were reproduced on ozalid skins.

12. Colombia has had a rather exceptional discovery pattern with three hyperbolic trends: an early one around the La Cira-Infantas discovery in the Magdalena Valley; a second in the 1960s with Orito, which opened up the southern Sub-Andean Basin extending into Ecuador; and a third with the Llanos discoveries highlighted by Canon Limon and later Cusiana.

13. The late discovery of giant fields, which may have more reserves than here estimated, distorts the picture somewhat. It might be better to use a bell-curve to model future

production than rely on the standard fixed depletion model (as discussed in Chapter 8). The wells face exceptional stress conditions and can cost as much as $40 M to drill.

14. Wallace Pratt, an American expert, put it well in 1952 when he said "Oil is found in the minds of men".

15. Knoepfel H., 1986, gives a first class discussion of energy issues.

16. One could include geothermal energy, which is renewable where it comes from volcanic rocks, but exhaustible where coming from a buried aquifer

17. This is an interesting case. Mr Brian Fleay has written a book "The decline of the age of oil", based on an identical reasoning to that developed here. It captured public imagination and led to strong opposition to the construction of a road called The Eastern Distributor, which had been planned as infrastructure for the Olympic Games in 2000. It may be one of the first examples of where concern about future oil supply has led to a major public outcry (see Macleay, 1996).

Chapter 2

OIL: CHEMISTRY, SOURCE AND TRAP

If you don't know much about where oil comes from, don't be ashamed. Until quite recently even professional explorers had only a rather hazy idea about the exact origin of oil. In the last Chapter, I described my experiences as a young geologist looking for oil in Trinidad and Colombia. I knew of the classic view that most of the oil in northern South America came from the *La Luna Formation*, a sequence of petroliferous Cretaceous shales and muddy limestones, but I had difficulty in understanding how this deep-seated source could charge the overlying Eocene reservoirs in the Middle Magdalena Valley without filling intervening sandstones. To overcome this difficulty, I proposed that the oil was generated in the shallow sequences themselves, despite their non-marine origins. It seemed a reasonable explanation but it was no more than a hypothesis. Now, geochemistry can specifically identify not only source but the date of generation. It has revolutionized the search for oil, permitting a much improved assessment of what the world's oil endowment actually is. It is of cardinal importance, and I will try to explain it in this chapter. It may be rather heavy going for the generalist, but it is important to understand at least something about the physics and chemistry of oil and gas at the outset.

Before coming to explain the origin of oil as now accepted by virtually all explorers, I should dispose of an alternative theory that has been advanced by Thomas Gold[1], an academic, who proposed that oil and gas come from the deep interior of the Earth. If this were so, petroleum would be present in almost infinite quantities. This theory apparently emanated from Russian scientists, and recently has been again advanced by J.F. Kenney[2]. When the Swedish government bowed to environmental pressures and suspended the production of nuclear energy in the wake of Chernobyl, it was persuaded at huge cost to test Gold's theory. Two wells were drilled to about 7000 m at Siljan, the site of a meteor crater in an area of granitic rocks, lacking[3] conventional prospects. Some indications of oil and asphalt were known in the vicinity, but are thought to be due to the local distillation of oil from Ordovician black shales, due to the heat of the meteor impact[4]. It is claimed that the wells encountered some indications of oil, but a degree of mystery surrounds the critical samples. If valid, which is far from certain, probably such indications either came from lubricants used in drilling the well, or from the shales into which the granite body was intruded. Few people take the hypothesis of an inorganic origin of oil seriously, but it has to be mentioned as it is one of the arguments advanced by those who dismiss resource constraints to oil production. I will leave it at that.

THE CHEMISTRY OF PETROLEUM

Organic chemistry is so termed because it is concerned with the essential element of living organisms; namely carbon. This is not the place to go into the subject in any depth: suffice it to say, that oil and gas, together with other related substances such as asphalt or bitumen, are hydrocarbons made up of carbon and hydrogen in various molecular combinations. In fact, we are concerned with a family of such molecules which can dissolve intimately in each other, depending partly on the ambient temperature and pressure: they are, so to speak, born, transformed, degraded and disappear, having in this respect a sort of life cycle that mirrors their organic origins. The term petroleum, meaning rock oil, is strictly speaking a misnomer given the organic origin of oil, but is nevertheless a useful generic term for oil, gas and related substances.

We are concerned with three hydrocarbon families:

1. Saturated Hydrocarbons

Saturated hydrocarbons are paraffins (or alkanes), which are quantitatively the most important, making up 50-60% of most oils. They form a linear molecular chain with the general formula C_nH_{2n+2}, and occur in three states:

gas	–	with C_1 to C_4
liquid	–	with C_5 to C_{15}
solid	–	with above C_{15}

So-called n-paraffins, with odd numbers of carbon atoms, are synthesized in living organisms, and such molecules found in oil are true biological markers inherited from the living organisms from which they were derived. C_{15}, C_{17} and C_{19} characterize microscopic organisms including algae, whereas molecules of above C_{21} typify plants. These chemical links give the game away, showing that oils come primarily from algae. Another group of molecules, the iso-alkanes with a branched structure, include pristane (C_{19}) and phytane (C_{70}), and is derived from chlorophyll in living organic material. They also demonstrate the link.

2. Unsaturated Hydrocarbons

Unsaturated hydrocarbons comprise the aromatics which have a ringed molecular structure. They are so named because of their pleasant smell, although the naphtheno-aromatic sub-family is commonly associated with sulphur compounds, giving them the exceedingly unpleasant smell of bad eggs. Benzene (C_6H_6) is the simplest aromatic hydrocarbon.

3. Resins and Asphaltenes

Resins and asphaltenes are complex compounds with high molecular weights, rich in nitrogen, oxygen, sulphur, nickel and vanadium. They are mainly the products of the chemical alteration of ordinary oils.

It is of course not as simple as this. For example, the carbon itself occurs in two isotopes ^{12}C and ^{13}C, the proportions of which can be used to help identify whether the oils were deposited in marine or non-marine conditions.

THE PHYSICAL PROPERTIES OF PETROLEUM

As I have explained, petroleum can occur as a solid, liquid or gas, partly depending on the ambient temperature and pressure, and each phase may contain dissolved elements of the others. Oil thus commonly contains gas; and gas contains liquids.

Natural Gas is divided into two types: *dry gas,* consisting mainly of methane, and *wet gas*, which contains liquids, such as propane and butane, and is normally found in close association with an oil accumulation. The isotopic composition of methane generally reflects the degree to which it has been subjected to rising temperature and pressure on burial. Some natural gas deposits contain hydrogen sulphide (sour gas), nitrogen and carbon dioxide, depending on depositional conditions and the effects of alteration. Small amounts of helium and argon are also sometimes present.

Methane is relatively soluble in water. Natural gas is highly compressible, such that its volume may be reduced by a factor of 200 to 300, which incidentally means that it can provide a valuable drive mechanism to expel associated oil from a reservoir.

Hydrates are special deposits of methane in an ice-like solid condition which are found in Arctic and deep oceanic environments. Some hopes have been entertained for exploiting such deposits, which may be very large, but they are unlikely to be fulfilled as the methane, being held in a solid matrix, has no opportunity to migrate and hence accumulate in commercial quantities.

Oil is a liquid hydrocarbon but generally contains varying amounts of gaseous and solid hydrocarbons in solution. These phases may separate naturally, and can be extracted by processing. Oils come with many different characteristics. Density, mainly reflecting the chemical composition, is one important property, which in the Anglo-Saxon world is traditionally measured under a scale set by the American Petroleum Institute, with most oils being in the range 15° to 45°API (0.9-0.7 S.G.). The heavier oils are rich in resins, asphaltenes and sulphur, whereas the lighter oils tend to contain dissolved gas. Heavy oils are generally dark brown or green in colour whereas the light oils may be almost as clear as a refined product. Another important property is viscosity, the inverse of fluidity, which generally increases with density, and decreases with the dissolved gas content and at higher temperature. It is measured in centipoise[5], and ranges from 1 cP to more than 10 000 cP. This property is related to pour point, which is linked to the paraffin content. High viscosity crudes, especially waxy ones of non-marine origin, become pasty and solid below about 10°C. The heat generated by a high through-put is needed, for example, to prevent the oil in the Alaskan pipeline from solidifying.

A third important property is solubility: namely the ability of the several fractions to mutually dissolve in each other. In particular, large amounts of gas can be dissolved in oil. It is measured as a gas-oil ratio (GOR) which may be as high as 6000 cubic feet per barrel (1000 m^3/t). The ratio varies inversely with density and rising pressure. Where conditions approach the bubble point, the gas separates to form a gas cap above the oil accumulation. The dissolved gas increases the volume of the liquid, and a so-called *Formation Volume Factor* has to be applied to convert volumes of oil in the reservoir to those at the surface where the gas comes out of solution.

Heavy oil is the term applied to oils with a gravity below variously 10°, 15° or 20° API . There is unfortunately no standard industry definition of the gravity threshold which is a cause of much confusion[6]. The sulphur content may be as high as ten percent. Heavy oils have various origins but most commonly are normal oils from which the light fractions have been removed by water leaching, oxidation or microbial degradation. Huge deposits of heavy oil and bitumen occur in Eastern Venezuela, Western Canada and Siberia, forming important resources for the future.

Solids comprise methane hydrates, already described, and a whole family of complex bitumens and asphalts[7], such for example as found in the famous Pitch Lake of Trinidad.

That, in a nutshell, describes the physical and chemical properties of petroleum: it is a slippery substance in more senses than one as we shall see as this account unfolds. Its diversity in a way reflects the diversity of the life from which it was derived.

Having briefly covered its properties, I will now turn to consider its formation in nature. It is another very complex subject, but one that has to be understood, at least in general terms, to gain an appreciation of the reasons why the world's endowment is so finite.

THE FORMATION OF OIL AND GAS

As those on diets soon come to find out, living organisms are largely made up of proteins, carbohydrates and lipids (fats). Lipids are abundant in algae, especially the *Botrycoccus* family, and in diatoms, and are also present in

Fig. 2-1. The Pitch Lake of Trinidad, a natural seepage that has long been a source of asphalt.

plants, being found in pollen, cutin, chlorophyll and caretonoids. The sunlit upper waters of seas and lakes support abundant life, especially micro-organisms, including these algae and bacteria. The surrounding lands are dominated by plant life. The organisms and plants have their life cycles ending in death. Their remains sink directly to the bed of the sea or lake in which they lived to form the basic ingredient of oil, while on land, leaves and plant remains are washed by rivers into lakes or seas and form a source of gas, see p. 110.

In the summer of 1988, the Italian tourist industry faced a serious setback. The beaches along the Adriatic Coast became covered with evil-smelling slimy masses; and offshore, the fishermen reported that their nets were being clogged with the same stuff. What had happened was that unusual weather conditions had led to a prolific flowering of algae which absorbed so much oxygen from the sea as to poison not only itself but marine life generally. Apart from what was washed onto the beaches to dismay the tourists, the organic debris eventually sank to the sea bed.

I am myself familiar with other examples of places where marine life proliferates excessively. Fishing in the Humboldt Current off Ecuador was such an experience. The current runs northwards along the coast of South America,

and is affected by offshore winds that drive the surface waters westwards to be replaced by up-welling deep cold water along the coast. This deep water is heavily mineralized, and provides important nutrients to support the proliferation of microorganisms on which the whole food chain is built. The grey sea with mist banks, caused by the cool water, was full of game fish, and the fins of numerous sharks could be seen circling our boat. The sky was as full of sea birds, living off vast shoals of anchovies. The sea and sky merged into one huge organic soup. The sediments being deposited beneath this current have been investigated[8], and do indeed form potential source-rocks for oil, although not in the life-span of *homo sapiens*. Note in passing that this same life stock was responsible for the great deposits of *guano* in Peru and Chile on which Europe depended for fertilizer prior to the advent of artificial fertilizer made from hydrocarbons and electrolytic methods. Likewise, the Norwegian fjords, in which we used to sail, sometimes turn milky white due to the proliferation of algae during the long hours of daylight in the northern summer. There were even reports of so-called "red tides" when algal flowerings took so much oxygen from the waters that they turned a reddish hue.

Looking into the geological past, it is clear that there were only periodic explosions of life in places where conditions were exceptionally conducive, as in the present day examples quoted. The critical factors were sea temperature due to latitude, global climate (warming) and local circumstances, as well as the supply of nutrients. In practice, abundant hydrocarbons were generated only in tropical latitudes: remembering that some areas that were previously in the tropics subsequently moved to higher latitudes under Plate Tectonic displacements. Most such movements were northerly which explains the paucity of oil in the Southern Hemisphere[9]. These brief and rare events are responsible for the oil trends as we know them, and we do now know almost all of them. Clearly, the abundance of the organic material was the first essential ingredient, but it was not in itself sufficient, for the organic debris had to be both preserved and concentrated before it could become a commercial source of hydrocarbons. The total organic content of a source-rock can exceed ten percent in ideal conditions. For these and other reasons to be explained later, only a very small fraction of the Earth's original endowment of biomass has been available to form oil and gas.

The fleshy parts of dead fish settle quickly to the sea bed, but the microscopic plankton, which forms the main source of oil, settle at a rate of no more than about 100 m per week. Only about two percent reaches the sea-bed of shallow seas, and perhaps one tenth of that survives in the deep oceans, because it is oxidized before it gets to the bottom. Furthermore, much of what does reach the seabed is destroyed by bottom-living organisms. It is only in the depths of stagnant marine troughs and deep lakes, where the oxygen content is low, that large amounts of the organic material can be preserved. It is also important that the material should be concentrated, as occurs in areas receiving little other sediment. Present day analogues are the Black, Baltic and Caspian Seas, the Gulf of California and Lake Maracaibo.

The world's most prolific oil province, at least so far as conventional oil is concerned, is the Middle East[10]. The organic material responsible for it was formed in warm Jurassic seas and accumulated in stagnant sink holes and lagoons within a broad carbonate platform that received only limited amounts of sediment washed in by rivers draining the surrounding low-lying lands. Another prolific province is the North Sea, where organic material accumulated in stagnant rifts towards the end of the Jurassic Period, 150 million years ago. A third example is provided by the rift lakes that developed along the west coast of Africa and Brasil, as the South Atlantic opened during the early Cretaceous.

The point is that these conditions were met only very rarely, both in time and place, which explains why prolific oil and gas deposits are restricted to only a few well defined trends.

The rate of sedimentation plays an important part of the process. If sedimentation is too rapid, the organic material becomes disseminated, whereas if it is too slow, its preservation is impaired. Clay minerals too are involved, helping to fix the organic material.

So far, I have spoken mainly of oil which is associated with restricted marine or lake environments characterised by an abundance of planktonic and algal organic debris. By contrast, gas is associated with more brackish environments, as found in deltas, where ligneous and humic material from plant life predominates. In some areas, the two environments overlap, giving both oil and gas source-rocks.

WALTER ZIEGLER
International explorer

Q Walter: you come from a family of well known geologists. Can you tell us a little of this background?

A *My father was a doctor in Winterthur in Switzerland. He was a keen naturalist, and instilled in all four of his children an early scientific interest, and an appreciation of nature, particularly botany and geology. He also had a philosophical streak, and taught us to first observe, and then to try and explain, and comprehend. He wanted logic not magic. It influenced our choice of career: but all the same it was strange that there should be three geologists in one generation.*

As children, we often went to our grandfather's summer house in the Jura, a beautiful range of beech clad mountains and green valleys. The shiny yellow-white limestone cliffs formed spectacular rock arches in the gaps where the rivers broke through. We later understood that they were structures termed anticlines, and still later that oil is trapped in anticlines. Ancient castles and ruins stood on these cliffs guarding the passages from one valley to another in this spectacular country. We couldn't fail to notice the alternating strata with the hard limestones standing out from the soft shales. We began to have an intuitive feel for geology, which was further encouraged when we found fossils: ammonites (Ammon's Horn), belemnites (Devil's fingers) and other shells, which our mother often pointed out.

A family friend, Freddy Senn, was an oil geologist, who impressed us all with his stories of travel in distant lands

including Burma and South America. When he died unexpectedly on his way to Morocco, his widow gave us a book from his library "Geologie der Schweiz". Its pages were full of cross-sections, maps and panoramas which allowed us to identify and understand the geology of the country we knew.

Naturally, we also went to the Alps and broadened our knowledge with minerals, igneous and metamorphic rocks,

Fig. 2-2. Walter Ziegler, an international oil explorer who took part in the North Sea discovery.

and the sight of the great overthrust nappes, exposed on the mountain side. My brother, Peter, won a school prize with his collection of minerals. I had a teacher called Dr Peter Walter with the prominent "beak" of the true scientist, who had a great interest in geology. During my last summer vacation, I hiked with him from the Helvetic Alps through the Aar and Gotthard Massifs to the Engadine, traversing a great section of Alpine geology. The results of my exams were less than impressive, but my mind was far away in the Alps. I remembered particularly a chance encounter at an altitude of 3402m on the Rheinwald Horn, where we met Wolfgang Leupold, a professor of geology at Zurich University, who gave us an impromptu lecture on the geology of the range we could see below us, stretching from the Po Valley in the West to the Tauern Mountains of Austria in the East. I was stimulated to learn more and understand how this great mountain chain came into being.

In the autumn of 1947, I started a general science course at the Swiss Federal Institute of Technology (ETH), but I did not enjoy it and was glad to be called up for Military Service. I was a driver and took the opportunity to learn as much as I could of the geology of the areas where I was stationed, especially the Tessin. I had a normal army experience for two or three years, alternating with university semesters, and ended up as an Artillery Lieutenant. It was the beginning of the Cold War, and we understood that we had to be ready to defend our country lest it suffered the same fate as Czechoslovakia where in the spring of 1948 a democratic government was overthrown by Russian-backed communists.

Gradually my interests at University concentrated on geology, and I was fortunate to fall under the influence of Professor Rudolf Staub, a great Alpine geologist, who instilled enthusiasm for the subject and taught us how to think, observe and interpret, and stepping beyond that to speculate about the unknown. Although I had inscribed in a course of Engineering Geology, I came to realise that working on foundations and tunnels would be rather restrictive, and resolved to try to become a petroleum geologist and see the world. To this end, I spent 2$\frac{1}{2}$ years on a higher degree, studying an area of flysch in the Grisson under Professor Staub. I enjoyed the good comraderie of the post-graduates under Capo Staub, as we affectionately called him. We learnt a lot from each other and at times had too much to drink, which didn't harm us either. I finished at the ETH in February 1955 at the great age of 27. I felt I needed to escape from the somewhat closed society of Swiss academia and the highly regulated social structure of the country. I yearned for freedom and adventure, not creeping socialism.

Q When you graduated, your brother followed the traditional path to Shell, but you went to Canada to join Esso, now Exxon, the world's largest oil company. Can you say something about these early days and how your career evolved?

A My first job was as a photogeologist with the Institute Francais du Petrole where I was engaged in mapping the Sahara under Daniel Trumpy, a famous retired Shell geologist. Never have I been so cold as that winter in Paris in 1955/56. Gas, which had to be piped from the Saar coalfields, was in short supply, and the pressure was barely sufficient for heating or cooking for days on end. France was then facing its retreat from Empire, having suffered the defeat of Dien Bien Phu in Viet Nam. There were frequent strikes, high inflation and a general malaise. Army trucks provided transport, but despite all the hardships, Paris still held a great attraction for me.

I was offered a job by Shell as a geophysicist, but that did not appeal to me, and instead I followed up a suggestion from friends in Canada, who told me that Imperial Oil, an Esso affiliate, was looking for geologists. I successfully applied.

I quit my job, and in due course set off on a great adventure: by train to Liverpool to board the Saxonia to Montreal and on to Calgary. A few days later I was sent on my first mission to join a field party of three geologists, a cook, helicopter pilot and mechanic working on the frontal ranges of the Rockies by the Athabasca River. The helicopter was a magic carpet that swept us up the mountain side to the outcrops we wanted to study, although often a hair raising experience when buffeted by mountain winds. However before long the carpet broke down and we had to hire a string of pack animals from a backs woodman called "Old John the Swede" who slept under his stinking saddle blanket. It was an exciting life of riding, climbing, measuring geological sections and living under canvas, and we saw lots of wildlife: moose, wapiti, Rocky Mountain sheep, goats, mules, deer, coyote, wolves and partridges. When the field season ended with freezing rain and snow flurries it was time to head back to base.

I then joined the research group in Calgary where amongst other things, they were studying how to measure the chemical alteration of hydrocarbon source-rocks. Diane Loranger had noticed the progressive loss of ornamentation on fossil ostracods with increasing depth of burial, and Frank Staplin had observed how fossil pollen became darker in colour. These lines of research were to provide critical means of determining where and when the source-rocks would give up their oil: it was a glimpse of a new understanding.

I continued to spend the summers on field studies full of colourful and interesting experiences, and I also widened my experience of basin studies, wellsite work and becoming generally more of an oilman. Then in the early 1960s came bad times. The industry was suffering one of its periodic downturns, and the geological department was reduced from 40 to 20. The Company had decided that it was not interested in gas, and preferred to import cheap Venezuelan oil rather than explore the foothills of the Rockies. It was to be a recurring theme.

But then a new play opened up with the discovery of Swan Hills and Judy Creek, where the objectives were hard-to-find reefs. We were back in business with all the excitements of a boom. I was promoted to Area Geologist in Edmonton, and now had for the first time a more commercial role, making deals and competing in lease sales. It was great fun.

Q Clearly you integrated into the Canadian environment, but still you were a European at heart with a wider perspective. Esso evidently recognized this, when they invited you to transfer to their international operations. What was your reaction?

A On January 2nd 1964 I was told that I was to be transferred to Esso's Geneva office to study Triassic reefs in Austria. It was totally unexpected, and I could not believe my good luck. I had greatly enjoyed my time in Canada, but the limitations of the Alberta Basin were already evident. The new frontiers of the Arctic and the offshore were only just awakening, and the thrust of exploration was turning overseas, especially to Libya. I welcomed the chance to broaden my horizons.

Q What were the highlights of this new assignment?

A The first was a cultural shock: I had almost forgotten how to speak French. But then another stroke of good luck was a reorganization that transformed what had been a research office into a European Frontier Exploration office. My new mission was to integrate a preliminary North Sea study into a regional context. I made numerous trips to Germany to collect information. Next came another transfer: this time to Spain where a new office to evaluate Africa had been established under Dave Kingston (see p.44), one of Esso's so-called "Rover-boy" geologists with whom I hit it off particularly well. I started collecting information with which to prepare the basic basin maps, and also made a brief trip to Morocco to investigate an exploration idea.

I was then recalled to Calgary for what turned out to be a splendid and stimulating period of work in a very congenial environment. I was able to catch up on the latest progress in geochemistry and the early steps towards seismic stratigraphy on which great strides were being made. But despite the enthusiasm, I think at the back of our minds we were beginning to see clearly the limits of exploration. There were fewer new ideas to be tried, and often we would find that many of the promising concepts that did develop in our minds turned out on further investigation to have been already tested.

Before long I was back overseas to be based in Spain on African studies. I began travelling in earnest with numerous trips all over Africa: Mozambique, South Africa, Tanzania, Kenya, Cameroons, Gabon, Rio Muni, and Fernando Po. I could see at first hand the full misery of post-colonial Africa: camps of starving refugee children, violence, stone-

walling petty bureaucrats, and the general degradation of a once fine people under ill-absorbed Western influences. Technically, I soon realized that much of Africa was effectively non-prospective apart from a few key areas along the Equatorial Atlantic Coast, where most of the possibilities were already controlled by other companies.

In March 1969, began my final assignment to Walton-on-Thames in England to act as Chief Geologist in a newly established East Atlantic Study Group. It was a bad experience: we drilled some forty dry holes along the coast of NW Africa at a cost of more than $100 million. At the same time, the Company found itself in the grip of restructuring as various affiliates were consolidated and absorbed into what eventually became Exxon, which at the time we unkindly dubbed as the "sign of the double cross". The ranks of middle management swelled with ill-experienced and sometimes abrasive individuals from the domestic organization in the United States. They had been successful in the Gulf Coast and now wanted to conquer the world, which they thought of as a great extension of Texas, marred only by grasping foreign governments. The Viet Nam War was in progress

Africa was assessed "to death" under the new administration, but the record shows that between 1972 and 1983 the Company had failed to find a single barrel of oil (save for Chad on a Conoco farm-in, which is still not in production) despite a search costing millions of dollars. Africa lacked the pervasive source-rocks of the Gulf Coast and so a "Gulf Coast" approach to exploration did not work. The Company had the expertise to understand but those in charge had grown up in a different environment and did not apply it rigorously. The thing rolled on under its own momentum at huge cost.

Later the Company did learn from this mistake, and instituted a programme of training to very high technical standards. Exxon courses became renowned for their excellence in the industry, but in many respects it was a case of closing the stable door after the horse had bolted. There was much less left to find.

Finally in the summer of 1983, the Europe-Africa regional office was packed onto a 747 and shipped to Houston where it reopened in diminished form.

Q The North Sea was opening up then, and you published one of the first syntheses of that area. Did Esso play a fundamental role there?

A In 1970, after the N.W.Africa fiasco, I joined a North Sea Study Group with the task of putting together a regional synthesis. We were in partnership with Shell as operator in the United Kingdom. The southern North Sea gas province had already been developed as an extension of the giant find at Groningen in Holland, and in the early 1970s some people thought that there was little further potential. However, when we were preparing for the UK Fourth Round of Licensing, Pat Kelly, one of our geophysicists, drew our attention to a regional seismic line that a contractor had shot east of the Shetlands. It showed "tilted fault blocks" below the Cretaceous, which reminded him of the El Morgan Field in the Gulf of Suez. Shell/Esso tested the idea with Well 211/29-1 in May 1971, and held the positive results in the greatest secrecy as a basis for bidding on surrounding blocks, many of which turned out to hold major oilfields. We did not have a close daily contact with Shell but followed its every action with close attention, contributing constructively as the ideas evolved.

During these days, I compiled the regional picture within a plate-tectonic framework, making analogies with other comparable settings around the world. I authored one of the first published accounts of North Sea geology.

When the oil crisis of 1974 hit, I was hustled off to evaluate coal potential around the World, the Company having decided to move into that business. In the spring of 1975, still another new study group was formed to evaluate the Circum-Atlantic in the hope that the eastern US shelf might provide much needed new oil supplies. I was called in to advise, but on the basis of my evaluation I was neither enthusiastic nor popular for explaining yet again that no amount of ingenuity can compensate for the lack of active source-rocks. Our team's advice was ignored and several expensive wells were drilled with predictable results. By now, we really understood the essential factors needed to make an oil prospect, and we had all the technology to provide the raw data. The sad fact was that there was much less left to find. The US East Coast was as poorly endowed as its counterpart in NW Africa, which we had already explored to our cost.

After other assignments in Libya and Africa, we faced in 1983 the final dismantling of the once might global Esso Exploration organization. The world evaluation had effectively been completed. It showed that there was nothing left to do beyond a mopping up operation and a single office was enough to handle the ever smaller projects that remained.

I was offered a senior position in this new organisation but preferred to stay in England, partly for family reasons. A chance meeting led to an offer to join Petrofina, a Belgian oil company, that was then seeking to expand. I accepted and after a period in the UK office moved to Brussels where I ended my working career. It was a new company with a very different character. We tried our best looking for by-passed ventures, but the world was the same and I could not find what was not there to be found.

Q So you had a very varied career in many parts of the world and you were able to observe the unfolding situation from a very privileged position in the world's largest oil company having access to a colossal database and call on the most advanced technical expertise. How did you assess the situation in resource terms?

A I regretfully came to the conclusion that it was a war of lost causes. There were thrilling moments as during the pioneering days in Canada and later in the North Sea when we opened up a new province, but much of it was a catalogue of frustration. We knew what we were doing and with the best will in the world and the best expertise available we simply could not make a purse out of a sow's ear. There are only about thirty provinces world wide as prolific as the North Sea, that is to say having an oil endowment of more than ten billion barrels, an amount itself enough only to supply the world for six months. The key to the North Sea was the prolific Upper Jurassic source-rocks. Every successful basin relies on a well defined source and an appropriate thermal history.

Q Time and again you have stressed the importance of source. What do we know about the world distribution of source-rocks.

A Each successful basin has its source system, by which I mean not only the rock itself, but the timing of generation and migration. It is a thermokinetic process whereby the rocks give up their oil and gas on critical exposure to heat. In addition, the preservation of the oil and gas formed is vitally important: all oilfields leak over time. So we had to search for structures that had been charged in the relatively recent geological past.

Q So how would you sum up the conclusions of a career spanning fifty years of exploration in many parts of the world?

A Above all I would say that it was great fun. I saw a lot, learned a lot, lived in many different countries and met a great number of varied and interesting people from many backgrounds. It was my good fortune to live during an epoch when petroleum geology became a

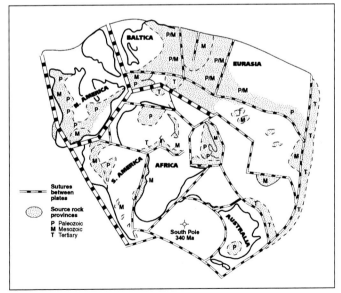

Fig. 2-3. The main oil and gas provinces of the world in their original setting along the sutures marking the break-up that led to the present continents.

science, subject to rigorous scientific discipline. It is well capable of answering questions about the availability of hydrocarbon resources on which the modern economy depends. Our studies have confirmed beyond any doubt that the globe has a decidedly finite potential for oil exploration. The implications are colossal. The World must finally come to terms with the fact that changes in the way it lives are imminent. It has no option but to adjust to resource limitations. "No more candy for the kids!" The game is nearly over.

Q One final question: can I ask you to use your worldwide experience to show us where the key provinces are: not a present day map but one showing the continents in the positions they then occupied.

A *Yes indeed, oil is concentrated into a few provinces, for well understood reasons, and we can plot where they are. I don't think that there are many, if any, new ones to find considering how extensively the world has now been explored, and considering that we now know so much more about the factors responsible.*

THE CONVERSION OF ORGANIC MATTER INTO OIL AND GAS

In a sense, the organic material has a life after death as it is gradually converted by chemical reactions into oil and gas. Once it has landed on the sea- or lake-bed, it is buried beneath layers of generally fine-grained sediment washed in by rivers from the surrounding lands. Physical and chemical reactions commence almost immediately, being primarily driven by bacteria and other micro-organisms that can continue to attack the sediments until they are buried to depths of several hundred metres. The main effect is to create a strongly reducing environment. The bacteria split the complex molecules releasing carbon dioxide and methane, leaving behind an insoluble residue known as *kerogen*. It occurs in three important types: *sapropel* yielding oil; *vitrinite* giving gas; and *inertinite* that yields neither. It is also classified into three categories known as Types I, II and III according to the path of chemical change. Types I and II with a high hydrogen/carbon ratio are oil prone, whereas Type III, with a low ratio, is gas prone. It is all rather complicated.

The temperature rises with burial below the sea-bed, and the resulting chemical activation tends to break down the complex molecules to more simple structures, commonly releasing methane in the process. Three stages of alteration are recognized: *diagenesis* at shallow depth; *catagenesis* at moderate depth, normally in the range of 2000 to 4500 m, when the bulk of the oil is generated; and *metagenesis*, at greater depth where the oil is cracked[11] to gas. Coaly material, called *vitrinite*, changes its reflectivity on alteration, which means it can be used as a sort of thermometer to indicate the degree to which the rocks have been "cooked" in terms of both temperature and the time to which they were exposed to it. *Vitrinite Reflectance* values thus allow us to track the course of alteration for both hydrocarbons and coals, as shown in Figure 2-4.

As temperature rises over the critical range, more and more *kerogen* is converted to oil, but as the process proceeds there is less and less left to convert. The term *oil window* is used to describe the depth range of generation, which generally does not extend over more than about 1000

m. The peak itself is even more important, as it takes place over a relatively short time span. It has a dynamism of itself, which I will consider more in relation to the migration of oil. It is critically important to determine the structural conditions that were present at the time of peak generation because they will primarily determine where the oil moved to and which traps were to be filled.

Stage	Vitrinite Reflectance	Hydrocarbons	Coal
Diagenesis	<0.5	Biogenic gas Early gas	Peat Lignite Sub-bituminous
Catagenesis	0.5-1.0 1.0-2.0	Oil Gas-Condensate	Bituminous Coals
Metagenesis	2.0-4.0	Methane	Anthracite
Metamorphism	>4.0		

Fig. 2-4. The alteration of organic material into hydrocarbons and coal.

OIL & GAS GENERATION

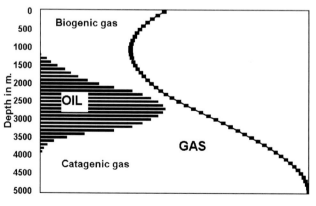

Fig. 2-5. The oil window: the depth at which oil is generated.

HOW OIL MIGRATES AND ACCUMULATES IN OIL FIELDS

I have explained how organic matter was converted to *kerogen,* which in turn yielded oil and gas on being buried to a critical depth, normally between 2500 and 4000 m. It was preserved only in certain environments in which, for various reasons, only fine-grained sediments were deposited. Such rocks, with their high organic content generally consist of dark coloured clays and muddy limestones, often having a petroliferous odour. They are called hydrocarbon *source-rocks.*

The expulsion of oil from the source-rock, under the circumstances described, is termed *primary migration,* whereas its subsequent movement into a reservoir is called *secondary migration.* Chemical reactions at the critical temperatures break down the large *kerogen* molecules to form the smaller molecules of oil and gas. This results in a moderate expansion of volume because the density of oil is 0.7-0.9 g/cm^3 compared with 1.15-1.35 g/cm^3 for *kerogen.* The differential in the case of gas is even greater[12]. This expansion leads to an increase in pressure, which is a critical factor in oil migration that needs to be explained.

At a given depth in a sedimentary basin, the weight of the overlying rocks is partly carried by the stress transmitted directly through the network of grains making up the rocks, known as *vertical stress,* and partly by the fluid pressure in the pore spaces. The generation of oil and gas, which enters the pore space, therefore increases the fluid pressure both absolutely and relatively to the *vertical stress.* The *kerogen* in the source-rock is commonly concentrated into a sequence of individually thin layers, which become fluid on being converted to oil. The pressure in such fluid intercalations rises until it exceeds that of the stress imposed by the overburden weight. Since the lateral rock stress is less than the vertical stress, the excess pore pressure leads to the formation of vertical fractures, along which the oil bleeds off to relieve the pressure, thereby closing the fluid layer so that the grains again come in contact to carry the overburden. If you sit on an air-cushion not strong enough to support your weight, you will experience something of the same sort of phenomenon.

Once the oil and gas have been forced out of the source-rock, the main driving force for further movement is buoyancy, because the density of oil and gas is less than the water that naturally fills the pore space. The strength of the force is determined both by the height of the oil column in the rocks and the counter capillary pressures either between the oil and rock grains directly, or between the oil and a thin film of water that coats the grains, as is usually the case. The capillary resistance to flow is related to the throat-size of the passages between the individual pores. The throat-size in an un-cemented sandstone is large enough to allow the oil through, in which case the sandstone is said to have adequate permeability. Rising damp in old buildings demonstrates how seemingly solid stone has both porosity and permeability, allowing water to gradually flow though it. So it is in underground reservoirs, save that the pressures

are much higher. The pore throat-size in shales is low, typically having a permeability of about a *nano-darcy* (10^{-9} D) – permeability being measured in units termed *darcys* – and in such cases the movement of oil is prevented. Micro-fractures, however, provide a pathway by which the oil can cut across such impermeable layers, many of which have normally to be breached in the course of its migration[13].

Knowledge of the process of rock fracturing is routinely gained from drilling operations. Ideally the drilling mud is weighted up to a level sufficient to exert enough pressure to preserve the borehole: if the mud weight at depth is too high, it will force its way into the formation, fracturing it in the process, whereas if it is too low, the formation may encroach on the borehole to grip the drill bit. The rocks at depth become quite plastic under the stresses to which they are exposed. One of the main challenges to efficient drilling is to properly match the mud-weight to the changing subsurface formation pressures: failure to do so can result in a dangerous blow-out or "stuck drill-pipe".

In many basins undergoing active subsidence, the critical fracture pressures are commonly found in and around the oilfields, suggesting that migration is in progress at the present time. In fact, in some instances, oil, or more commonly gas, from depth can to some extent continuously replenish the reservoirs. In other cases, where the original basins have been subjected to subsequent uplift, the formation pressures are much lower, approximating to the hydrostatic head of the water in the formation. In indurated sequences, geological faults can form migration paths for oil, but in cases where the sediments were still un-compacted at the time of faulting, the faults are smeared with clay, and form seals rather than conduits for oil.

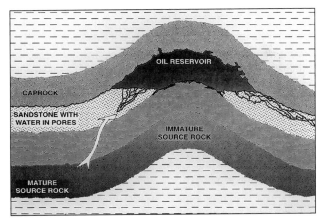

Fig. 2-6. Oil migrates from the source-rock into an anticlinal trap (after Bjørlikke).

In cases where there are no intervening reservoirs and traps, oil and gas may migrate vertically along the fracture systems created in the process of migration to escape at the surface, giving rise to seepages which are found in many petroliferous basins. In other cases, the migrating hydrocarbons encounter a conduit such as a porous sandstone, and then flow through it laterally and upwards under the influence of buoyancy.

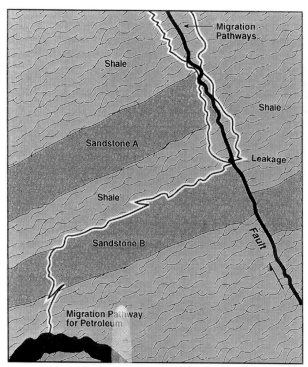

Fig. 2-7. Oil may also be trapped by faults, some of which leak (after Bjørlikke).

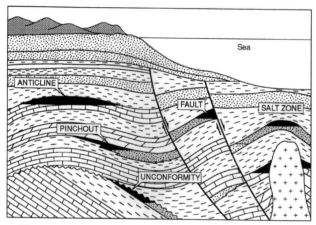

Fig. 2-8 A stylized illustration of diverse oil traps.

Again, if this conduit leads directly to the surface, the oil and gas will escape. If however the rocks in the basin have been folded and faulted, the hydrocarbons will accumulate at the highest point of such traps. There is an infinite variety of circumstances: sometimes the trap may lie directly above the source, as for example in the case of the Ekofisk Fields of Norway. In other cases, it may lie far away, perhaps at the margins of a basin as in Eastern Venezuela, where oil is found at shallow depth far from where it was generated. The catchment area for traps is colourfully referred to as the "kitchen", namely where the source-rocks are cooked to give up their oil and gas. The traps themselves are subject to various pressure regimes as they become charged with oil and gas.

Figure 2-8 is a cartoon of the different types of traps that can hold oil in a sedimentary basin bordering a continent. In the real world, it is unusual to have so many different types of trap in close proximity.

The degree to which the traps can hold their charge depends on the sealing quality of the overlying rocks, and the pressure differential. Salt and anhydrite are particularly efficient seals, and their widespread occurrence in the Middle East is one of the factors responsible for its rich endowment. Shales and clays can also serve as seals that can hold the hydrocarbon charge until a critical pressure is reached. When that is passed, the seals fracture, as described above, and the hydrocarbons vent from the weakest point in the trap, either to collect in a shallower trap, if one is present, or to escape at the surface. Since gas tends to segregate from the oil and collect at the top of the trap, the gas escapes from an imperfect seal more easily than does oil. No seal has complete integrity, and some

degree of seepage is inevitable over time. It explains why the occurrence of oilfields becomes progressively rarer in older rocks. It is still another factor that limits the resource. The process of oilfield destruction is a very important one that is not always appreciated[14]. At shallow depth, the petroleum is prone to leakage, biodegradation and weathering. At great depth, it is cracked to gas. However, oil that has leaked from one accumulation may in some circumstances collect at shallower depth. For example, the oil in the recent West of Shetlands discoveries in the United Kingdom evidently migrated early into a larger structure termed a "holding tank" from which it later leaked and remigrated into other structures that had formed in the meantime[15]. Much is naturally lost in the course of each re-migration, but it does explain some otherwise anomalous discoveries. It is estimated that the median age of the world's oilfields is 35 Ma (Oligocene), which is quite recent in geological terms. Older fields are preserved only where they lie in quiet tectonic settings and where they are endowed with very effective seals, normally salt.

In summary then, oil and gas are hydrocarbons derived from organic material: oil coming primarily from algae that lived in seas and lakes; and gas from plants whose remains were washed into them. Only a small fraction of this bio-mass was preserved from oxidation and destruction, and it accumulated in only a few places where conditions were right. Such organic material as was preserved was buried below sediments that were subsequently washed into the seas and lakes, and it was heated on burial by the Earth's heat flow. Chemical reactions converted it to a material known as *kerogen*, which in turn yielded oil and gas on further burial. Peak generation was reached at a certain depth of burial and lasted for a comparatively short period of geological time. The oil from this short-lived charge filled such structures as were in communication with it. The organic material and any hydrocarbons formed are progressively destroyed when depressed beneath the *oil window*. Only a very small fraction of the oil generated, perhaps about one-millionth, actually finds its way into exploitable oilfields.

This may seem all very complicated for the non-specialist, and indeed there are many more factors than covered in this summary. But it is important to have at least an inkling of what is involved to understand why it is such a finite natural resource.

NOTES

(For references, see Bibliography)

1. *See* Gold T., 1988.

2. Kenney, J.F., 1996, describes a Russian theory that oil comes from primordial material deep in the Earth, suggesting that the international industry is mistaken in its understanding and should now look in crystalline rocks and beneath volcanoes. Most experienced explorers would dismiss the idea. In any event, we should not rely on such an untested hypothesis for future supplies. Not surprisingly, Kenney's views are promoted by Odell, an economist who dismisses resource constraints to production.

3. Krayushkin, V.A. *et al.,*1994, provides the original paper behind Kenney's paper.

4. Vlierboom, F.W. *et al.* 1986 gives an explanation of the Siljan drilling in Sweden.

5. In the S.I. system, it is measured in mPa.s or milliPascal-second.

6. The official (UNITAR) classification terms oil with a gravity below 10° API *Extra Heavy Oil.*

7. Bitumen (tar) is a term applied to oils with a viscosity above 10 Pa.s (10 000 Cp), and asphalt is a term for bitumen with a viscosity above 20 Pa.s.

8. Demaison, G.J. and G.T. Moore, 1988, have investigated the sediments being deposited by the Humboldt current. Demaison emanated from the French school of geochemistry but worked for Chevron in California.

9. Klemme, H.D. and G.F. Ulmishek, 1991, provide a very valuable insight into the world's source rock realms, which forms the basis for the US Geological Survey assessment.

10. In fact, the most prolific provinces, if oil that has been subsequently degraded is included, are in northern South America; Western Canada and Siberia, where huge tar sand and heavy oil deposits occur. A very effective seal is as important to the Middle East as is a source.

11. The term "cracked" means that the bonds holding the atoms together in oil molecules are broken down at high temperatures to yield simpler gas molecules.

12. Bjørlykke, K., 1995, explains the complex process of oil migration in a very readable paper.

13. Heum, O.R., 1996, provides further information on modern concepts of oil migration.

14. Macgregor, D.S., 1996, has published an excellent discussion of this subject including a listing of giant fields.

15. see an excellent paper by Doré and Lundin, 1996, whose Fig. 14 explains the West-of-Shetlands petroleum system without specifically naming it.

Chapter 3
THE PIONEERING EPOCH

The last chapter was rather heavy going technical stuff, but important reading all the same if one is to gain an appreciation of the physical constraints to oil production. There are not great underground caverns full of oil waiting to be tapped. It accumulates in the minute pore space of rocks under complex geological circumstances, which only rarely come together in the right combination. One weak link in the geological chain converts a promising prospect into an expensive dry hole. In the last chapter, I explained how flush oil generation occurs when the critical temperature is reached in the source-rocks, and how it progressively consumes the amount of organic raw material available to convert. The result is a classic bell-shaped curve of generation, starting at zero, rising to a peak and then falling back to zero. This is a pattern which will recur in this story in different contexts: it is well to get used to it, for it is immensely important.

The world is likely to reach peak production around the end of the Century, only a few years away. Our experience in most countries has been of the left-hand side of the curve, when production was rising. We can learn much from it before we face its mirror image when production makes its inevitable decline.

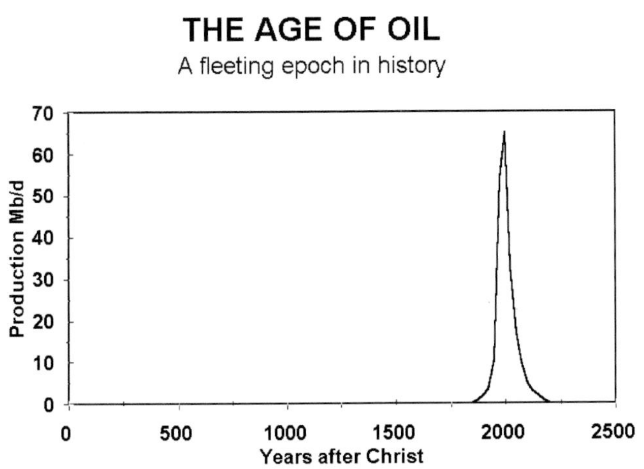

THE AGE OF OIL
A fleeting epoch in history

Fig. 3-1. The age of oil: a fleeting epoch in history.

While many of the details are fuzzy, no one can doubt the generality of the picture of the oil age in Figure 3-1[1].

Considered in greater detail, it can be usefully divided into four epochs: *pioneering* to 1950; *growth* to 1970; *transition* to 2000 and then *decline*.

In this chapter, I will describe the Pioneering Epoch when most of what we know about oil was worked out and when most of the technology needed to extract it was developed. There have been important subsequent advances, but the back was broken during the pioneering epoch.

Oil and gas from surface seepages have been known since long before the time of Christ. Moses' basket of reeds was caulked with tar; Nebuchadnezzar's fiery furnace and the burning bush were, it may be assumed, located on gas seepages; and the eternal flames that were worshipped in antiquity were flammable oil impregnated shales near Baku on the shores of the Caspian[2]. I remember in the 1960s seeing Papuan natives in the highlands of New Guinea daub their bodies with oil collected from seepages when they gathered for their ritual *sing-sing* dances, a custom that had being going on from time immemorial.

It did not take much to move from skimming oil off the pools into which the natural seepages ran, to start digging shallow pits to extract it more efficiently. It was used primarily for medicinal purposes. The early Burmese were the most advanced in this regard, using bamboo pipes to case shallow wells and transport the oil. The Romans and Chinese were also exploiting oil before the time of Christ. There has also been a very long history of constructing wells for water and salt brine; salt having been a very valuable commodity in the Middle Ages and before. The early Chinese are credited with the development of a form of rig for drilling wells, consisting of a heavy stone on the end of a rope that was repeatedly raised and dropped, slowly punching a hole into the earth. It was the precursor of the *Cable-Tool* with which the early oil wells were drilled.

A trade in oil had been established both in Europe and the United States long before wells were drilled specifically for it[3]: there was much demand for domestic illuminants derived from lard, whale-oil, camphene, as well as oil from seepages and coal workings. The distillation of cannel coals in Scotland in the last century was particularly noteworthy as it involved the process of refining. The growing demand for this lamp-oil, coupled with the decline in whale oil due to over-whaling, were the stimulants for the search for oil. The technique for drilling wells was already well established in North America in the production of brine, from which to extract salt needed for preserving meat. There were even cases where such wells encountered gas, which was used to fuel the salt works.

The kerosene lamp was a great revolution in the way people lived, adding a useable evening to the working day, especially in rural areas. A second and greater revolution came on July 3rd 1882, when Carl Benz in Germany powered the first automobile with the internal combustion engine, which had been invented a few years before by Nicholas Otto. At first, it used carburetted benzene distilled from coal but soon turned to gasoline refined from crude oil. The automobile developed an unquenchable thirst for oil.

Fig. 3-2. In 1882, Carl Benz powered the first automobile with the internal combustion engine.

Courtesy of Mercedes Benz

THE BIRTH OF THE OIL INDUSTRY

The birth of the oil industry is generally attributed to the famous well drilled for oil in 1859 by the self-styled Colonel Edwin. L. Drake at Titusville, Pennsylvania, although it is now claimed that F.N. Semyenov actually got there first with a well on the Apsheron Peninsular, near Baku in Russia, eleven years before[4]. The first *production* statistics in Romania go back to 1854. It does not really matter who wears the crown, for in any event the oil industry grew rapidly in the succeeding years in both Pennsylvania and on the shores of the Caspian[5]. Much has been written on the early history of the oil industry, and it does indeed make colourful reading. The oil pioneer, Beeby-Thompson[6], arriving at Baku in 1898, for example, wistfully notes that:

"The Caucasian women, about whose beauty I had heard so much, were disappointing, for all the most lovely creatures had long ago been sold to Turkish harems, leaving the least attractive for reproduction".

Fig. 3-3. In 1859, Drake's well in Pennsylvania launched the American oil industry.

From Pennsylvania Historical & Museum Commission Drake Well Museum Collection, Titusville Pennsylvania

Fig. 3-4. Early oilfields at Baku, another important source of oil.

Here, I will be content to summarise only the highlights of this exciting epoch, concentrating mainly on the discovery of oil; and the companies and people that were responsible.

Drake's well, which encountered oil at a depth of 67 feet[7] in an Upper Devonian sandstone, led to the first great oil exploration boom as the shallow reservoirs were tapped by an army of pioneers and speculators who descended on the oil lands of the Appalachian Basin of the United States. Stills were erected nearby to make kerosene, which within the remarkably short span of two years was already being exported to Europe. Fortunes were made, and as many were lost. Prices fluctuated wildly: from its outset, the industry has been plagued by "boom or bust". The reason is the special depletion pattern of oil, which flows rapidly under its own pressure from the wellbore as soon as it is tapped in a manner very different from the labourious process of mining, for example, coal. Depletion is a theme that will reappear frequently in the following pages. The early explorers drilled by guesswork, although soon began to develop an empirical understanding of geology, identifying the trends where drilling succeeded. The State of Pennsylvania did appoint a geologist to investigate; and his report of 1865 observed that oil tended to accumulate in anticlines, where the strata are folded into an arch, and went on to state that

"Petroleum is a bituminous liquid resulting from the decomposition of marine and land plants" [8].

In other words, the essential geological constraints had already been understood in America within six years of the first discoveries in the country. It did not take them long to figure out the basics[9].

Looked at with to-day's eyes, it is rather surprising that these ancient rocks, deposited over 350 million years ago, should still be oil-bearing so close to the surface. No seals are perfect, and this was a long period of time over which the oil could have seeped away. Probably the explanation is that, although the source of the oil is ancient, generation was not achieved until much later after long exposure to low temperatures. The Appalachian Basin is an interesting example of depletion. The oil boom was already over by 1900 by which time 183 oilfields had been found, yielding an ultimate recovery of 1.33 billion barrels[10]. It is in fact quite a small province, notwithstanding its early importance, as its total contribution would be enough to supply the world's present demand for less than a month. It is noteworthy that it still has reported reserves of 28 million barrels, showing how production declines exponentially, such that old fields continue to produce a few barrels a day for a very long time during the tail-end of their depletion, and how ever smaller fields continue to be found even in mature areas. To-day, we hear a lot about technological progress, with much play made of the importance of horizontal wells and multi-lateral wells, in which several branches are drilled from a single initial borehole. It is sanguine therefore to learn that both techniques were already being used in Pennsylvania before the turn of the Century[11], albeit with by no means to-day's technical sophistication.

STANDARD OIL

In the same year as Drake drilled his well, a man by the name of J.D. Rockefeller went into partnership with a newly arrived British immigrant, Maurice Clark, to establish a trading company in Cleveland, Ohio. It was an opportune moment with the Civil War of 1861-65 creating a demand for goods of all sorts, and it was an opportune place connected by two railways and the Great Lakes navigation system. The new firm soon turned to trade in kerosene which before long led it into the refinery business. In this way started the great Standard Oil Company which grew to become the world's largest corporation, being run on ruthless business lines. It was the precursor of the modern company with its bureaucracy, driven solely by the motive of return on investment. It was a marketing company: a market it sought to control by fair means and foul. It succeeded in doing so by placing a stranglehold on oil transport both by securing the pipelines and negotiating special rebates from the railways. The wild fluctuations in oil price were anathema to its orderly plans. Even in these early years of the industry, a need for regulation had already arisen: Standard Oil was in fact exercising a function no different from that subsequently applied by the Achnacarry Agreement[12], the Texas Railroad Commission[13] or OPEC.

Fig. 3-5. J.D. Rockefeller, ruthless founder of Standard Oil, and philanthropist.

Courtesy of Exxon Corporation

Fig 3-6. Henry Flagler, Rockefeller's associate who had a key role in the creation of Standard Oil.

Courtesy of Exxon Corporation

Standard was reluctant to enter the hurly-burly of exploration, although it eventually did so in the 1880s when another new oil province was found in Indiana. Its motive in doing so was to protect its existing market from competition. By then, the Pennsylvania fields were beginning to decline, the first example of the depletion of this natural resource in an oil province. By 1885, the State Geologist of Pennsylvania had already stated:

"the amazing exhibition of oil is only a temporary and vanishing phenomenon – one which young men will come to see come to its natural end"

He was both right and wrong in his prognosis: he was right about the area he knew but wrong insofar as he did not know how much oil would be found in new areas. Now a century later, we have a much better idea of that issue, and can confidently repeat his words on a global scale.

Standard Oil's ruthless capitalism was much reviled by the independent oil producers who regarded it as a creeping octopus that would eventually ensnare and devour them. It was particularly disliked in Texas and the southern States, still smarting from the Yankee victory in the Civil War. Pressure against it grew, until in 1911 the government was forced to break it up under anti-trust legislation, a democratic response to an overweening monopoly, probably unrivalled save for the unswerving centralism of

the Soviet Union. After the break-up, some of Standard's daughters, Esso, Chevron, Mobil, Amoco, Conoco, Sohio and Arco, to name the largest of the thirty-seven, grew to become some of the world's most important oil companies in their own right.

Although they did practice successful exploration throughout the world, in some cases pioneering new projects in the best traditions of the explorer, it can probably be said that they owed most of their growth to their great

Standard Companies in 1911	External merger & liquidation	Today
SOC (California) SOC (Kentucky)		CHEVRON
Swan & Finch	liquidated 1965	
SOC (New York) Vacuum Oil Co		MOBIL
Waters Pierce	dissolved 1940	
Borne, Scrymser Co		Borne Chemical
Galena Signal Oil Co Cumberland Pipe Line Co Southern Pipe Line Co	Ashland 1917	Ashland Oil Co
Chesebrough Mfg Co		Chesebrough-Ponds
The Crescent Pipe Line Co	liquidated 1925	
Prairie Oil & Gas Co Atlantic Refining Co	Sinclair 1917 Richfield 1955	ARCO
SOC (Kansas) SOC (Indiana) SOC (Nebraska)	Pan American 1930	AMOCO
Washington Oil Co	liquidated 1976	
SW Pennsylvania Pipe Lines National Transit Co South Penn Oil Co Eureka Pipe Line Co		Pennzoil
Union Tank Car Co		Trans Union
Colonial Oil Co	liquidated	
Indiana Pipe Line Co Buckeye Pipe Line Co Northern Pipe Line Co New York Transit Co	Pennsylvania Co '52	Penn Central
Anglo-American Corp SOC (New Jersey)		EXXON
Continental Oil Co.	Maryland 1917 Du Pont	Du Pont (Conoco)
SOC (Ohio) Solar Refining	BP 1938	BP
The Ohio Oil Co	US Steel	US Steel (Marathon)

Fig 3-7. Standard's daughters.

financial strength, inherited from Rockefeller's empire which allowed them to swallow their weaker competitors. I think that the major companies have always been traders at heart, with making money being their sole objective: nothing wrong with that of course – save when one comes to consider the depletion of a finite resource that should perhaps be governed by more sophisticated criteria to better respect the value of this irreplaceable resource to Mankind.

Standard Oil's control was centred on the eastern States, but new oil lands were being explored in the south. California came in first with several important discoveries before the turn of the Century. Unocal was the dominant producer, and has managed to preserve its independence until today. The complex geology of California where the oil occurred in strongly folded and faulted Tertiary rocks, partly affected by the famous San Andreas Fault, prompted a greater attention to scientific geology here than elsewhere, and the companies in California took on increasing numbers of professional geologists to guide their efforts. Later, Standard Oil did move in, and its affiliate, Standard of California, now Chevron, has had an exceptional reputation in exploration, in due course bringing in Saudi Arabia.

Oil was found in Texas when a water well at Corsicana unexpectedly encountered oil in 1893. It was followed in 1901 by a spectacular blow-out at Spindletop near Beaumont, in which seventy-five thousand barrels a day gushed high into the sky. It was a mighty roar that heralded another oil boom, opening up a new province. It transformed the economic life of the United States, and indirectly contributed to its world political power. Discovery followed discovery as the new trends were drilled up, but eventually it peaked in the 1930s. Today, only a few million barrels are found each year and they in very small fields. The total endowment of Texas, again resulting mainly from the early discoveries, is about 60 billion barrels, about the same as the North Sea. The discovery at Spindletop spawned two more major companies, Texaco and Gulf Oil, the latter now taken over by Chevron, following an unsuccessful assault by a clear thinking corporate raider, Boone Pickens[14]. He correctly realized that Gulf's past was worth much more to its shareholders than its future. It is a reality that most other major oil companies now face, with varying degrees of frankness.

TEXACO - a hard nosed company

The Spindletop discovery well was drilled by a one-armed, self-educated mechanic, named Patillo Higgins, who was backed by Captain A.F. Lucas, an immigrant from Yugoslavia. They eventually sold out to what became Gulf Oil. Another man on the scene was Joe Cullinan, known as "Buckskin Joe" for his abrasive manner. He set up the Texas Fuel Company in Beaumont to trade in oil and oilfield equipment. He was backed by New York and Chicago investors, managed by a retiring German immigrant, by the name of Arnold Schlaet[15], who exerted a considerable influence on the Company. In 1906, the name Texaco with its logo of a green "T" overlying a red Texas Star was registered. Buckskin Joe knew his job and ran the business with an autocratic style that did not endear himself to his investors who eventually ran him off. Some say that the Company has retained a characteristically autocratic manner ever since, which perhaps owes something to its founder.

As the business grew, Texaco moved its headquarters to New York, and expanded its upstream operations throughout the United States, being more successful than many of its competitors. Because much of the US market was controlled by Standard, Texaco paid especial attention to overseas markets. In 1936, its global marketing strategy prompted it to join forces with Chevron in the Eastern Hemisphere, creating the Caltex Group. With hindsight, this

can be seen as one of those rare transcendental corporate decisions with incredible and unforeseen consequences transforming a company. As a result, it gained access to Chevron's rights in super-rich Saudi Arabia, becoming a founder member of Aramco. Caltex also found the Minas

Fig. 3-8. Joe Cullanin, "Buckskin Joe", (1860-1937) founder of Texaco.

Courtesy of Texaco

Fig. 3-9. Spindletop on fire: the field that opened the southern United States.

Field in Sumatra, the largest in Indonesia, which could not, however, be developed until after the Second World War. The oil supply from Caltex would meet Texaco's needs for a long time to come, which probably explains its own rather indifferent performance as an international explorer. It had no need to find more, save for strategic reasons to reduce its dependence on Saudi Arabia.

The mate of the first tanker to leave Spindletop with Texaco oil was Torkild Rieber, a tough Norwegian seaman, born in Voss near Bergen in 1882. This man built up the Company's tanker fleet, and by 1935 had risen to be Chairman of the Board. He was a ruthless businessman, intent on building a market in Europe, where he admired the efficiency of the growing Fascist movements, which he supported. He was by no means alone in such a view: more European businessmen than later cared to admit it shared this respect for power and decisiveness. In Spain, he supplied Franco's forces in contravention of the US Neutrality Law, and did so on credit, being eventually rewarded with a dominant share of the Spanish market for much of Franco's life. This connection brought him into contact with leading German Nazis with whom he struck deals to swap tankers built in Germany for oil, which continued to be supplied after the Second World War broke out. His close ties with the Germans were exploited by the latter's intelligence service who established a presence in Texaco's New York office. This group sent intelligence about convoy sailings, disguised as corporate patent numbers, through the Company's cable office to Germany, and sought also to influence American politicians to the German cause. The plot was unmasked by British counter-intelligence, and the Captain was eventually forced from office after America came into the war[16]. Texaco's black and red livery and its red star does seem to have a slightly Germanic touch to it even now.

Fig. 3-10. Captain Rieber (b.1882 in Norway).

Texaco, like most other American companies, was operating in a cut-throat environment in the United States. Oil production was rising, and competition for market was fierce. It occupied most of the Company's attention. Growth was achieved primarily by buying up competitors, and for a period many of the major companies had large portfolios of partly owned affiliates. The deals were often complex and contentious, and some of the directors of swallowed companies remained on the board of the affiliates representing minority interests. Litigation was the order of the day. Texaco's half interest in Caltex was the jewel in its crown. Robert E. King, based in New York, was its well respected Exploration Manager. The corporate history[17] makes no mention of Texaco's own Exploration Manager; perhaps, there was no such title. In the post-war years, Texaco in its own right achieved one of its few international exploration successes when it found the Orito Field in Colombia in the 1960s, which led to a string of finds in neighbouring Ecuador, in a joint venture with Gulf.

It also had a modest success in the UK North Sea. Otherwise, it seems to have followed a policy of obtaining its reserves by acquiring other companies including Seabord in Venezuela, Trinidad Leaseholds, and Getty: the latter landing it in an expensive law suit with Pennzoil. Later tax

Fig. 3-11. Augustus C. Long, (b.1904) a post-war Chairman, with an autocratic style.

regulations led to the consolidation of affiliate interests, and Caltex was subsequently absorbed by its parents.

An interesting development in the marketing area was the Company's decision to sell in 1985 a one-half interest in its refining and marking organization in 33 US states to a Saudi company, after Aramco was expropriated[18]. It was a noteworthy link between a producing government and a downstream organization that made sense for both parties. It is surprising that it did not set a precedent for other major companies. Perhaps only the hard-nosed Texaco was willing to let a cuckoo into the nest[19]. It would not be surprising if there was a strong Saudi shareholding in Texaco.

In a recent article[20], the Company explains a new strategy of concentrating on a few key fields, returning in fact to its US homeland where it aims to buy out its partners and conduct highly efficient operations with a slimmed-down staff. In effect, it wants to behave like an independent producer, as in earlier years, rather than as a global explorer.

The climactic moments in Texaco's life were its spectacular birth at Spindletop which opened up Texas; its decision in 1935 to join forces with Chevron in the Eastern Hemisphere that delivered Saudi Arabia; followed in turn by the loss through sequestration of that country's production in 1979. It seems never to have quite had its heart in international exploration, although from time to time has made rather transitory forays. The most successful of these was opening up the Sub-Andean basins of southern Colombia and Ecuador, which involved some stirring exploits in the Amazon jungles. It seems to be the weakest of the former sisters, and some analysts wonder about its future.

OTHER EARLY OIL PATCHES

The early development of the oil industry in the United States had a lasting world influence: the large American companies expanded overseas taking their business and technical culture with them. In technical terms, the industry's American roots has left its legacy, including for example its units of measurement: the traditional well-casing sizes of $9\frac{5}{8}''$ and $13\frac{3}{8}''$ are still in almost universal use as are such colourful drilling terms as *roughneck, rathole, kelly bushing,* not to mention a piece of equipment delightfully known as a *Donkey's Dick.* But the United States was not by any means the only pioneering oil country. During the last Century, there were already developments in many places including: Baku in Russia; Borneo; Burma; Sumatra; Romania; Poland; Trinidad; Peru; and Mexico. Of these, the most important was Baku on the shores of the Caspian, a backward and poorly administered territory on the southern fringe of the Russian Empire. Oil extracted from hand dug pits had been a state

monopoly, but in the early 1870s, the area was opened up to private capital. One of the first entrepreneurs to arrive on the scene was Ludwig Nobel, a member of an inventive Swedish family which made a fortune out of dynamite. It is now remembered by the Nobel Prize, which it endowed.

The world's first tanker, the *Zoroaster,* was developed by the Nobels to transport oil on the Caspian, in the absence of sufficient oak to make traditional barrels. The Rothschilds Bank later came in to finance a railway to Batumi on the Black Sea opening an export route to the West. They in turn were followed by Shell, which started exporting Baku kerosene in tankers to Europe and the East. That company later merged with Royal Dutch which had been pioneering oil production in the Dutch East Indies (now Indonesia) to become the giant Royal Dutch/Shell or simply Shell as it is known today.

The Baku oilfields lie in a Tertiary basin in front of the Caucasus. The geology is characterised by complex folds and faults, and multiple reservoirs.

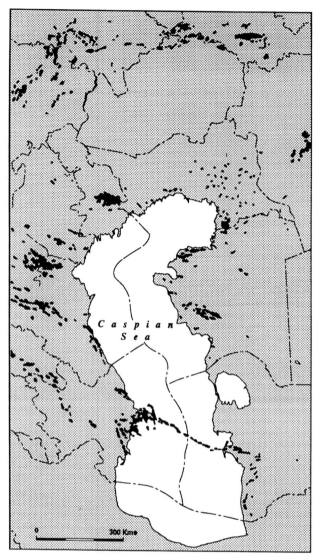

Fig. 3-12. Map of Caspian: one of the first and last major oil plays.

Fig. 3-13. A sailing tanker: the bark "Brilliant" that carried a cargo of 30 000 barrels of oil.

Courtesy of Exxon Corporation

Seepages of both oil and gas were abundant. A peculiarity of the geology was the presence of numerous so-called mud-volcanoes, as also found in Romania, Colombia

Fig. 3-14. Marcus Samuel, a London merchant, who built Shell Oil.

Shell International Photographic Services, London

Fig. 3-15. Henri Deterding, dynamic head of Royal Dutch, which merged with Shell in 1907.

Shell International Photographic Services, London

and Trinidad. They are mounds of mud, up to several hundred feet in height that form over active gas seepages. They sometimes explode and catch fire. Beeby-Thompson[21] describes how he commenced geological investigations of the area, but no doubt much was found by hit-or-miss. It was evidently easier to hit than to miss in this prolific area that in 1900 produced as much as 75 million barrels from 1700 wells less than 1000 feet deep. It was a violent place of banditry and strife, with appalling operating conditions. No less a figure than Joseph Stalin was a workers' leader, masterminding strikes and disturbances in Baku in the early years of the Century. Such pressures spread and culminated in the Bolshevik rising of 1917, which, to put it mildly, transformed the world's political scenery. It was not to be the last occasion on which oil shaped human destiny: the most important of which is about to come. The Bolshevik Revolution effectively brought to a close the first Caspian oil boom. I say the *first* advisedly, for today Western and Russian oil companies are endeavouring to bring about a second Caspian oil boom, if and when they can settle the pipeline export route and many contractual issues. They aim to produce the offshore fields that were not developed by the Soviets. Indeed, this second bite at the Caspian cherry may offer the world the last hope of bringing major new

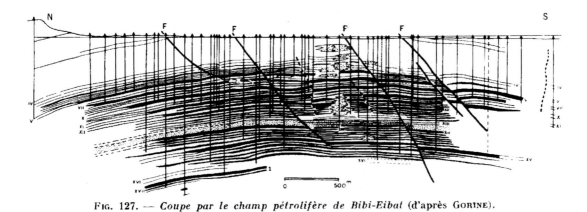

Fig. 127. — *Coupe par le champ pétrolifère de Bibi-Eibat* (d'après GORINE).
1 = Horizons de sables pétrolifères (IV-XVII) ; 2 = Brèches et anciens volcans de boue enfouis ; F = Failles.

Fig.3-16. The Bibi-Eibat Field near Baku, characterized by multiple thin reservoirs.

production onto the market. But there are doubts that much will ever leave the Asian mainland once demand in the Former Soviet Union recovers. It is noteworthy that some of the Western pioneers in this project have got cold feet, selling out to the Japanese, who are very conscious of their strategic oil supply[22]. The Caspian is not of course by any means *new* in resource terms, being the site of some of earliest discoveries, and its late potential arises solely because it was isolated for political reasons. It is sometimes claimed that the Caspian may be another Saudi Arabia, but such claims are likely to prove greatly exaggerated[23].

The greatest oil province of all, the Middle East, was also attracting attention as the last Century drew to a close, but to do business there was difficult. Most of the area was controlled by the Ottoman Empire with its decadent and corrupt Sultan in a harem surrounded by eunuchs. The rest was in the hands of the Shah of Persia whose authority barely extended beyond his own capital.

The Germans became interested in building a railway from Berlin to Baghdad as part of a foreign policy initiative aimed to[24] catch up with the colonial expansion of France and Britain in other parts of the world. As a land power, it recognized the military mobility afforded by railways, which it conceived would be more effective than the slower British sea-power. It secured to this end a concession in Anatolia and Mesopotamia, which included mineral rights for twenty kilometres on either side of the track, presumably as a source of building stone. Its engineers soon reported the numerous oil seepages which they encountered in the Mosul area of what is now Iraq. Although the Sultan was alerted to the possibilities, and retracted the rights from the German railway company, he was too idle and ill-informed to do much about it.

At about the same time, the head of Persian Customs, General Antoine Kitabgi, hearing of the growing oil interest in the vicinity, resolved to see if he could let an oil concession in his country. After one or two false starts, he managed in 1900 to bring it to the attention of William Knox D'Arcy, an entrepreneur who had just returned to London from Australia where he had made a fortune in

Fig. 3-17. William Knox D'Arcy (1849-1917) whose initiative in Iran founded what is now BP.

Photograph courtesy of British Petroleum

gold-mining.

He saw the possibilities, and eventually secured the rights: a £20,000 signature bonus being the main inducement to the impoverished Shah. Drilling commenced in 1902 under appalling conditions, with summer temperatures of over 110 °F., but success was slow in coming. D'Arcy was becoming overextended, but was encouraged when the British government started to take an interest in his project. Britain had always sought to deter Russian expansion into the Middle East to protect its trade route to India and the East, and oil was now a new factor. D'Arcy's immediate problems were, however, partly solved when the old established Burmah Oil Company, based in Glasgow, agreed to take a share in his company.

In January 1908, after six long years of travail and

disappointment, the third well was spudded at Masjid-i-Sulaiman (The Mosque of Solomon) in the Zagros foothills: it was pretty much the last throw of the dice. By May, the well was down to 1000 feet without results, and a cable to suspend operations was received from London. The local manager, G.B. Reynolds, however, decided to continue until he received written confirmation. His initiative was rewarded at 4.30 a.m. on the morning of May 26th, when the well blew out throwing a jet of oil fifty feet into the air[25].

Fig. 3-18. The well that opened the Middle East in 1908 and changed the World.

Photograph courtesy of British Petroleum

In World resource terms, it was a climactic event. As Dr Ziegler explains (p. 25), the World contains no more than about thirty significant petroleum systems, that is to say provinces with the unique set of geological circumstances to yield prolific oil. This discovery in 1908 found the largest[26]. In terms of such petroleum systems, as opposed to individual fields, half the world's oil had now been found, although that would not of course become known until much later. It was undoubtedly a turning point in history.

Another turning point of a different sort was about to unfold in 1914: the First World War. Britain's last major naval engagement had been the Battle of Trafalgar in 1805, a critical action in the Napoleonic wars. At the height of Empire, the British Navy was the corner-stone of Britain's power, but by the turn of the Century it had become more of a symbol, with polished brass, holystoned decks and smartly dressed crews, than an efficient fighting machine. It was just this pageantry that so impressed the Kaiser, himself

an honorary admiral in the British Navy, when he came to take part in his favourite sport of yacht racing at Cowes[27]. "Why?", he asked himself, "does Germany not have such impressive battleships, dressed over all in flags and illuminated by night, to stand guardians for forthcoming yacht races at Kiel". From this harmless vanity began the Anglo-German arms race, as each new German warship had to be matched by a British one to maintain the balance of power. Gradually the emphasis changed from the pomp and splendour of the Marine band on the quarter deck to actually making the thing a lethal weapon, able to out-speed and out-gun its competitor. The maverick Admiral Fisher was dedicated to this transformation. He realized that his ships would have to convert to oil fuel to obtain the performance he expected, but his proposals met resistance. For the first time, but certainly not the last, as this book will explain, the issue was *security of oil supply*: Britain had no oil of its own, and was reluctant to rely on American oil or even Shell oil with its Dutch connection. What it needed was its very own supply that it could control. Winston Churchill, then the First Lord of the Admiralty, concluded that Persian oil was the answer for two reasons: first, to supply the Navy; and second, to strengthen a British presence in the Middle East to deter the threat of German or Russian expansion in that area. The government took up a fifty-one percent interest in the Anglo-Persian Oil Company, later to become British Petroleum, or BP, royal assent being granted six days before war broke out[28].

The War opened with cavalry charges, as plumed Uhlan lancers galloped into action, but it ended with tanks driven by internal combustion engines running on fuel refined from crude oil[29]. Oil became the great new driving force of the world, changing the meaning of the term, horse-power. But in terms of oil resources, perhaps the most significant feature of the First World War was that Turkey backed the losing side. Had the country been an ally or neutral, events would have turned out very differently[30]. As it was, Britain had a motive to encourage Arab nationalism[31], which effectively resulted in the creation of the state of Saudi Arabia out of a tribal desert. It was to have far reaching economic and political consequences that have still to be played out, as this book will discuss in later chapters. The defeat of the Turks led to the break up of the Ottoman Empire into new administrations, of which Iraq, Kuwait and Saudi Arabia are the most important. Not far below the surface, was the division of the region's oil rights to the three victorious allies, Britain, France and the United States. As I will discuss later, Turkey may yet again come to exert its considerable influence in the area, perhaps even renewing its old alliance with Germany.

The first solution for the division of oil rights was to share them. This was achieved by the formation during the 1920s of the Iraq Petroleum Company, owned by Shell, BP, Compagnie Francais des Pétroles (CFP, now Total), Mobil and Esso, not to forget the legendary Calouste Gulbenkian (Mr 5%) who put the deal together (see p. 44.) This group had what is called an Area of Mutual Interest (AMI) agreement that prohibited independent activities by the partners in the area of the former Ottoman Empire. It became known as the Red-Line Agreement covering all the productive Middle East territories outside Iran and Kuwait[32], and was the cause of bitter conflict for a long time to come.

Chevron, which was not restricted by the Red-Line Agreement, took up rights in Bahrain in 1929, to be later joined by Texaco (see p. 35), and struck oil there two years later. This find, coming from Tertiary sandstones at fairly shallow depth, was itself comparatively modest, but it was nevertheless of immense importance, for Bahrain lay only a few miles off the coast of Saudi Arabia. Up to that point, interest in oil had been concentrated on the huge folded structures of the Zagros Foothills in Iran and Iraq that were obvious surface features visible for miles around.

Fig. 3-19. An anticline in Iran, an obvious oil prospect.

Photograph courtesy of British Petroleum

Many geologists, seeking analogues for this familiar type of prospect, were then sceptical of the platform province to the west of the Persian Gulf, where the strata were largely obscured below sand dunes and, where seen, were flat-lying or, at most, shallow dipping. At first sight, it seemed to lack adequate structure to provide large traps for oil. So, the discovery of oil on Bahrain on the edge of this new province carried immense implications, which were at once recognized. Chevron began negotiating for rights in Saudi Arabia, partly through an eccentric and disaffected Englishman, by the name of Harry St. John Philby. He was trading in Jidda, and was, remarkably enough, no less than the father of the infamous British double-agent, Kim Philby. King Ibn Saud, himself a British protégé from the War, was desperately short of money, and Chevron clinched the deal in 1933 with delivery of thirty-five thousand gold sovereigns that were shipped to Arabia in

seven boxes aboard a P&O liner. It was a substantial and risky investment at the time, for no one could have imagined that Saudi Arabia would become the world's most prodigious oil province, with an ultimate endowment of about 300 billion barrels, sixteen percent of the world's total. Chevron, when it later found that it lacked the resources to develop the area single handed, brought in Texaco (see p. 35), followed in 1947 by Mobil and Esso, the latter two in flagrant disregard for the famous Red-Line Agreement. This grouping formed the Arabian-American Oil Company (Aramco), the emphasis being on the second word. In pre-war days, Britain under its imperial mantle had successfully exerted an almost exclusive influence throughout the Middle East, but in its weakened and socialist post-war state had loosened its grip in favour of the United States. Ibn Saud, absolute ruler of a feudal and primitive country that was little more than his private estate, effectively became an American satrap. The further evolution of this remarkable and extraordinary situation has yet to unfold with or without the House of Saud[33]. I will return to the issue in later chapters.

While rights to Saudi Arabia were being negotiated, BP and Gulf turned attention to Kuwait, which lay also on the western shore of the Persian Gulf and outside the Red-Line Agreement. They eventually decided to join forces rather than compete for the territory, and signed a lease for it in 1933. It completed the primary carve-up of the Middle East.

Although by far the most important, the Middle East was by no means the only oil territory being explored and developed. Most progress was in the Western Hemisphere, especially in the United States itself which was already becoming a fairly mature province, but also in Venezuela and Mexico, where impressive finds were made. Shell, which had rather missed out on the carve-up of the Middle East, took up a strong position in the Western Hemisphere, competing successfully with the major American companies. Generally smaller scale operations were also taking place in many other countries. By 1935, twenty-five were in production, of which the seven largest are shown in Figure 3-20. The United States was producing sixty-four percent of the World's needs. The Middle East was barely represented: Iran, the largest producer, was in seventh place with only two percent. The other countries were, in decreasing order: India (incl. Pakistan); Poland; Peru; Colombia; Argentina; Trinidad; Japan; Sarawak; Brunei; Iraq; Canada; Germany; Egypt; Sakhalin; Ecuador; France; Italy; Czechoslovakia and Bolivia[34]. What a different world it was !

Seven major companies, comprising Shell, BP, Esso, Mobil, Chevron, Texaco and Gulf, later dubbed the "Seven Sisters" by Enrico Mattei, the Italian oilman, had already brought world supply under their control. With most oil coming from the United States, security of the supply was not a serious issue, although one not altogether without concern. The Soviet Union was closed to foreign companies, and those with rights from the pre-war days in Baku were formally expropriated in 1928. Mexico also

USA	18 Mb	64%
USSR	3.4	12%
Mexico	1.9	7%
Venezuela	1.2	4%
Dutch East Indies	0.7	2%
Rumania	0.7	2%
Iran	0.6	2%

Fig. 3-20. Oil production in 1935.

ousted the foreign companies ten years later as a nationalist movement, somewhat akin to the Fascist influences of contemporary Europe, gained political ascendancy, believing perhaps with some justification that foreign influences were becoming excessive. Its oil industry was placed in the hands of a state enterprise PEMEX, the first example of the state oil company that was later to be copied widely. These expropriations were harbingers of what was to come.

CALOUSTE GULBENKIAN
oil man par excellence
"If you can't bite the hand that feeds you, kiss it"

This immortal dictum, with which he is credited, sums up one of the most remarkable oil men of all time and also underlines a common feature of oil deals: the joint venture between several companies, each having an undivided interest in a concession. The management of undivided interests calls for compromise and negotiating skill, as families who inherit property know to their cost, and Gulbenkian needed both to hold onto his 5% in his brainchild and life's work, the Iraq Petroleum Company.

Calouste Gulbenkian[35] was the son of a prominent Armenian businessman in Istanbul, who had made a fortune as an importer of Russian kerosene. He was educated in France and England, where in 1887, at age nineteen, he secured a first-class degree in engineering at London University. Although he had some aspirations to further studies, evidently having serious intellectual leanings, his father sent him off to Baku to learn about oil, the foundation of the family fortune. On returning, he wrote several well received articles on his experiences in *Revue des Deux Mondes,* a respected French journal, which established him as something of an oil expert in the Turkish capital. When the government became aware that oil seepages had been encountered in Mesopotamia (now Iraq) by a German group, backed by the Deutsche Bank, which was surveying a proposed railway, it asked him to investigate. He produced a comprehensive report, which fired his imagination to somehow secure rights to this oil himself. It was a finally successful quest that was to occupy him for many years.

But in 1896, the Sultan embarked on a ferocious campaign of what would now be called ethnic cleansing, in which large numbers of Armenians were massacred. The Gulbenkian family was forced to flee to Egypt. A chance meeting on the boat with a fellow refugee, Alexander Mantashov, who was an influential figure in Russian oil, introduced him to the right circles in Cairo. They included Sir Evelyn Baring of the then famous, but recently discredited, British family bank. This contact proved useful when he moved to London in the following year to represent Russian oil interests. He soon became a close associate of Henri Deterding, the head of Shell.

Fig. 3-21. Calouste Sakis Gulbenkian, oilman.

Gulbenkian Foundation

In 1910, the Sultan was overthrown by the Young Turks who were bent on modernizing the country. They invited a group of London financiers and Gulbenkian to establish a national bank. Gulbenkian then persuaded the bank to form an alliance with the Deutsche Bank and Shell to establish the Turkish Petroleum Company in London, Gulbenkian retaining 15% for himself. Later, British government pressure brought about the entry of BP, reducing his stake to 5%. A concession for the Mosul oil lands was granted on June 28th 1914, a few weeks before war broke out, whereupon the Deutsche Bank's interest was taken over by the British Custodian of Enemy Property.

With the advance of Allied forces, the future of the Ottoman Empire was settled by France and Britain under the so-called Sykes-Picot Agreement of 1916. It provided, amongst other clauses, that the Mosul area with its oil potential should fall within the French sphere of influence, but the British eventually recovered control provided that France could participate in the oil development. A further complexity was introduced when Woodrow Wilson, the American President, proclaimed at the peace treaty the worthy principle of self-determination, which was not what France and Britain exactly had in mind for the Middle East. Britain in particular reneged on the promises it had made for wartime expediency to create an Arab nation. This American pressure eventually led to the somewhat artificial creation of the state of Iraq. However altruistic the proposal of self-determination was as a general principle, it was not long before American oil companies with the support of their government were pressing for a share of Mesopotamian oil. Difficult negotiations continued over many years, involving issues as fundamental to Gulbenkian as the validity of the Turkish Petroleum Company's rights. He followed every step with immense care, preferring the written word of the telegram to the spoken one at a meeting, and somehow contrived to hold on to his five percent. In 1927, drilling commenced six miles north-west of Kirkuk, and at 3.00 a.m. on October 25th, the well blew out at a depth of only 1500 feet, flowing at some 95 000 barrels a day. Even so, one more year was to elapse before Gulbenkian was ready to sign the final agreement for what became the Iraq Petroleum Company, to be owned by Shell, BP, and CFP of France, with 23.75% each; Mobil and Esso with 11.875% each[36], and Gulbenkian with 5%. The numbers themselves speak of who did, and who did not, give ground. It had taken him thirty-five years to get there, but he made it by dint of attention to detail and remarkable tenacity.

By now, even without Iraq oil, he was a fabulously wealthy man living variously in the Ritz Hotels of London and Paris, attended by a string of very young mistresses, whom he apparently regarded as a medical necessity. Apart from business, his interest was his art collection.

Fig.3-22. The discovery well near Kirkuk, October 1927.

Photograph courtesy of British Petroleum

Although his Iraq venture was his most spectacular success, he was also instrumental in getting Shell into Mexico and Venezuela, which were not inconsiderable feats in their own right. He worked closely with Henri Deterding until their relations soured – over their shared admiration of Lydia Pavlova, the wife of a Russian general, who eventually became the second Mrs Deterding. This was one of his few negotiations that failed.

When the Second World War came, he moved from Paris to Portugal, where he died in 1955 at the age of 85, leaving a vast fortune, a remarkable art collection and a charitable organisation, the Gulbenkian Foundation. He regarded himself as an architect – not of buildings but of immaculate and elegant business constructions and negotiation.

In the same way as we may speculate about the different course of history had Turkey not joined the losing side in the War, we may be sure that the world of oil would have been very different had not Calouste Gulbenkian dedicated his life to it.

PIONEERING EXPLORATION

Exploration and production during the Pioneering Epoch was onshore, where in fact most of the world's oil lies. It began with very primitive technology and understanding, but progress was remarkably rapid. The rotary rig was introduced to replace the cable tool; well logging was developed to identify the formations and their oil content; and seismic surveys were brought into wide use to investigate geological structure at depth. All were established well before the end of the epoch.

It was the great age of field geology. Pioneering geologists on foot and muleback scoured the world for prospects, working under often arduous conditions far from base[37]. They had to be men of initiative, as well as rugged individualists. They searched, first, for seepages, and then for promising structures in the vicinity. But most of all they looked at the rocks to see if they had the essential characteristics and structural style to make an oil province. They developed a nose for what worked and what did not, and they mainly got it right. Kingston[38] describes how Esso had a small group of such explorers that roamed the world from 1940 to 1960, searching out every nook and cranny. He comments on the ending of this chapter:

The field studies and overseas travel came to an end, and the roving assignments ceased [in response to the move offshore]. *During the 1970s and early 1980s, the price of oil rose significantly, and the oil industry went on an exploration binge. Basins that we had repeatedly rejected in the past (we called them "dogs") were leased and drilled, mostly with negative results ... Some of our reconnaissance field work led to acreage acquisition, and to drilling, a few of which resulted in major oil discoveries in Europe, North Africa, and South America. Very often our regional studies resulted in negative recommendations – to stay out of basins because they did not look "oily" from a geological standpoint.*

It is a telling comment. It is not so difficult to determine if a new area holds promise or not, and the pioneering geologists were not often mistaken in their assessment.

In the 18th Century, there was a belief in the existence of a habitable "Great Southern Continent". When James Cook[39] in 1769 sailed into the Southern Ocean and confirmed that it did not exist, the World realized that all

Fig. 3-24. J.V.Harrison, who mapped Iran with Norman Falcon, has the rugged countenance of the pioneer explorer.

the continents had been found. Later voyages of discovery found and mapped the smaller islands. Now satellite imagery can identify the smallest rock. Oil exploration has followed a similar path: the "continents" and most of the larger "islands" were all found during the pioneering epoch.

The international fraternity was a small and closely knit one, despite the fact that individually they were working for competing oil companies (see p. 45).

I take 1950 to mark the end of the Pioneering Epoch, a convenient mid-century milestone, but it could as well be the Second World War. Oil was of course a critical factor to both sides. Germany was supplied by Romania and large quantities of synthetic oil distilled from coal (rising from 28% in 1939 to a peak of 62% in 1944), whereas the Allies were supplied largely from the United States.

Fig. 3-25. Geological Field work: an almost lost exploration technique, which found most of the World's oil.

Photograph courtesy of British Petroleum

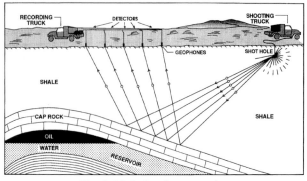

Fig. 3-23 Seismic surveying

The Coming Oil Crisis

	Production kb/d	Cumulative Production	Reserves	Yet-to-Find
USA	5407	41.3	40.0	123.7
Venezuela	1498	5.5	38.2	38.2
FSU	758	6.7	23.2	210.1
Iran	664	2.4	31.8	85.9
Saudi Arabia	547	1.0	131.6	167.4
Kuwait	344	0.3	60.5	29.3
Mexico	198	2.5	3.9	45.6
Iraq	140	0.5	28.6	85.9
Indonesia	136	1.3	10.9	17.9
Colombia	93	0.5	1.3	8.2
WORLD	10477	67	394	1339

Fig. 3-26. The important producing countries in 1950. Note how much had been produced in the USA.

In terms of the resources, about 460 billion barrels of oil, which is about one-quarter of the world's ultimate endowment, were discovered during the Pioneering Epoch. Many of the critical giant fields had been found. But, no more than 67 billion had been produced by the end of the period, leaving plenty of reserves to be produced in the future. Production had been inexorably rising: to 3.8 million barrels a day by 1930 and 10.4 Mb/d by 1950.

The world was still perceived to be a large place, and few people doubted that the discovery of oil and gas would continue unabated. The War had ended with a feeling of optimism for growing economic prosperity and a new world order of justice and respect. Oil, by now fully controlled by the major international companies, was expected to fuel the wheels of industry. Things have not quite worked out that way[40].

A Selection from the Role of Honour of International Pioneering Geologists

They found most of the world's oil basins with technology no more advanced than the hand lens and the hammer. The country of their most notable contributions is shown in parenthesis

E. Lehner (Trinidad)
H.G. Kugler (Trinidad)
D. Trumpy (Ecuador)
W.E. Humphrey (Mexico)
N.J. Sander (Saudi Arabia)
V. Oppenheim (Latin America)
J.V. Harrison (Iran)
N.L. Falcon (Iran)
H.D. Hedberg (Venezuela)
M. Steinekke (Saudi Arabia)
H.H. Renz (Venezuela)
O. Renz (Venezuela)
J.W. Bausch van Bertsbergh (Venezuela)
L.G. Weeks (Australia)
H.R. Tainsh (Burma)
M.K. Hubbert (USA)
J. Ruthven Pike (worldwide)
A. Gannser (Venezuela)
E. Rod (Venezuela)
F.B. Notestein (Colombia)
T.A. Link (Brasil)
W.C. Hatfield (Colombia)
H. Loser (Peru)
J.W. Harrington (Latin America)
H.V. Dunnington (Iraq)

G.M. Lees (Iran)
De Boeckh H (Iran)
P.E. Kent (UK)
F.E. Wellings (Iraq)
J.D. Moody (worldwide)
F.C.P. Slinger (Iran)
A.N. Thomas (Iran, Canada, Libya)
R.A. Brankamp (Saudi Arabia)
S. Elder (Iran)
H.P. Schaub (Venezuela)
J. Weeda (Borneo)
E.H. Cunningham Craig (Trinidad)
R. Arnold (Venezuela)
G.A. Macready (Venezuela)
T.W. Barrington (Venezuela)
V. Benavides (Peru)
S. Papp (Hungary)
L. Eötvös (Hungary)
V. Zsigmondy (Hungary)
F. Bóhm (Hungary)
F. Pávai-Vajna (Hungary)
G. Csíky (Hungary)
H. Kirk (Middle East)
G. Flores (Cuba, Sicily, Mozambique)
R.E. King (World)

W.E. Aitken (Colombia)
C.C. Wilson (Trinidad)
W. Saville (Colombia)
H. Widmer (Borneo)
H. Harrington (L. America)
A. Beeby-Thompson (Worldwide)
W. Link (Brasil)
G. Noble (Iraq)
G.M. Mugoci (Romania)
I. Popescu-Voitesti (Romania)
G. Macovei (Romania)
G. Cobalcescu (Romania)
W.O. Leutenegger (worldwide)

Charlie Hares (USA)
Tom Wilson (Alaska)
L. Mrazac (Romania)
G. Grigoras (Poland)
Desheng Lee (China)
I. Lukasiewicz (Poland)
H. de Cizancourt (Poland)
W.D. Gill (Pakistan)
P. Bokov (Bulgaria)
V. Dank (Hungary)
T. Buday (Czechia)

and many unknown warriors

NOTES
(For references, see Bibliography)

1. See also an interesting article by Haldorsen 1996.

2. Bilkadi, Z., 1996, has published a beautifully illustrated book "Babylon to Baku" covering references to oil in antiquity: the earliest record being 3000 BC.

3. Fuller, J.G.C., 1993, gives a good account with historical insights.

4. Narimanov, A. and A. Palaz, 1994, point out that wells had been drilled for oil in the Baku area before the famous well of Col. Drake in Pennsylvania.

5. In fact, Romania boasts of being the country with the oldest record of oil production: 275 tonnes in 1857. A foreign oil company had been registered there already in 1864. Production comes from the foothills of the Carpathians which were also investigated in early years in neighbouring countries including Poland (Popescu 1994, Jacob. 1938)

6. Beeby-Thompson, A., 1961, gives a marvellous account of the early days of exploration and production around the world.

7. Yergin, D., 1991, has written the seminal oil history which is essential reading; while Pees, S.T., 1989, fills in the detailed memorabilia of the early Pennsylvanian fields.

8. Beers, F.W. 1865 quoted in Pees S.T., 1989.

9. Possibly the only useful contribution of Kenney's paper on the inorganic origin of oil was the record of an observation by Academician Mikhailo V. Lomonosov in 1757, who stated *Rock oil originates as tiny bodies of animals buried in the sediments which, under the influence of increased temperature and pressure acting during an unimaginably long period of time, transform into rock oil*. He got it right in 1757: who said that the Russians were not good explorers!

10. Energy Information Administration, 1990, provides useful data on the USA.

11. Pees, S.T., 1989.

12. An agreement between the companies to define their spheres of interest. It did not survive for long.

13. This organization which controlled the movement of oil by rail was used to prorate production to support prices.

14. Pickens T. Boone, 1987, has written a candid and illuminating book about his assault upon a moribund Gulf Oil, with not always complimentary observations about the senior management of oil companies.

15. James, M., 1953, authored the official, and possibly expurgated, history of Texaco.

16. Farago, L, 1973, in a book on wartime espionage reveals the role of Texaco.

17. James, M., 1953.

18. Apparently this situation arose from circumstances relating to the Pennzoil law suit. Texaco's suppliers began withholding credit, fearing for the Company's future, and it turned to Saudi Arabia for help. The latter provided the much needed credit, but the *quid pro quo* was access to Texaco's downstream.

19. At the time of writing it is reported that US Shell plans to join this venture with Saudi Arabia in a highly significant move.

20. Cazalot, C.P., 1996, in a rare interview explains Texaco's policy today, which seems to be to concentrate on US opportunities.

21. Beeby-Thompson, A. 1961.

22. World Oil, June 1996, reports the sale of Pennzoil and

McDermot's interest in the Azeri-Chiraq-Guneshli fields to the Japanese company ITOCHU for $132 M.

23. I don't know much about the offshore potential, but have seen reports that "reserves" in the order of 20 billion barrels may be ready for development from already partly identified fields. If confirmed it would be equivalent to about one-third of the North Sea, providing for less than one year's world supply. Others who have seen the seismic data conclude that the large structures offshore in the North Caspian lack an adequate salt cover, and may not deliver as hoped. Time will tell.

24. British Petroleum, 1959, provides a delightful pictorial history of the company.

25. British Petroleum, 1959, *idem*.

26. In some regards one could distinguish the Zagros foothills as a somewhat different petroleum system from the rest of the Persian Gulf and Arabian area, but that is probably splitting hairs at least in this context.

27. Massey, R.K., 1991 writes about the events leading up to the First World War.

28. In a sense, BP became the first state-controlled oil company, although the State did not normally intervene as other than an investor. BP had difficulties entering Latin America because of this state-ownership which was not allowed under the constitution of several Latin American countries.

29. Clemenceau, the French premier in the war is quoted by Zischka as saying *"Une goutte de pétrole vaut une goutte de sang"* (a drop of oil is worth a drop of blood).

30. See Fromkin, D., 1989, for an excellent account of the factors that led to Turkey's alliance with Germany, which were evidently much misunderstood by the British Government at the time. British policy had hitherto supported the maintenance of the Ottoman Empire to act as a buffer against Russian imperialism.

31. Lawrence, T.E., 1935, a former archeologist, led an Arab rising against the Turks, promising recognition for an Arab state, a promise the Allies reneged upon in the Peace Treaty.

32. Kuwait had of course been part of the Ottoman Empire, although as a trading port it had more external links, especially with Britain. It was excluded from the Red-Line Agreement by Gulbenkian, on the famous occasion when he defined the area of interest with a red-pencil at a meeting. He probably did so to avoid difficulties with Britain.

33. Aburish, S.K., 1994, gives a revealing account of the corruption of a family that controls the world's supply of oil.

34. See Sell, 1938, for a listing of production data.

35. Yergin, D., 1991; Tugendhat, C. and A. Hamilton, 1975, both give lengthy accounts of the life of Gulbenkian, which is also covered in an autobiography *Pantaraxia*. Other accounts are by his colourful son, Nubar, *Portrait in Oil*, and a biography by Ralph Hewins.

36. Other American companies were originally involved but dropped out.

37. Blakey, S., 1985, 1991 has published two well illustrated books on pioneering exploration

38. Kingston, D. in Hatley, A.G., 1995, describes the life of a roving field geologist.

39. Boorstin, D.J., 1983 has written an excellent book of scientific discovery and exploration.

40. Kennedy, P., 1995, highlights the dangers of the global economy in a very perceptive work.

Chapter 4

GROWTH AND TRANSITION

GROWTH 1950-1970

In 1950, world oil production stood at 10 million barrels a day (Mb/d), but within twenty years it had risen to 45 Mb/d, a staggering near five-fold increase. I call it the *Growth Epoch*. At first sight, you would be forgiven for thinking that this must have been a golden age for oil. But in fact there were darkening shadows, and the rapid growth was in a sense a reaction to a new uncertainty and insecurity. The companies had no intrinsic reason to flood the world with cheap oil, which, had they been assured of their future, would have been contrary to their long term interests. There are several reasons why they began to dig into capital, which I will explore in this chapter.

As was noted in the last chapter, the Soviet Union had formally expropriated the foreign owners of its oil in 1928, and Mexico did the same ten years later. They were ripples on the pond compared with the waves that enveloped the industry only one year into the *Growth Epoch.*

For several years prior to 1950, BP had been in dispute with Iran over the level of government take. The Iranians found it inequitable that they should receive royalties of £90 million while the Company registered a profit of £250 million, half of which went to the British government, holding fifty-one percent of the shares. The well-worn cries of foreign exploitation rang in the Iranian Parliament and in the streets. The most outspoken and impassioned voice was that of Mohammed Mossadegh, a frail-looking, seventy-year old, land-owning aristocrat, who harangued Parliament with theatrical tears and faints. He was often interviewed, stretched out in pyjamas on his iron bedstead, seemingly close to collapse. The situation deteriorated when the Prime Minister, who proposed moderation, was assassinated, followed soon afterwards by the Minister of Education, who suffered the same fate. Events were spiralling out of control. In the face of these popular pressures, Parliament passed a law, nationalizing BP's Iranian affiliate, and on April 28th 1951 elected Mossadegh as Prime Minister to implement the decision. He did. By September, a British warship, with the band playing "Colonel Bogey", had evacuated the last British nationals from Abadan, the Company's base since 1908.

It was a far cry from the gunboat diplomacy of an earlier epoch. A war-weakened Britain no longer had the stomach for Empire or the enforcement of contract. It was a retreat much welcomed by the American government, perhaps for misguided idealistic anti-colonialist reasons, but as likely with an eye to oil. A few years later, when the Iranian crisis was resolved, American companies found themselves usurping, through their stake in the Consortium which had replaced BP, much of the position once held exclusively by BP.

At the time, this was seen as little more than an unfortunate chapter in a changing post-war world. After all, the socialist government in Britain was then bent on nationalising almost everything in sight. So, who should deny the Iranians the right to do the same thing? But in fact it had far-reaching implications, and led to shifts of attitude and policy, some of whose consequences are yet to be played out.

Let us reflect for a moment on BP's position before the Iranian nationalization. It had a concession that ran for sixty years from 1901. It relied on Iran for most of the oil it needed to supply its European and imperial markets that were gradually growing. It could easily balance supply and demand, and make long term plans, both for the duration of the concession and with the reasonable expectation that it could be extended. The Company knew what resources it had, and could plan an orderly exploitation so as to conserve them in a reasonable way. Such an attitude likewise influenced the American companies in their home country, where they owned the mineral rights outright. It had been an epoch of stability – even complacency one could argue – but at least it provided no motive for squandering resources.

All that changed after 1951. It was a searing experience for BP, which reacted by launching into a vigorous and remarkably successful exploration campaign around the world to find new supplies: an effort, culminating in Alaska and the North Sea. The fate of BP in Iran did not pass unobserved by the other major companies who relied heavily on Middle East oil. They realised too that they were becoming increasingly unwelcome tenants of unfriendly landlords: an ironic contrast with the socialist attitude to property in Europe where the tenant was favoured at the expense of the landlord. The companies concluded that they had only two priorities: to produce as fast as possible while they still owned the rights; and to find more sources of oil to lessen their dependence on a single supply.

Fig. 4-1. Cheap imported oil replaced indigenous coal for electricity generation.

Photo: Jørgen Schytee/Still Pictures

Fig. 4-2. Releasing a huge explosive charge for calibrating the refraction seismic survey that found Hassi Messaoud, Africa's largest field.

Photo: Jean Laherrère

To produce as fast as possible meant that they had to find new markets; and the main challenge of the epoch was to do just that, dumping cheap oil on the world. It soon created an energy dependent society, driving to work from suburbia and buying consumer durables that could be transported around the world at minimal cost by cheap oil. Strawberries became available everywhere on every day of the year. The European market in particular was opened up, even to the extent of fuelling electricity generation by cheap oil imports at the expense of indigenous coal.

To find new sources of cheap oil was a greater challenge. It was obvious already by then that nowhere in the world compared with the abundant Middle East. Nevertheless, for strategic reasons as opposed to strictly economic ones, exploration was stepped up. Attention turned to Africa, a continent that had not been widely explored before. BP and Shell in joint ventures took up pioneering positions in East and West Africa, the latter soon yielding important discoveries in Nigeria. French companies turned to Algeria, where they were rewarded by the giant Hassi Messaoud discovery in 1956. A third prolific new basin in Libya was developed by both European and American companies, with major discoveries in 1958 and 1959.

Other useful finds were made in India by the State oil company, later culminating with the Bombay High field in 1974; as well as in Indonesia; Australia and Latin America. Meanwhile, behind the Iron Curtain, the systematic exploration of the Soviets was being rewarded by major discoveries of oil and gas.

In 1968, towards the end of the epoch, another frontier was crossed with the discovery of the giant Prudhoe Bay Field in Alaska, which was of enormous importance to the United States, giving it a second lease of life after discovery onshore in the Lower 48 States had peaked. The field was found by Arco who had leased the crest of the structure which contained the gas-cap, but BP's lesser bid yielded the flanks where most of the oil was. Prudhoe Bay did much to compensate BP for the loss of Iran, and further cemented the Company's reputation as a first-class explorer. It was no mean feat for the Company, a newcomer from Europe, to walk off with most of North America's largest oilfield.

These new areas were onshore, but attention also began to turn offshore. In fact, offshore extensions to fields had already been attacked by drilling deviated holes from the land, as in Trinidad and Peru. Some shallow water prospects had been drilled from steel platforms erected on the sea-bed, as in Lake Maracaibo in Venezuela and the Gulf of Mexico.

What was needed was a mobile floating rig that could be used for truly exploration purposes where the cost of a fixed

Fig. 4-4. The *Ocean Traveller* rig drilled the first five wells in the stormy waters of the Norwegian North Sea.

Fig.4-3. Drilling in Venezuela.

Photo: Shell International Photographic Services, London

platform could not be justified. Engineering work in this direction had already been put in hand, and the first such floating rig, the *Breton Rig 20*, designed by John T. Hayward, was put into operation in the Gulf of Mexico in 1949. It consisted of no more than a barge with a drilling rig mounted upon it: the innovations being in the sea-bed wellheads and the connections with the barge above. This technology could be used in waters up to about 100 m in depth, but was very susceptible to wave conditions.

Further development led to the concept of the semi-submersible rig which was based on the ingenious idea of building the rig on two submerged pontoons that floated below the wave base, providing a stable platform irrespective of surface conditions[1]. The first such rig, *Blue Water No 1,* came into operation in 1962. This new technology widened the scope of exploration to the continental shelves of the world within water depths of about 200 m. The fleet of such rigs rapidly grew so that during the 1960s semi-submersible drilling was undertaken in Borneo, Iran, Canada, West Africa, the United Kingdom, Norway and New Zealand. Designs were continuously improved such that, by the end of the epoch, wells were being routinely drilled in the stormy North Sea.

Two alternative technologies deserve mention. One was the jack-up, in which the platform, having arrived on location, put down long retractable legs that sat on the sea-bed, and jacked itself up above the waves. The Zapata Company of Houston dominated this market. It was managed by George Bush, the former American President, who gained thereby a particular insight into the oil business that was later to influence American foreign policy. The second alternative was the drill ship in which a rig was mounted midships on a conventional vessel, which was held in place either by anchors or thrusters. Global Marine of California pioneered this approach to meet the deeper water needs of that State.

Much play is made of technological progress, which some hold will provide a solution to the coming supply crisis. The technology of the semi-submersible rig did indeed make a major contribution by bringing the reserves of the world's continental shelves into reach for the first time, but I think it will prove to be the last such technological breakthrough having a significant global impact. Most subsequent technology has succeeded mainly in increasing production rate and so accelerating depletion, without adding much in terms of reserves. It is an issue to which I will return.

Five years after the BP debacle in Iran came another nail in the sanctity of contract: the Suez Crisis. Anthony Eden, the new British Prime Minister, having previously resigned office in protest at the appeasement that led to the Second World War, was resolved to react forcibly to Egypt's nationalization of the Suez Canal. An Anglo-French force was mobilized to defend the canal from a contrived Israeli attack. But America saw it as neo-colonialism and undermined the effort, such that the troops had to withdraw ignominiously a few weeks later. It was plain for all to see that countries could renege on contracts with impunity: likewise the socialists of Europe had scant regard from the sanctity of private ownership. Whereas France and Britain had had long experience of overseas political administration and exerting their influence around the world, America, which was now in the driving seat, had only two policies: economic hegemony and curtailment of the perceived Communist threat, itself not without significance for its burgeoning arms industry. The fruits of this abnegation of the enforcement of international order were soon to be felt, especially by BP as the flag-bearer of an ex-colonial power. It was expelled from both Libya and Nigeria for political problems related to its mother country.

Production rose rapidly during the *Growth Epoch,* mainly because fields in the Middle East, especially those in Kuwait and Saudi Arabia which had already been found

PRODUCTION GROWTH
1950 & 1970

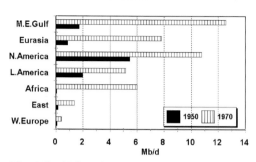

Fig. 4-5. 1950 and 1970 production by region.

Gb	Production Gb/a	Cumulative Production	Reserves	Yet-to-Find
USA	3.5	96	45	64
FSU	2.6	32	106	102
Iran	1.4	14	79	27
Venezuela	1.4	26	34	22
Saudi Arabia	1.3	13	237	50
Kuwait	1.0	12	71	7
WORLD	16.8	193	949	608

Fig. 4-6. The most important producers in 1970.

before and during the War, were now brought into production. Yet again, the classic depletion pattern manifested itself, as production in the giant fields rose rapidly to peak levels. It was the same problem of flush production as had so concerned Standard Oil during the early days.

Figure 4-5 shows the respective production levels of the major producers at the beginning and at the end of the period.

Discovery too peaked as shown in Figure 4-7 although this was not evident at the time. In particular, the discovery of the critical giant fields peaked as shown in Figure 4-8.

DISCOVERY
By decade

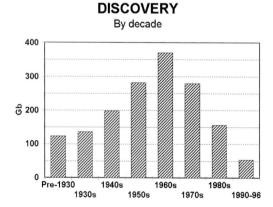

Fig. 4-7. Oil discovery by decade.

GIANT FIELDS
Initial reserves by discovery year

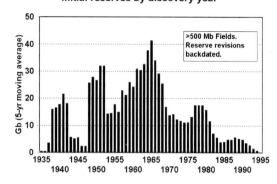

Fig. 4-8. Peak giant field discovery.

The growth of oil production from new sources around the world held down the price of oil, and hence the revenues accruing to the principal producers. Their populations were growing, and their economies were expanding. They needed more money. Since the main source of such money for most of them was oil revenue, they set about to see how they could increase their cut. Such moves led to the creation of the Organization of Petroleum Exporting Countries (OPEC) in 1960. It was a move in which Venezuela joined the Gulf producers to try to exert more influence. I will return to the pricing of oil and the evolution of OPEC in a later chapter.

The period of growth drew to a close around 1970, and gave way to what can be termed the *Transition*.

TRANSITION 1970-2000

Many of the companies had been pumping as if there was no tomorrow: as it turned out, for many of them there wasn't. Around 1970, radical changes began to be felt in many ways, leading to a new epoch, which I term the *Transition,* before *Decline* sets in around the end of the Century. During the previous twenty years, production had been growing at an average of seventeen percent a year. But, over the two decades from 1970, it fell to respectively three percent and almost zero. It is today rising again at above two percent a year, with increasing demand in the Far East, but the total growth over the *Transition Epoch* as a whole is unlikely to exceed two percent per year.

This fall in demand was brought about by a combination of factors. They included improvements in engine efficiency, as exemplified by the jet engine, but more than that, there was a sort of saturation in the industrial countries. You can only drive one car at a time, even if that is stationary in a traffic jam. The pattern differed from one region to another with the recent increase in Asia and Australasia being offset by an anomalous fall in the Former Soviet Union, due to its difficult economic circumstances after the collapse of Communism (see Figure 4-9). On a per capita basis, consumption ranged from about three tons in North America, to 1.5 tons in Western Europe and 0.25 in the rest of the world.

OIL CONSUMPTION
1995

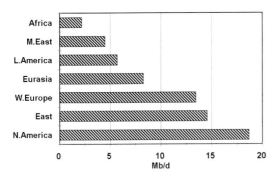

Fig. 4-9. Consumption by area (after BP).

Gasoline has been the dominant product in the United States, and its consumption is growing fast in both South East Asia and the developing countries. Fuel oil use, by contrast, has been in decline, especially in Western Europe.

As noted, the share of production coming from the five main Middle East producers had been rising rapidly during the preceding *Growth Epoch*, such that by 1973 they were supplying thirty-six percent of the world's needs. The world was dangerously vulnerable to this single source, as it was soon to find out.

This is not the place to review the Arab-Israeli conflict in more than a few words. In pre-war days, Britain had some sympathy for the idea for the creation of a Jewish homeland in Palestine[2], which was then a British protectorate, provided that it did not adversely affect the indigenous population. It had not bargained for the large-scale immigration in the aftermath of the holocaust that brought the matter to a head. Jewish terrorist pressure finally led the British to leave in 1948, and the State of Israel was proclaimed, soon to be the recipient of massive aid from the United States, which was being influenced by a strong Jewish lobby. The Palestinians and neighbouring Arab States were not exactly consulted, and it is not surprising that they objected to what was tantamount to an invasion and an exercise in what would now be called ethnic-cleansing. Whatever claims the Jewish people have had to reclaiming a homeland they lost 2000 years ago, it is obvious that the indigenous Arab community have become the victims of what transpired. A deep-seated conflict has developed ever since and shows little sign of abating.

In Reading, England, far from the Middle East, I recently met a taxi driver, clad in loose-fitting white clothes, who described himself as a Pakistani Islamic Fundamentalist. He confided that, given the chance, he would be very willing to strap Semtex explosive to his waist and destroy Israelis: a strong reaction for a man living in gentle Berkshire, having had no direct connection with the conflict. He further confided that whereas the Israelis had modern American weapons, his group had time and people: an ominous message.

On October 6th 1973, a combined Egyptian-Syrian attack was launched on Israel in what became known as the Yom Kippur war. The idea of using oil supply as a weapon had been around for several years, and on October 17th, the Arab oil ministers, meeting in Kuwait, decided to use it. They resolved to reduce oil production at five percent a month, and to totally cut off supplies to America and the Netherlands[3]. By December, oil prices had risen to $17/b, up from about $2, around which level they had hovered for most of the Century. It was what came to be known as the First Oil Shock. By definition, a shock is a short-lived affair, and the embargo was relaxed by the end of the year. Oil prices did not however fall back to their previous levels but fluctuated in the low 'teens.

Fig. 4-10. Sheik Yamani, oil minister of Saudi Arabia. The world listened to his every word.

Photo courtesy of Sheik Z. Yamani

A second oil shock was not slow in coming. It arrived six years later in 1979 when the Shah of Iran fell from power to be replaced by the Islamic Fundamentalist, the Ayatollah Khomeini, who returned from exile in France to form a new government bent on reintroducing Islamic values in the country. Panic buying drove oil prices up to almost $50/b.

These two shocks were warning signals, like the minor tremors that presage an earthquake. They prompted many people to become aware of oil resource constraints for the first time. Could the world continue to sustain economic growth indefinitely was a question that was being increasingly asked. It was, and remains, a very valid question, but the fears then raised proved to be groundless so far as the immediate future was concerned. The new productive areas that had been found during the proceeding *Growth Epoch,* especially in Alaska and the North Sea,

Fig. 4-11. The Ayatollah Khomeini, who returned Iran to Islam.

Photo: Associated Press

distorts most rational behaviour. Previously, the companies had been paying tax at 50% on the basis of "posted price", which was substantially above the price actually being realized in the market. In fact, the host government take was closer to 80%. However, by virtue of the double tax agreements which were in force, the companies were able to claim the tax paid to the producers as a charge against the tax payable to the consumer governments. In effect, the consumer taxpayer was paying the tax to the producing governments. This happy hidden subsidy ended after sequestration, when the producing governments had to sell their oil on the open market. It really is remarkable to see how the companies actually made so much of their money by juggling their tax obligations.

With the expropriations, the major oil companies lost ownership of most of their sources of supply, such that by the mid-1990s their share of world reserves was no more than about seven percent. The Seven Sisters, now down to Six on the demise of Gulf Oil, had to put their finery back in the cupboard, and get down to cleaning the steps. The concept of property is a deep one, ranking close to life itself as the expression "life and property" confirms. People who own something tend to try to protect and preserve it; tenants have a different attitude, wanting to make the most of their rights while they have them; and traders have an even shorter-term view, being interested only in the moment of transaction. The oil companies have passed through each of these stages. Now, they are effectively only marketers; and will soon face the further constraint of declining markets. It is no wonder that they are short-term in attitude.

The Middle East share continued to slide as production grew in other areas, such that it had fallen to sixteen percent by 1985. Prices collapsed to a low of $6/b in what became known as the "Oil Glut". It was not really a glut, because the new sources were small in comparison with the established Middle East: it was simply flush production produced at maximum rate by those with a very short view of the future. It was also partially driven by discount economics that depreciated future earnings under the rules of Discounted Cash Flow (DCF) and Rate of Return (ROR). Much of the new production came from offshore and remote areas that incurred high front end investments: the companies had every incentive under conventional economic principles to produce as fast as possible.

I remember an occasion when I was advising the Bulgarian Government at the time it was trying to attract Western oil companies to explore the country soon after the fall of Communism. The official concerned was reading one of the applications for a concession, and turned to me with a puzzled expression "What means please DCF?" he asked. I replied "Us capitalists like to make money fast". He shrugged his shoulders and confided that such an idea was rather obscene to him. He had a point so far as exploiting a finite resource is concerned.

But then the trend changed again. It was a fundamental discontinuity as important as the shocks that preceded it, although much less spectacular: many people did not even notice. *The share from the Middle East began to rise.* This

were now coming on stream, and flush production from the giant fields, found early in the exploration process, was beginning to flood the market. It was a repetition of the same old problem that had plagued the industry since its birth. It is important to remember that these new provinces had already been found, and were not a response to the oil shocks, even if higher prices did accelerate development. As a result of this new production, the share of world supply from the five Middle East producers fell, and was down to about thirty percent by the time of the second oil shock, which was accordingly a short-lived affair.

Meanwhile, there was another important development that transformed the industry during the early years of the *Transition:* in a word, sequestration. OPEC had substantially failed to regulate the business either by imposed posted price or later quota, and the producing governments now turned towards sequestration: Iraq in 1972; Kuwait in 1975; Venezuela one year later; and finally Saudi Arabia itself in 1979. Furthermore, State companies, often with privileged rights, arose almost everywhere under the socialist movements of the day, except of course in the capitalist United States. In Britain, rights previously freely negotiated were clawed back by what was cynically described as a voluntary transfer to give to a new state-owned British National Oil Company (BNOC); and in 1973, Norway established Statoil which, as we shall see, has become almost a state within a state.

Sequestration did not work as its proponents had hoped: they had forgotten one vital ingredient, namely tax which time it is set to continue to do so, as now there are no new provinces even in sight to counter the Middle East dominance. It has nothing to do with politics or economics, and everything to do with the fact that the Persian Gulf area was uniquely endowed with both oil source-rocks laid down in the warm Jurassic seas and with tight seals from the salt deposited when the seas subsequently dried up.

This general trend of rising share is controlled by geological factors, but there was a political factor that triggered the fall of oil price in 1986. According to the remarkable revelations of the book *Victory*[4] by Peter Schweizer, King Fahd dropped the price in an orchestrated effort by the United States Intelligence Services to undermine the Soviet economy, which relied on oil exports for foreign exchange. The Gulf War was perhaps one of the consequences.

Saddam's Strategic Oil Reserve: A Sinister Speculation

We are so used to hearing about the mistakes of governments and the ineptitude of their policies that it has become difficult to credit that the unfolding of events could reflect a deliberate policy, immaculately carried out. The image of the intelligence community is equally tarnished with their defectors, double agents and spooks eavesdropping by night: the stuff of a Le Carré novel.

Perhaps we are mistaken to be so disparaging. To read Peter Schweizer's book *Victory*[5] would certainly suggest so. It is a remarkable book, not only for its contents, but that it should see the light of day. How credible is it? Internally it seems to be, with its numerous references to cited interviews and now released Presidential Executive Orders.

Fig. 4-12. Peter Schweizer, author of a remarkable book.

Photo: Atlantic Monthly

Externally too it seems plausible, judging from the comments of, for example, Hodel[6] who was President Reagan's Energy Secretary at the end of the Cold War, and the explicit confirmation by Allen, the chief foreign policy adviser[7].

The thrust of the book is that President Reagan and Mrs Thatcher were not content to coexist with the Soviets but resolved to bring down the regime by economic means. The Reagan Administration set up an inner secret cabinet to orchestrate the endeavour through the offices of the CIA. There was little public money available for it, so it had to be paid for by covert means. Enter King Fahd of Saudi Arabia, who was not short of cash, and certainly had a motive to prevent Soviet influence spreading to the Middle East or his Kingdom, already endangered by dissident movements based in neighbouring Communist Yemen. It transpires that he cooperated in the purchase of arms to send to the freedom fighters in Afghanistan who began to undermine Soviet military credibility. Star Wars was then proposed as an immensely expensive undertaking which the Soviets would feel obliged to match even if it bankrupted them, especially in terms of foreign exchange that was earned by the sale of oil. Enter King Fahd a second time in 1985, to be persuaded to open the oil valve and flood the world with oil, causing oil price to collapse and, with it, critical Soviet earnings of foreign exchange. That is what the book says anyway. The fall in oil price also bankrupted a large section of the American oil business, and explains why the Energy Secretary of the day, who had been working hard to try to come to the industry's rescue – so vital to America, by then importing almost 50% of its needs – feels so resentful that he was not informed of what was happening.

As discussed in Chapter 3, Saudi Arabia became the cornerstone of American oil interests in the Middle East, when Chevron successfully leased it in 1933. It contains the Ghawar oilfield which holds about 5% of the world's ultimate endowment. The northern end (Ain Dar) was found in 1948 but the full extent was not determined until 1950

Fig. 4-13. King Fahd of Saudi Arabia, who helped the American initiative.

Photo: Associated Press

with the drilling of a well 200 km to the south at Haradh[8]. The country owns almost a quarter of the world's remaining conventional oil. Once populated by a few tribesmen, now eight million people live there, of whom half are non-Saudi nationals, doing most of the work, but utterly dependent on royal patronage. Aburish[9] in his book paints a disturbing picture of this strange regime with its "princely allocation" of oil for the benefit of the privileged, and their free, reserved seats in the national airline. There are now more tanks in the desert than drivers: each tank having been bought on terms yielding a princely commission. The United States supports this helpful monarchy, and according to *Victory,* has provided various surveillance and other facilities to protect it from the growing number of dissidents at home and abroad. The recent bombings in Riyadh and Dahran were aimed at this American presence. In addition to the dissidents, concerned about progress in the country itself, is the wider issue of the guardianship of the holy shrines of Medina and Mecca, which many fundamentalists feel are not well served by the profligate House of Saud.

But the Cold War victory carried its price: oil price. That stayed down, and King Fahd began to find it increasingly difficult to keep up the flow of royal patronage to the huge number of dependents and claimants. His problem may have worried his friends in Washington and Langley, Virginia. They needed him to survive and continue to supply America with oil at a reasonable price: not so low as to further offend the domestic oil lobby, but not so high as to burden the world economy and the US balance of

payments. It is said that Schlesinger, the former US Energy Secretary, when asked recently about US oil policy, replied "that King Fahd should live for ever". OPEC was proving ineffectual in prorating oil, and some oil somewhere had to be taken out of the market. This leads to the conspiracy theory which bears retelling although I have no idea as to its validity.

Having so successfully ended the Cold War by covert means, perhaps attention now turned to finding a new way to prorate oil production without putting all the burden on poor old Fahd. Casting around for a candidate to perform this useful role, eyes may have fallen on President Saddam Hussein of Iraq. Here was a bellicose man who had been engaged in a long war with Iran, having been encouraged and partly armed for the purpose by America. He had, with justification, accused his neighbour Kuwait of cheating on OPEC quota, which had the effect of lowering oil price and thus denying Iraq its just revenue. He also thought Kuwait was sucking oil from his side of the Rumaila field, straddling the boundary. He was upset and wanted to teach his neighbour a lesson, but that was not the whole story as he had long wanted additional access to the Persian Gulf. The Tigris and Euphrates flow through Iraq providing natural communications, but their exit to the sea is blocked by the State of Kuwait, all the boundaries having been artificially created by the British after the First World War. This same issue partly explained his previous conflict with Iran. As he planned his invasion, presumably over a period of time, he must have tried to assess his chances of getting away with it. We do not know what soundings he made, nor what responses may have been secretly conveyed to him. He evidently did not suspect that he was being manipulated, if that was the case. In any event, he satisfied himself, mobilized his army, and headed south in 1990.

It is hard to imagine that this movement passed unnoticed by the surveillance satellites that presumably keep this sensitive area under constant observation. It was therefore strange that April Glaspie, the US Ambassador,

Fig. 4-14. Saddam Hussein of Iraq, here in less than bellicose mood: was he duped?

Photo: Associated Press

The Coming Oil Crisis

on the eve of the invasion made a statement to the effect that the conflict with Kuwait was of no great concern to the United States[10]. She was later called before a Senate investigating committee to explain her words, but could only offer the excuse that in the difficult circumstances of Iraq she, as a woman, was confined to her Embassy and did not know what was happening. Tell that to the marines! They meanwhile had been given another message: namely to prepare to "defend" Saudi Arabia. While all of this was going on, British arms-making machinery was being supplied to Iraq through Jordan with the knowledge of the government, as now revealed by the Matrix Churchill trial[11]. Furthermore, there are reports that Saudi Arabia had been contributing to Iraq's nuclear weapons programme[12].

What followed is well known. Desert Storm was organized by the United States under a UN mandate, with loyal support from Britain and several other countries, and led to a huge military build-up in Saudi Arabia. General Schwarzkopf[13] led his troops north to liberate Kuwait, not before about two billion barrels of oil (equivalent to the reserves of a major North Sea field) went up in smoke as the retreating Iraqis fired the wells. He entered Iraqi territory and was soon within a few miles of Baghdad. But, to everyone's surprise, was then ordered to halt and return, leaving the potential threat of Saddam in place.

The Gulf War was well covered by CNN; and was widely hailed as a great victory for the United States, restoring faith in its military prowess after Viet Nam. Yellow ribbons fluttered in flag-flying America, and bumper stickers rallied the nation in support of its heroes. Politicians were seen haranguing the troops in the sand dunes. It was all good stuff and remarkable for the near absence of allied casualties: many of such as did occur were caused by road accidents or were self-inflicted in the confusion of war. We should not overlook, however, the so-called Gulf Syndrome, affecting many ex-servicemen, who were exposed to the release of sarin nerve gas as a result of the explosion of an Iraqi munitions dump at the end of the war.

It may indeed have been another *Victory,* but not quite as generally perceived. In a few weeks it was all over. Saddam proved remarkably resilient, remaining in full

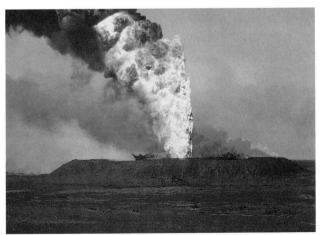

Fig. 4-15. Two billion barrels of oil went up in smoke in Kuwait wells when Iraq fired the wells.

control of his country. King Fahd was more firmly on his throne than ever, surrounded by a US military presence. All was well. Then the UN, led by the United States, imposed an oil export embargo on Iraq ostensibly to persuade Saddam to pay reparations and stop his nuclear and chemical warfare programmes. In this way, Iraqi oil was taken off the market; Fahd increased production, and the United States was able to import large quantities at low to moderate prices. With Iraq production removed, it became easier for OPEC to manage its quota arrangements as it was now close to capacity anyway.

At last, an effective solution to world proration had been found: Saddam would provide a huge strategic reserve for the future at no cost and without interfering with anyone's commercial interests. It was too good to be true, but how to maintain the arrangement?

As the war ended, things returned quickly to normality. The Iraqis managed to survive the hardships well enough, and seemed to support their leader. The flames at Kuwait were put out, without having done as much damage as was at first feared. But normality was the last thing that the policy needed. A perceived threat was required to justify the continuance of the embargo and the protection of Fahd. It did not prove difficult to preserve the state of tension:

– first, there were the Marsh Arabs, a Shiite community with Iranian links, against whom Saddam moved, triggering an Allied military response;

– next, incidents with the Kurds also prompted Allied reactions, in which a helicopter carrying UN officials was mistakenly shot-down by Allied fire;

– next came an alleged plot to assassinate President Bush in Kuwait for which some Iraqi culprits were quickly found. That triggered the firing of a missile from the Red Sea on a ministry in Baghdad;

– next, an undaunted Saddam conducted some military manoeuvres, which were quickly interpreted as a prelude to a new invasion of Kuwait, and red alerts were sounded in the Allied camp.

– next a new face appeared on the scene. Turkey marched into northern Iraq to harass the Kurds, to the consternation of the world with the sole exception of Saddam, who surprisingly seemed to have no difficulty with this invasion of his country. It sounds like the beginning of another story.

How long can "Saddam's Strategic Reserve" be maintained? The need for it will not be for very long. Oil demand has been almost stagnant for many years, but is now expected to grow, driven largely by the East and the burgeoning population of the Third World. Production in most non-OPEC countries is past or close to peak. Soon the boot may be on the other foot, with calls for Saddam to release his oil. Perhaps reason will soon have to be found to show that he is not such a bad chap after all. Unlike Fahd, he is not maintained on a throne by US marines. Iran, also vilified by US policy, has no motive to give its oil away and may bury the hatchet with its former enemy and press for higher oil price.

The oil market now relies heavily on the futures market. With the fear of a return of Iraq production overhanging the market, many refiners have preferred paper to physical oil in their tanks, stocks of which have sunk to a twenty-year low. Prices began to firm in the winter of 1995/6, and there was growing popular concern in America about soaring gasoline prices. Then, surprise, surprise, the UN talked of relaxing the embargo for "humanitarian" reasons, and oil prices fell back temporarily.

In the summer of 1996, America launched a virtually unprovoked attack of 44 cruise missiles on southern Iraq. It was ostensibly a reaction to Saddam Hussein's support for the Kurdish Democratic Party, which controls the lucrative customs posts on the Turkish border and which had invited Saddam to help them oust their rival the Kurdish Patriotic Union. The latter were successfully removed from the provincial capital of Arbil near the giant oilfield of Kirkuk, with in fact minimal intervention by Iraqi troops. Refugees were welcomed by Iran with which the KPU evidently had some ties. The missiles had virtually no effect, and the action was widely condemned by all but the United States' faithful ally, Britain. It is too soon to analyse this extraordinary move. It is variously attributed to Clinton's election campaign; a diversion to allow CIA agents to escape from Arbil; a miscalculation; or a continuation of previous policy of maintaining tension in the area to justify the continued support of King Fahd. One can imagine two other scenarios: King Fahd is under greater threat from his dissidents than is known, and the United States needs justification to maintain its grip on the country; or the United States realizes that it will soon need Iraq's oil, and that Saddam Hussein is likely to be a less generous seller than was King Fahd. Perhaps policy has changed on this realization, and now they actually do want to dispose of him. It was in any event playing with fire, and the world's oil supply. For a while, it looked as if the title of this book would be already out of date by the time it is published as oil prices rose to levels not seen for many years during the autumn of 1996. The embargo was then unexpectedly relaxed, despite the missiles, allowing Iraq to export four billion dollars' worth a year. Prices have not so far fallen markedly, confirming the underlying growing scarcity. It is too soon to evaluate the real significance of these events.

So much for the conspiracy theory: I don't know what to make of it.

The *Transition* has witnessed important scientific and technological progress all round. Great advances were made in geology, as for example by the growing understanding of *Plate Tectonics* and their significance. Geological processes were now much better known. As a consequence, it has become easier to understand the circumstances responsible for oil accumulation, and hence to predict them.

New exploration tools were developed. The advent of the computer chip brought in digital seismic, which radically improved its resolution. Geologists could now see the structure of the oil zones with remarkable clarity. So-called 3D surveys were used to still further enhance resolution, and now come 4D surveys, which are permanently installed over oil fields and allow the actual movement of oil in the reservoir to be observed. The work station has brought amazing computing power to the interpretation process. In parallel with this have come great advances in well-logging so as to determine accurately where the oil and gas zones lie, along with much other useful information.

The greatest breakthrough of all in the context of resource assessment was the advances in geochemistry in the 1980s that for the first time showed where and when oil and gas were generated, as discussed in Chapter 2. Geochemistry also tellingly explains why large tracts are barren, countering the argument of those who claim much oil remains to be found in statistically under-drilled areas. There are usually sound geological reasons why they are under-drilled.

Drilling and production technology advanced all round, as did offshore construction. Highly deviated wells could now reach far out from the platforms to chase thin reservoirs; multi-lateral wells, in which several branches are drilled from a single borehole, increased the extraction rate; and sub-sea completions and floating production facilities allowed smaller accumulations to be tapped. However,

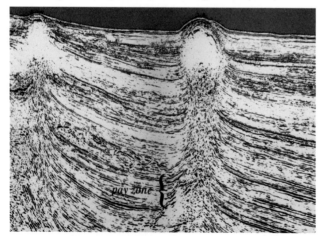

Fig. 4-16. Seismic section: a sort of X-ray of the Earth that allows geologists to study the oil zones. Here, a salt-plug in the deep Gulf of Mexico gives rise to a small trap on its flank.

The Coming Oil Crisis

most of these developments were specific to the offshore and especially the smaller fields. They have much less impact on the old onshore fields which hold most of the world's reserves.

Fig. 4-17. The drilling bit, on which everything depends.

Photo courtesy of Exxon Corporation

All of this represents an amazing achievement for which the engineers can be justifiably proud. But what has been the consequence? Production rate has been increased but less has been found. The short explanation is that there is much less left to find.

Lastly, there has been a radical shift in the way the industry conducts its upstream business. In early years, the major companies did everything; then the contracting business grew in a competitive market, such that the major companies found it expedient to contract most of the work related to seismic surveys, drilling and construction. When oil prices collapsed in the mid 1980s, as already described, the major companies started wholesale purges of their technical staffs, hiring many back again as, or though, contractors, but free of long-term career commitments to them. Lastly, came the so called *Alliance* in which a group of contractors cooperate closely with the oil company to undertake a project.

It is a story of the declining role of the major oil company. In a sense it reflects the ever-diminishing technical risks, consequent upon the technological progress and knowledge. There is clearly less and less of a practical role for an entrepreneurial oil company when the precise parameters of a project can be defined in advance. The major companies are increasingly being reduced to being bankers and investors. Their role in coordinating the contractors' work may also be undermined with the arrival of consultants specializing in that sphere. There is a flaw in this, however, insofar as they are also eating the seed corn of technical expertise. Whereas contractors can mop up skilled technicians who were trained in the major companies, it is much more difficult to picture the hard-

pressed contractor being able to provide the same breadth of experience. So once this generation of expertise has gone, there may not be much to replace it. Perhaps it wont matter anyway if production declines and exploration dries up as this study suggests is inevitable.

Before leaving this theme, it is worth commenting that the State-owned companies, who control most of the world's reserves, have little technical incentive to open their doors to the international companies, when contractor alliances can do the job as well or better. All the companies really offer is money; and you would think that there are other ways of finding that without surrendering national control of the resource. It will in any event be much easier to find the money once oil prices rise. It can be claimed that the major oil companies offer downstream markets to the producers, but as these markets become increasingly hungry, access to them will not command a premium. There is little difficulty in persuading a starving man to eat.

At the time of writing in late 1996, there are still three more years to go until the end of the *Transition*. For many people, there is not a supply cloud on the horizon. The man in the street still drives to work in crowded streets; air line fares have never been lower. Even in the quiet of rural France where I write, there is the nearly perpetual thump of distant tractors at work, burning diesel oil. The European Commission in its Green Paper on *Energy to 2020*[14] admits to growing oil demand, but does not even ask the simple and obvious question of how much there is going to be available.

In the following pages, I will examine how dangerous this complacency is. I think the World is about to face, not another shock if that means a short and sharp interruption in supply, but a fundamental discontinuity when conventional oil production peaks and begins an irreversible decline. It is time to take stock of the resources.

NOTES
(For references, see Bibliography)

1. Nerheim, G. and B.S. Utne, 1992 give an excellent account of the Smedvig Company, which began in shipping but successfully became a drilling contractor.

2. Lloyd George, the British Premier in the First World War, liked the idea from the standpoint of a homeland for the Children of Israel in romantic biblical terms and not as a new Judaic secular state, see Fromkin.

3. Yergin, D., 1991.

4. Schweizer, P., 1994 gives a remarkable account of the secret policy of the United States to bring down the Soviet Union by economic means and with the help of the oil weapon applied by King Fahd of Saudi Arabia.

5. Schweizer, P., 1994.

6. Nation, L., 1995 reports on the dismay of Hodel, the former US Energy Secretary, to discover that he was not privy to

his government's policy. He is knowledgable enough to express concern about the coming oil crisis.

7.   Allen, R.V., 1996, the Foreign Policy Adviser at the time, explicitly confirms the validity of Schweizer's book.

8.   see Perrodon, A., 1985 for a description of the details.

9.   Aburish, S.K., 1994.

10. According to Roberts (1995 p. 315) Glaspie did have several meetings with Saddam, in which she said that the United States did not take sides in disputes but respected sovereignty. Roberts states that Saddam thought he had a "green light" to press his claim against Kuwait, but explains it as a failure of the language of diplomacy.

11. Scott Report., 1994, a judicial enquiry, details the British Government's connivance with the export of arms making equipment to Iraq.

12. Colvin, M., 1994, reports that Saudi Arabia contributed to Iraq's nuclear programme.

13. General Schwarzkopf, who led Desert Storm but turned back at the gates of Baghdad, does not explain the reasons for this strategic retreat.

14. European Commission, 1995, issued a Green Paper on Europe's energy policy to 2020, oblivious of the resource constraints to oil production. Is it blind? Or is it politically more advantageous to react to a crisis rather than anticipate one?

Chapter 5

INSIGHT AND INFORMATION

"Put a dime in the slot, and magic fingers will console you": we did, but nothing happened save an imperceptible vibration of the bed. We had arrived in the spring of 1968 in San Francisco to stop over in a motel after a long flight across the Pacific from Australia. It was our first experience of the technological and, indeed, the commercial prowess of the New World. America then seemed an exciting place of opportunity; and we had arrived as "landed immigrants", following the well-trodden path of Europeans seeking a new life.

In Chapter 1, I described how I had fallen in love with Colombia and had become deeply absorbed in its geology. I felt that I was on the brink of great discoveries into the nature of mountain building and continental structure. The geological revolution of Plate Tectonics was then breaking; and I felt that my Andean studies could somehow contribute[1]. So, after nine months in Australia and New Guinea to which BP had transferred me after the golden Colombian days, I had quit to try something new in America. My hope was that I would somehow find my way back to Latin America. I had lined up interviews with both the State University of New York, and Amoco.

We continued to New York where we stayed with our good friends from Colombia, the Barrys: Richard had recently joined Mobil in New York to work in a new subject called Operations Research.

The interview with the University was not a success: the faculty terrorised me over lunch with questions on academic subjects which had long since passed from my mind. Surprisingly, they did offer me an Assistant Professorship, but I lacked the confidence to return to academia. I realized that I did not know how to teach freshmen, whatever they were. I thought I would try Amoco before accepting. When a few days later I presented myself at their office at 555 Fifth Avenue, I felt much more at home.

It was not in fact then Amoco, but the American International Oil Company, the international arm of the Standard Oil Company of Indiana, one of Rockefeller's own, based in Chicago.

I liked the office immediately. I was vastly impressed by Bill Humphrey, the Vice-President of Exploration. He was a short man with clear blue eyes and brushed back hair, who had had much experience

of Latin America, and we were soon discussing the details of the age of the *Santa Rosa Formation* in Colombia. We seemed to hit it off. I was also impressed by the Regional Geologist, Nestor Sander, a silver haired Californian, who sat with his distance glasses perched on his forehead, in a darkened room surrounded by books. He told me of his early experiences in Arabia, and of the time he had spent in Paris, where he had married a French girl, Georgia. He was a civilized and knowledgable man. There too, was the Chief Geologist, Ward O'Malley, a charming Irish American, whose father had been a well-known journalist on the New Yorker. It was a small and delightful office; and when they offered me the job of regional geologist for Latin America, I thought that I had landed on my feet.

We explored the beautiful countryside of New England with its white clapboard houses, reminiscent of Scandinavia, and found a house to rent near Greens Farms, close to Westport. I remember that it belonged to a splendid

Fig. 5-1. Dr Nestor Sander, who ran the regional evaluation.

New Yorker named Zerlene Joffe, the wife of a crocodile skin importer of all things, who confided that "she did not function under stress", and saw her psychiatrist frequently in the best traditions of the ladies of that city.

Commuting into New York on the bankrupt Newhaven Railroad was an experience in itself. "Sure glad they don't run an airline" was how one traveller succinctly put it as we walked along the track to a relief train, a common occurrence. The bar was full of Wall Street brokers, and there was a special elite carriage called the "Southport Club Car" hooked on the back for distinguished gentlemen from Southport, who played cards and drank gin-and-tonic served by a Negro steward.

But this new life was not to be. Within a few months of our joining, the Company announced that it was going to consolidate the international company with the domestic organization in Chicago. So before long, we found ourselves driving through the acrid atmosphere of the steel mills of Gary, Indiana, the gateway to Chicago, not having been reassured by the numerous billboards through the prairies, advertising *chewing* tobacco.

One of the first tasks given to the new consolidated organization was to try to determine just how much oil the world had as a basis for defining its new strategy. The project was placed under the direction of Bob Blanton, a chain-smoking rather stressed man, with a clipped southern accent. I found myself trying to assess the potential of Latin America. As I look back, I realise that it was not a very sophisticated analysis.

The first step was to search out the maps and reports on each country, put out by Petroconsultants of Geneva. The maps showed each basin, together with its oilfields and exploration wells. The next step was to look up the reserve and production data, as published each year by the Oil and Gas Journal.

We evolved a standard approach which involved calculating the area and rock-volume of each basin; determining how much had been found by how many wells; and then making some intelligent guess for the undiscovered potential. In those days, there were no spreadsheets or even electronic calculators, and so we did not even try the various statistical techniques that are now easily applied. Instead, it was simply pragmatic geological judgment. Unfortunately, I don't remember what the global total was, but there have not been too many surprises in Latin America, save for the deep water discoveries off Brasil. Having previously recommended the Llanos as prime territory, the subsequent giant discoveries of Canon Limon and Cusiana were broadly anticipated.

Until then, although I had been aware of the limitations of the areas in which I had worked, I had always thought that the world was a large place with plenty of scope for new discovery. Now, for the first time, I began to realize that the prospects worldwide were indeed as limited as in the places I knew: distant fields turned out to be no greener. Even in those days, when much less was known than now, many areas did not smell right for oil: and indeed they have proved to be barren. As the implications of this new insight

seeped in, I began to look at life differently. I realized that the mindless consumerism, nowhere more evident than in the country where I was now living, was insupportable. I began to have a doomsday view of the future. We were, I came to realize, on the edge of an abyss.

It absolutely changed my attitude to exploration. I realized that the few prospects that actually met the criteria to succeed had a transcendental value; and I realized that most of the others would fail. I began subconsciously to appreciate the 20:80 rule: 20% of the people own 80% of the wealth; 20% of the patients use 80% of the state health funds; and only 20% of the prospects succeed, if that. This polarity was not understood by the Company; whose management systems tended to make everything look the same under the strait-jacket of hypothetical economic evaluation (see Chapter 10). As a consequence, it missed out on the few prime prospects, where the concession terms were usually tough, and frittered away its effort on the 80% that lacked the essential ingredients.

My career in exploration was to continue for twenty years, but this insight never left me. Circumstances, especially arising from government licensing policy, often forced me to apply for acreage in which I had little confidence as part of a sort of mad competitive game. I hardly ever expressed an objective opinion to my own management, and when I did was accused of pessimism. They were substantially blind to what they were doing, but often it did not matter because the cost was deductible from high marginal tax. I will describe this situation more fully in Chapter 12, covering the Norwegian experience.

We did not like living in Chicago. The climate was awful: icy cold in winter when winds blew off the frozen Great Lakes; and sweaty hot in summer. When we asked Mrs Muther, the real estate agent, to find us a charming house in the shade of Evanston University, she sniffed with the comment *"Very definitely integrated"*, before coming to the point with the question *"how much do you earn?"* That information placed us, with few options, in Forest Avenue, Wilmette, a northern suburb. It was a pleasant old house in a tree-lined street, where the neighbours complained that there were weeds on our lawn. It was not a colourful or interesting place in which to live, although the job was excellent.

For two years, I worked in the small new ventures team, concerned mainly with Latin America, but often appraising opportunities in other parts of the world as they came in: Portugal, Korea, Svalbard, Guatemala, Peru, Borneo. I frequently found myself writing persuasive memoranda for Bill Humphrey's signature, the Vice-President of Exploration, who preferred to travel and search out opportunities and contacts, rather than feed the corporate bureaucracy. I tried always to end them on a Churchillian note with a call to arms, hoping to galvanize the tired group with enthusiasm: *"I urge the Members of the Committee to authorize our negotiators to set forth and open discussions on this exciting new opportunity..."*.

Eventually, I managed to persuade the Company to go into Ecuador to explore a promising area in the Amazon

Fig. 5-2. Ecuador, which enjoyed an oil boom in the late 1960s. We lived at Farsalia, at 8500 feet altitude outside Quito, in a beautiful house surrounded by eucalyptus woods and with a view of seven snow-capped volcanic peaks.

Fig. 5-3. Discussing negotiations with Juan-Carlos, the intermediary.

Fig.5-4. I look more urbane than previously.

headwaters, and I got myself transferred there as Chief Geologist. Since I spoke Spanish, I soon found myself doing less geology and more negotiating. Everything from permission to use jungle airstrips to labour contracts and applying for new concessions involved labyrinthine negotiation. It would be wrong to describe Ecuador as a corrupt country, but it was necessary to find ways to pay for services received: a subtle but valid distinction. I became a specialist in this murky but intriguing business: finding the intermediary who performed; and judging the style and the amount required to make the wheels turn in our direction.

It was not Ecuador that was corrupted, but myself: geology was abandoned in favour of more facile amusements in management. I was 38 years old, and mapping the Cretaceous unconformity had become stale.

When this venture ended and a recall to Chicago beckoned, I quit and returned to London, becoming manager of a small Texas company, The Shenandoah Oil Corporation of Fort Worth, whose principal shareholders were no less than Julie Andrews and James Stewart. Now, instead of the Ecuadorean generals, I had to deal with the socialist ministers who were deciding how to allocate North Sea oil rights. I had leant my lessons well and knew what to do, when signing up as a partner the owners of the Daily Mirror Newspaper, which supported the Labour Party. This chapter lasted several years and was followed by others that need not concern us.

I will pick up the story of when I was Executive Vice-President of a Belgian oil company in Norway in the late 1980s.

Whereas the wheels in Ecuador were greased by money, those in Norway were turned by "Research". Norway had a passion for research; and the oil companies' applications for promising licences were awarded partly on the degree to which they had contributed to Norwegian research. It became a racket as the companies postured about the claims for their particular projects, few of which resulted in anything particularly useful.

When wondering what project I could find to support our claims for preferential treatment, I fell upon the idea of sponsoring a research project into the world's endowment of oil, which I thought would be of singular value to the Norwegian government in planning its strategy. I aired the idea with Farouk al-Kasim, an Iraqi who was effectively running the Norwegian Petroleum Directorate, and had an enthusiastic response. Initially, I was not able to do much myself on the project, but it began to take shape. Later, when I had become the victim of a palace revolution, and was locked up in solitary confinement in an outhouse with nothing to do while the Company plucked up courage to show me the door, I was able to dedicate myself to the study full time.

Fig. 5-5. Farouk al-Kasim, of the Norwegian Petroleum Directorate who supported research into oil resources.

Photo courtesy of Farouk al-Kasim

Fig. 5-6. A cartoon of Buzz Ivanhoe, an expert who showed us the way.

The first step in the project was to invite Buzz Ivanhoe to come and visit us. I had read an article by him in the Oil and Gas Journal about resource depletion, which immediately struck a chord. It was evident that he shared my underlying concern. He had done some interesting work, plotting discovery against footage drilled[2] and investigating distribution patterns. I invited him to Norway to give us the benefit of his advice and knowledge

He was an interesting man, who had been an adviser to Occidental, but was now retired (see p. 84). He had been investigating the subject for a long time and had cooperated with the legendary M. King Hubbert, who in 1956 had correctly predicted when production in the United States would peak, developing the so-called Hubbert Curve[3].

He alerted us to the work being done by the United States Geological Survey on the subject. We accordingly sent someone to the United States to meet the team, which was led by Chuck Masters[4], and to obtain their material.

Next, we secured the database on historical oil reserves and production from the Oil and Gas Journal, and a listing of giant fields from Petroconsultants.

I continued this work with the Directorate after my precipitate "retirement", working with an enthusiastic team who soon saw the importance of the work. We started building the database, and began to investigate various analytical methods. I began to systematically collect references.

At an early stage of the study, we spotted the anomalous increases in reserves reported by several OPEC countries in 1988, and we began to collect data on published estimates of the *Ultimate Recovery*. I wrote an article for Noroil[5] at that time, which coined the term "political reserves" for the anomalous increases. The expression caught on. I remember sitting up all night in Stavanger with my rather primitive computer, trying to calculate depletion curves, using a programme designed to calculate mortgage rates.

We had enjoyed life in Norway and had really no wish to leave. But the prospect of retiring there with the high cost of living and the punitive tax regime, made the future seem fraught with hazard. Furthermore, having been at the centre of the oil business for many years, it now seemed strange to be a nobody. It takes only a few weeks to become out of touch.

For some years we had been taking vacations in rural France, and so we thought we would move there, eventually buying a beautiful old stone house near the sleepy village of Milhac, where I may be one of the younger inhabitants. Rural France is a magnificent place where people with smiling faces still till the land, as they have for centuries. It will be survival territory after *The Coming Oil Crisis*.

We have not regretted the move, especially since the advent of fax and the personal computer makes it possible to stay in touch. I set up my office in the loft surrounded by ancient beams, heaps of papers on the floor, and an old copy machine. I share it with a herd of spiders.

The Norwegian authorities kindly allowed me to publish the results of the study I had made with them as *The Golden Century of Oil 1950-2050 – the depletion of a resource*[6]. Although with hindsight, it seems a rather primitive analysis, it did capture most of the essentials.

To that point, my estimates of the undiscovered potential were based on my own intuitive judgment and knowledge of the underlying geology. The next step forward came when I was invited to the Total oil company in Paris to discuss the subject. There, I met Jean Laherrère, a specialist, also recently retired, who had read the *Golden Century*. That meeting in turn led to an invitation from Petroconsultants to develop the subject further using their authoritative and comprehensive database. Jean Laherrère began to explain some of the statistical techniques he had developed to refine old fashioned judgment. Since I am not very mathematical, it took me some time to grasp the *parabolic fractal* and the *shifted logistic curve,* but gradually I came to understand them and see the strength of the techniques. I will cover these things in a later chapter.

As will be discussed further, the great difficulty in any study of this subject is to secure reliable data on what has been discovered so far, as a basis for extrapolating future discovery. Petroconsultants had been accumulating this information for many years, and access to its database has proved invaluable.

This brings the story almost up to date. I continue to write articles on the subject, and am increasingly being asked to speak at conferences. I don't do it very well, but the subject itself usually does command attention. Many people intuitively understand the general position without necessarily having exactly defined it in their minds. In the next few chapters, I will put some meat on the bone, explaining the conclusions of the study.

AMOCO :
King of the Mid-West

Standard Oil, J.D. Rockefeller's remarkable empire, was broken up by anti-trust legislation in 1911 into thirty-seven independent companies with their own boards of directors, although Rockfeller interests did retain a substantial shareholding in all of them. One such daughter was the Standard Oil Company (Indiana), or simply, Standard of Indiana, based in Chicago. It inherited the company's premier refinery at Whiting, Indiana, seventeen miles from the centre of Chicago, just across the state line into Indiana. It was here that William M. Burton conducted successful research into new ways of refining petroleum to remove the sulphur content, and to use catalysts and other methods to optimize the production of a range of products, especially gasoline.

When it commenced its independent existence, the enterprise was strictly a refining and marketing company, having retained the brand name *Standard,* under which it sold products throughout the Mid West. It had no

Fig. 5-7. Horse-drawn Standard tanker.

Photo courtesy of Amoco

production of its own, securing its supply from other pipeline and production companies, notably the Prairie Oil Co., the Midwest Oil Co., and later its competitor, Sinclair. From time to time, shortages of crude caused its refineries to be partly shut down. To overcome this problem, the Company decided to move into the upstream and secure its own sources of oil. In 1918, a Col. Robert W. Stewart became the Chief Executive, and set about acquiring production. He had started his career as a lawyer in Dakota, but soon rose to prominence in the business world. He is described as intelligent, shrewd, 6ft tall, 250 lbs in weight and endowed with tremendous physical strength and a commanding personality. He was in control for eleven years, during which time he acquired numerous producing properties and other companies, of which the most important was Pan American. It was a company that had been built by Edward Doheny, a Californian mining prospector turned oilman[7], having substantial rights in Mexico and Venezuela, as well as a sales arrangement with the American Oil Co of Baltimore. It was well on the way to becoming a major international oil company.

Despite these achievements in securing production, Stewart's career ended on a bad note when he was deposed by the Rockefellers who thought he had had his hand in a dubious slush fund connected with the infamous Teapot Dome scandal, in which the Secretary for the Interior was corrupted.

In 1930, Standard decided to sell Pan American's major assets in Venezuela and Mexico to Esso. It was facing the great Depression and the threat that import levies might be imposed to protect US producers. It was also nervous that a clause in the 1917 Constitution of Mexico, providing for the expropriation of foreign assets, might be invoked, as indeed

Fig. 5-8. Robert W. Stewart, an imaginative force in the early days who tried to build up production for the company.

Photo courtesy of Amoco

In 1948, it put its toe again into international waters, taking up abortive rights in Colombia, but withdrew a year later. It was not until 1957 that it began to make a serious

Fig. 5-9. Bill Humphrey, a charismatic geologist who combed the world for prospects.

Photo courtesy of Amoco

it was eight years later. However logical at the time, it was one of those transcendental decisions that affect the destiny of a company: with it, Standard of Indiana surrendered the chance of becoming a leader in the world of oil, which it was then close to grasping. With the loss of these foreign holdings, the Company became a strictly domestic enterprise, serving primarily the Mid West. The sale of Pan American's properties gave rise to protracted litigation with the Blausteins who owned the American Oil Company, who relied on it for their oil supply. The dispute was not settled until 1954, when Standard acquired American and put the Blausteins on the board. There had been earlier disputes with Esso over the use of the Standard brand name, and the Company then adopted the name, "American", from its acquisition, which was later contracted to Amoco.

The Company built up its domestic production in the inter-war years by a long sequence of complex acquisitions, in which all sorts of residual minority interests survived, providing the lawyers with a field day. It operated upstream variously under the names of Pan American and Stanolind. By 1952, all of these endeavours had resulted in it securing production of 235,000 b/d (or 3.5% of the US total) about half of its marketing needs. It was not in fact a great deal: being less production than provided by a single North Sea field.

attempt to expand overseas, creating an arm in New York with the name of The American International Oil Co (AIOC) to which it recruited two dynamic executives, Chris Dome and Bill Humphrey

Within the span of a few years, these two men had secured rights in Trinidad, Argentina, Venezuela, Colombia, Egypt, Indonesia, Iran, the Netherlands, United Kingdom and Norway, which forty years later still provide about sixty percent of the Company's oil supply. They were men who knew their job, and worked on their own initiative with a minimal staff, most of whom were hand picked and came with international experience from other companies. In Iran, they broke ranks with the industry offering the government a higher take than was normal. In Britain, they teamed up with the influential British Gas Council (predecessor of British Gas), and secured a key position in the southern North Sea gas province. They understood the international environment, and knew they needed friends at court.

But Humphrey and Dome had too much independence and initiative for the head office in Chicago, who probably resented their success. I have often noted that nothing in the corporate jungle upsets a head office so much as a successful affiliate. Dome was thrown out, later being killed piloting his own stunt plane. The New York office was closed, and the international arm was integrated with a domestic organization in Chicago, later moving to Houston. The staff was expanded enormously, and a massive

The Coming Oil Crisis

bureaucratic structure of committees was erected. The predictable consequence was that the Company hardly found any more oil, and generally withered away in the areas such as the North Sea where it had been a respected pioneer. Its latest major engagement has been to secure a strong position in the Former Soviet Union, but it remains to be seen if that will turn out to be successful in view of the intractable political, fiscal and contractual problems, which can only be expected to deteriorate as domestic demand recovers. Otherwise, its global exploration efforts have had dismal results apart from some gas discoveries.

A telling article[8] by its Exploration and Production Manager explains how the Company has devised a complex economic formula to make money out of exploration, heavy with all the buzz words of "Impact", "Creativity", "Focus", "Vision", but it says it all when it reveals that it is premised on an estimate that world Ultimate recovery of conventional oil is an unrealistic four trillion barrels, more than double the number proposed here, demonstrating a poor grasp of the true position. Unfortunately, no amount of "creativity" can find what is not there to be found.

Amoco at heart is a company of the Mid-West, where it has had a long and successful role as a marketer and refiner. Its achievements have been primarily in the refining domain where it invented several important new processes. It was also a successful post-war explorer in the United States and Canada, at one time having more acreage under lease than any other company, a policy smacking more of quantity than quality. But by then most of the world's major fields had been found: it had missed the boat. Thanks to its marketing expertise, it remains nevertheless one of the largest corporations in the United States. Rumours circulate from time to time of a possible merger with Phillips Petroleum. It would be a logical move, consistent both with the Company's past acquisition policy and with a future that this study charts for the world's larger oil companies.

Amoco was a good and generous company to work for; and one that was far more welcoming and open to foreigners than were some of its European counterparts. I am grateful for the time I spent with it.

DATA BASE

In this chapter, I have summarised how this study of the world's oil endowment evolved. As much as anything, it sprang from an early very pragmatic evaluation made for Amoco in 1969. As I will explain later, the great difficulty to be faced by anyone attempting such an exercise is to find valid data on how much has been discovered so far and when. To know that is clearly the first step in extrapolating what remains to be found. A cornerstone is the simple and obvious recognition that there is an *Ultimate Recovery;* and that production will one day end because it is a finite resource, formed and preserved only rarely in a few places in the Earth's long geological history. Once this mental barrier of there being an *Ultimate* is crossed, plausible estimates of future production can be made, even if the numbers are not very precise and even if there are disagreements as to the details.

There are four principal sources of information:

Petroconsultants

This firm was set up in 1955 by Harry Wassall[9] to provide an industry news service with information on concessions, drilling, production, reserves and much besides. It began in Latin America but later provided worldwide coverage from its headquarters in Geneva. It maintains close contact with the oil companies and governments, and has now assembled a colossal database, which it provides to the industry. Its reports form a sort of ready reference, and the first port of call when evaluating a new area. In particular, its maps showing wells, fields and concessions, are an invaluable source of information.

In addition to data on production and reserves, it carries important information on drilling, which is especially critical to the evaluation. Its data relate to individual fields, with reserve revisions backdated, as if known at the discovery date. The reserves are properly classed as "median probability" reserves, meaning that the risks that the actual number will be above or below the estimate are equally matched. This largely overcomes the distortions deriving from apparent "reserve growth", discussed more fully in Chapter 6. The database does not cover North America, but otherwise is indispensable for any serious analysis of the position.

Oil & Gas Journal

The Oil & Gas Journal has for many years published data on reserves and production by country. The data come from a questionnaire distributed to governments and others, and is reported as received. In recent years, much of the information reported by governments in many countries has become unreliable, partly for political reasons. The reserves are supposedly *proved reserves,* although in reality they are often far from that. The numbers relate to crude oil and condensate.

BP Statistical Review of World Energy

BP publishes a booklet each year, which contains a wealth of energy statistics. However, it is important to recognize that in the case of oil reserves, it simply reproduces the numbers from the Oil & Gas Journal, adding natural gas liquids (NGLs). Many analysts assume that these numbers have the tacit support of a prestigious company with considerable knowledge, but that is not the case. Since the reserve numbers are erroneous, and the concept of Reserve to Production Ratios, which are depicted in the booklet as a measure of security of supply, is grossly flawed, it is unfortunate that the Chairman in the Introduction should describe it as a "bible" intended to "improve our understanding of the world energy scene and so make better-informed decisions". In fact, it does just the opposite.

World Oil

World Oil publishes comparable reserve and production data to that of the Oil & Gas Journal. The numbers are however different in many cases, being more reliable for several OPEC countries.

United States Geological Survey

The USGS has published an assessment of world reserves and undiscovered potential in successive meetings of the World Petroleum Congress[10]. The study is in fact based on Petroconsultants' material. It is important to recognize that the USGS has its own definitions, which need to be decoded before the data can be put to good use[11]. As an arm of the United States government, the organization has to be circumspect on what it says on this sensitive issue.

NOTES

(For references, see Bibliography)

1. See Stoneley, R., 1969, formerly Professor of Petroleum Geology at Imperial College, who incorporated some of the results in a global study.

2. See Ivanhoe, L.F., 1985: Buzz Ivanhoe was one of the first to investigate the size distribution of oilfields and the patterns of discovery. He remains a leading authority on the subject.

3. Hubbert, M.K., 1956, is the father of production forecasting based on an appreciation of the underlying resource, having correctly predicted when the United States would peak almost twenty years before it did. He was an eminent scientist who wrote on many other topics. He is immortalised in the "Hubbert Curve", the details of which he did not publish until he was 80 years old.

4. Masters, C.H., 1987, 1991, 1993, 1994, has published valuable resource estimates at successive World Petroleum Congresses. He skilfully found a way to meet the political needs of his principals in the USGS, while maintaining his professional integrity. His definitions need to be decoded before the numbers, which are High Case numbers, can be used.

5. see Campbell, C.J., 1989: one of his first articles on this subject.

6. Campbell, C.J., 1991: a book resulting from a study for the Norwegian Petroleum Directorate. While the study has evolved greatly since then, the *Golden Century* curves do provide a depletion profile linked to the estimated ultimate recovery, not found elsewhere in the public domain.

7. see Knowles, R.S., 1959 for a colourful description of the early oil days.

8. Schollnberger, W.E. 1996, describes Amoco's economic analysis system, but lets slip that it is premised on near "infinite" resources, quoted at 4 trillion barrels. With this unrealistic premise, the system is unlikely to work.

9. See a booklet dedicated to Harry Wassall, published by Petroconsultants in 1996, which describes the entrepreneurial drive of this imaginative man, and also a Dedication at the beginning of this book.

10. Masters, C.H., 1994.

11. Campbell, C.J., 1995, explains the USGS definition: in effect the reported Ultimate of 2300 Gb reduces to about 1800 Gb if the numbers are redefined in terms of *median probability* reserves.

Chapter 6

HOW MUCH OIL HAS BEEN FOUND

In the earlier chapters, I have discussed both the origin of oil in geological terms and the history of the industry. Now, I approach the heart of the matter, and turn to how much has been found, how much is yet to find, and, most important of all, how much remains to produce, not only in quantity but when. It is, however, first of all, necessary to decide what we are talking about, for, as explained, petroleum is a slippery substance that comes in many different forms, each having its own characteristics and depletion pattern. It is essential at the outset to distinguish *conventional* from *non-conventional* oil and gas, but that is easier said than done.

CONVENTIONAL OIL

As already discussed, the unfettered production of oil, as in an uncontrolled blow-out, rises rapidly to a peak and then declines exponentially until the reservoir is exhausted. To understand this, it is necessary to think of the reservoir. The oil lies in pore-space within the reservoir rock under enormous pressure. When the wellbore taps the reservoir, the pressure in the adjoining pores is sufficient to force the oil into the well through which it flows to erupt at the surface. In an uncontrolled well, or blow-out, there is a plume of escaping oil several hundred feet high (see Figure 3-22). As the pressure in the immediate vicinity of the wellbore is depleted, so oil farther and farther away has to be tapped, but instead of flowing through a pipe several inches in diameter it has to make its way through a network of interconnecting pores. The pore-throats restrict the flow, as do the capillary pressures at the interface of the oil and the rock grains, or the film of water that commonly surrounds them. It is obvious therefore that the flow will be progressively reduced at a rate reflecting the declining differential pressure between the formation and the wellbore, as well as the constrictions to the flow. It is clearly easier for light oil to flow than for heavy viscous oil. The pressure itself is due variously to the simple expansion of the oil that is compressed in the reservoir; the effects of an expanding gas-cap above the oil; the effects of gas coming out of solution within the oil itself; or a water drive due to the encroachment of water under pressure from below the oil contact. To begin with, the oil flows to the surface under its own pressure, at any rate in most fields, but as the natural flow rate begins to taper off, water or gas may have to be injected into the reservoir to help maintain

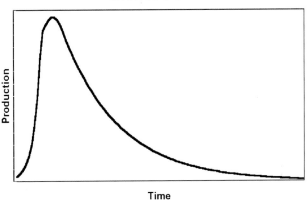

BLOWOUT

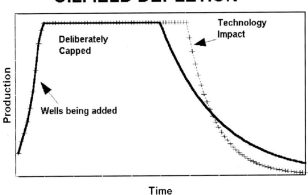

OILFIELD DEPLETION

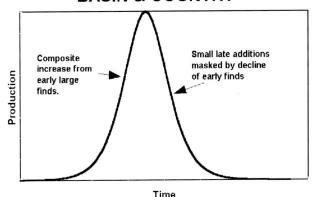

BASIN & COUNTRY

Fig. 6-1. Depletion profiles.

the pressure. In fact, in most modern fields, such stimulation commences early in the life of the field. The procedures are known as gas injection and waterflood, and are collectively termed secondary recovery. Eventually the well has to be pumped.

Most of the production to date has relied on primary and secondary recovery techniques, and it is likely that most of the oil to be produced over the next two decades will also do so too. This may be termed *conventional oil*, which has its own characteristic depletion pattern: starting at zero, rising rapidly to a peak or managed plateau, and then declining exponentially to exhaustion. The plateau of production in a field results from the deliberate choking back of production to optimize recovery or for other reasons. I will later turn to *non-conventional* oil, including particularly heavy oil, which has a very different depletion pattern, rising slowly to a long low plateau of production. In short, *conventional* oil production is cheap, easy and fast, whereas *non-conventional* oil production is expensive, difficult and slow.

About two million barrels a day of *non-conventional* oil[1] are already in production, but in practice it is not easy to draw a firm line between the two categories because the statistics are not always sufficiently accurate to do so. That said, it is of cardinal importance to make the distinction even if the boundary is a little fuzzy. What concerns the World most is when conventional production peaks, and shortages begin to appear. Since peak production is primarily driven by *conventional* oil, that is the stuff we have to think about most. I will accordingly confine my attention to this category of oil in the next few chapters, and all the statistics I mention will relate to it unless otherwise specified. Later, I will come to *non-conventional* oil which is likely to be an important, but very different, resource in the future.

We are unfortunately not quite out of the woods yet, because there is another confusing element: *Natural Gas Liquids* or NGL. As already explained in Chapter 2, gas at high pressure in the reservoir commonly contains dissolved hydrocarbons that condense into a liquid phase on being brought to the surface, being known as *Condensate*. Furthermore, additional amounts of liquids, mainly butane and propane, can be extracted from gas by processing. These substances are in many cases fed directly into the oil pipeline and are not distinguished in the records of production. Ideally, they should be treated separately since they are related to the gas and not the oil domain. If they are lumped together with oil, they distort the discovery and distribution patterns affecting the assessment of the undiscovered potential. Peru is a case in point: a large part of its liquid reserves relate to the giant Camisea gas-condensate field, which, if included, would distort the assessment of its oil endowment.

CUMULATIVE PRODUCTION

Apart from the difficulties of defining what is meant by the term *Conventional oil*, the concept of *cumulative*

production is a straightforward one: it is simply the total amount produced in the area under consideration as of the reference date.

Naturally, the records in early years were not particularly accurate in many places, although the amounts involved were by to-day's standards not large enough to be significant. It is also uncertain how much credence can be placed on the statistics in countries where the oil industry is controlled by a national company and is producing oil for the domestic market. Many countries do not report production by field, sometimes with production from several fields being metered together. Territorial changes also affect the statistics: for example, the *cumulative production* of Israel is in some databases exaggerated by the inclusions of oil from temporarily occupied Egyptian fields. The treatment of Kuwait war-loss, when about two billion barrels – an amount equal to the size of a large North Sea field – went up in smoke in the Gulf War is another issue. In principle, it should be treated as production insofar as it depleted the *reserves* by like amount, but how it is treated in the statistics is sometimes obscure. There were of course other war losses in Romania, Indonesia, Burma and the Soviet Union during the Second World War, which are hidden in the pages of history. There are also leakages, especially in the FSU, and operating losses. We may note in passing that gas statistics are more difficult because of the amounts flared or reinjected. Lastly, is the issue of NGL as already discussed. As we will see, however, the problems of knowing how much has been produced are minuscule in comparison with those of arriving at valid numbers for *reserves:* the amount yet to produce from known fields.

While on the subject of *cumulative production*, we may note first that it eats into the reserves: as the first goes up, the second comes down, other things being equal. Secondly, *cumulative production* when it ends is what is termed *ultimate recovery*. Obviously, it is production which is all that really matters to us: how much was generated and how much is held irretrievably in the ground are themselves irrelevant except to the degree that they influence production. Production is what makes money: and a short-term view of it is natural.

Figure 6-2. shows production together with *cumulative production* by region and the major countries as of the end of 1996. It is always important to stress the reference date.

It is noteworthy that the United States has produced more oil by far than any other country, and indeed more than all but two of the World's eight regions. It is not difficult to appreciate on this basis alone that the country is at an advanced stage of depletion. The importance of the Former Soviet Union is also evident.

The World as a whole had produced some 784 billion barrels as of the end of 1996: the first of the key elements to address when considering depletion. It amounts to about forty-three percent of the *ultimate recovery*, as we shall come to see. In other words, the World will soon have produced half of its *conventional* oil endowment: that says a lot about the future.

Produced by country	Production kb/d 1996	Cumulative Production Gb
USA	6478	173
FSU	7044	128
Saudi Arabia	7481	77
Venezuela	2955	48
Iran	3675	47
Kuwait	1818	27
Iraq	600	23
Mexico	2854	23
China	3127	21
Libya	1403	20
By Region		
N. America	8297	193
ME Gulf	16264	193
Eurasia	10347	156
L. America	8637	95
Africa	6720	63
East	3886	33
W. Europe	6207	29
ME Other	2844	19
World	63486	784

Figure 6-2. Production and Cumulative Production (excluding NGL). The regions are: ME Gulf is Abu Dhabi, Iran, Iraq, Kuwait, Neutral Zone and Saudi Arabia; N. America is USA and Canada; Eurasia is FSU, China and E. Europe.

RESERVES

Whereas oil may be readily metered at the surface when it is produced, there is no direct way by which to measure the amount of oil in the reservoir far underground. The term *reserves* is an unsatisfactory one in many ways. In normal parlance, it means something sure and fixed, such as financial reserves or reserves of troops in a battle, whereas in the case of oil, it is no more than an estimate, and often a loosely defined one at that.

When a prospect is evaluated prior to drilling, estimates are made of how much oil the trap may contain, namely what is called its *oil-in-place*. From this starting point, a certain assumed recovery factor is applied to estimate the *reserves,* meaning by definition how much will be produced[2]. Much is held irretrievably in the reservoir. If the first borehole, termed a New Field Wildcat (NFW) is successful, a new estimate will be made with the new information about the reservoir obtained from the well. The estimated reserves are termed *Initial* (or *Original*) *Reserves,* being the amount in the trap before production commences. They are progressively reduced by production. The reserves are what is estimated as remaining to produce at the reference date[3]: namely the *Initial Reserves* less the *Cumulative Production*. When production ends, as one day it will, the *Ultimate Recovery* of the field is equivalent to the *Initial Reserves* subject to such revisions as have had to

be made.

I will come back in a moment to discuss the confidence attaching to the estimates of the reserves of a discovery of oil or gas, but first will consider the more difficult issue of how to describe the estimates of what remains to be found in new discoveries yet to be made. They are clearly not *reserves,* which by definition refers to something known, or rather, known within a band of probability. What we mean are estimates of what may become *reserves,* if found. There is not a good word for this category of "reserves". I evade the issue by referring to it as the *"yet-to-find"* or the *"undiscovered"*.

It leads us to another difficult term: *"resources"*. In coal mining, for example, we can picture the deposit as a whole, as defined perhaps by core-drilling, which could be said to hold *resources* of a certain quantity, and to distinguish that from the amounts that are actually being exploited from the adits and shafts, which could be termed *reserves*. The procedure works less well with oil, essentially because it is fluid and very profitable, meaning that, with some important exceptions I will come to, most of the oil in known deposits in production qualifies almost immediately as fully exploitable *reserves,* leaving little to fill the *resources* category. Some classifications, treat the estimates of the *yet-to-find* as *resources,* but I find that misleading, implying that they are there to be found, which is less than sure. The term *resources* is sometimes also applied to the amounts of oil that may become available as a consequence of radically changed economic or technical circumstances, but they are very much more dubious estimates than applicable to the unexploited part of a coalfield. Another usage of *resources* makes it synonymous with *non-conventional* oil, which I discussed at the beginning of the chapter. On balance, I think it is better to avoid the term *resources,* except to describe the material in very general terms, as in the expression "natural resources". In any event, it is very important to avoid the misconception that the world is endowed with almost limitless *"resources"* of oil and gas that will be inexorably converted to *reserve* status by technology or investment – a view mistakenly held by some economists[4] and institutions who should know better.

Returning to consider further the nature of the estimates of *reserves,* it is obvious that they, as all estimates, are subject to degrees of uncertainty. The *initial reserves* of a prospect are determined on the basis of seismic surveys, that so to speak provide an X-ray of the structure, whose volume may be calculated. Certain assumptions are made about the reservoir's character and thickness, as well as the extent to which the trap is full, to provide a basis for calculating the *oil-in-place*. A recovery factor is then applied to calculate the *reserves* after volumetric adjustment for expansion at surface conditions of temperature and pressure. (*Reserves* by definition are recoverable, so it is tautologous to speak of *recoverable reserves,* as is often done).

A wealth of new information becomes available when the first well on the prospect has been drilled. It permits a more accurate estimate of the likely *reserves,* but the distribution of the reservoir within the prospect is still uncertain. It may thicken and change in quality in a particular direction, reflecting variations in the sedimentary environment in which the reservoir rocks were laid down; the prospect may be found to be divided up into separate fault-compartments, each with its own characteristics; or there may be more than one reservoir, each having its own variable distribution and characteristics.

It is therefore normal to drill several appraisal wells to delineate the field before embarking on a development, especially offshore where expensive facilities have to be installed. It is only then that sound estimates of the *reserves* can be made.

The explorers should try to estimate the amount of the reserves likely to be recovered, namely those having a median probability of occurrence, but in practice they tend towards optimism under the pressures of "selling" their idea. In other words, they commonly apply a probability of no more than 30 or 40% to their reserve estimates. But the engineers, who take over on a discovery, take a much more cautious line, looking for what is absolutely sure. They need "proved reserves", having say a 90% probability, which understate what will actually be produced. They don't build bridges with a fifty percent chance of standing up: they want to be quite sure.

The uncertainties have led to the practice of classifying reserves as *Proved (or Proven), Probable* and *Possible,* with meanings the words imply. A more modern, and better, method, is to apply statistical probabilities to the several parameters, and to recognize a *High Case* (with a 5-10% probability), a *Low Case* (with a 90-95% probability) and *Median* or *Mean Cases.* The *Mean Case* is statistically the more correct expected value, but the *Median (P50) Case,* is more commonly used, meaning generally that the risks of the actual recovery proving higher or lower than the estimate are equally matched. There is, however, no general agreement on which method should be used, and a vigorous debate occupies the technical fraternity[5]. A hybrid definition, whereby *Proved & Probable Reserves* together are defined as having a 50% Probability, is a compromise in increasing use[6].

Much of the uncertainty regarding the size of the field will have been removed by the time the decision to develop it is taken, especially offshore or in remote locations where substantial investments are at stake. However, some uncertainty will always remain until the field is finally abandoned.

In practice, more and more emphasis turns to the actual performance of the wells as the depletion proceeds, as there is by then no particular need to consider the notion of how much oil might be in-place or what the recovery factor is. It becomes simply a pragmatic matter of how much may be actually extracted from the wells. Commonly, the early estimates of *oil-in-place* remain on the files, and are compared with the new estimates of what the wells will actually deliver. This often indicates an *apparent* improvement in recovery, which is commonly attributed to

advances in technology or improved management, for which medals are awarded. But it is clear that genuine revisions to reserves based on increased knowledge should be as often down as up. Obviously, modern technology is far in advance of what was available fifty years ago when many of the large fields were found, and that has had an impact on reserves. But it is a mistake to extrapolate that improvement to the future, recognizing that the technology is now extremely advanced and efficient.

All of these strictly technical uncertainties affect the estimation of reserves, but by and large the procedures are straightforward and well understood, with the appropriate degree of uncertainty being quantified in technical terms.

The reporting of reserves is an entirely different matter. In the absence of strict, enforceable, audit-able and universal norms for definition and determination, there is scope for considerable latitude in the reporting of reserves, whether by companies or governments. The reporting of reserves is effectively a political act: subject to under- or over-statement depending on the circumstances and motives of the reporting entity. We try to exaggerate our assets when going to the bank to borrow money and understate them when meeting the tax man.

Major companies tend to understate *initial reserves* for a variety of motives: they may desire to be less visible; they may wish to smooth their asset appreciation so that the occasional discovery supports the lean intervening years; there may be stock-market or management implications; there may be debt considerations where reserves are collateral; there are tax implications in countries such as the United States where there is a depletion allowance encouraging understatement; and there is the laudable tendency of general caution, with the risks attendant upon overstating normally being greater than those of understatement. In practice, the major companies tend to treat *reserves* as a form of inventory: to be held as low as prudent management dictates, irrespective of what is actually there[7].

Smaller companies sometimes overstate with their eye on the stock market. The different approaches are well illustrated by the early reports of the size of the Cusiana Field in Colombia: One partner, an "independent" company reported ten billion barrels; BP (whose shortly to be fired Chairman was in desperate need of good news) reported three billion; and Total (whose technical people like to preserve some good news with which to satisfy their management on rainy days) reported one billion[8].

Another important influence is the Securities and Exchange Commission in the United States, which imposed strict rules to control the activities of unscrupulous promoters, when the bulk of the reserves were onshore. For financial purposes, only *Proved Reserves* being in close proximity to a producing well were acceptable. It meant that the reserves of the fields in question have been subject to progressive upward revision as they were drilled up, and as more and more of their reserves qualified under the strict SEC definitions. I will return to this issue in relation to the subject of "reserve growth", which is much misunderstood.

	Abu Dhabi	Dubai	Iran	Iraq	Kuwait	Neutral Zone	Saudi Arabia	Venezuela
1980	28.00	1.40	58.0	31.0	65.4	6.1	163.3	17.9
1981	29.0	1.4	57.5	30.0	65.9	6.0	165.0	18.0
1982	30.6	1.3	57.0	29.7	64.5	5.9	164.6	20.3
1983	30.5	1.4	55.3	41.0	64.2	5.7	162.4	21.5
1984	30.4	1.4	51.0	43.0	63.9	5.6	166.0	24.9
1985	30.5	1.4	48.5	44.5	_90.0_	5.4	169.0	25.9
1986	31.0	1.4	47.9	44.1	89.8	5.4	168.8	25.6
1987	31.0	1.4	48.8	47.1	91.9	5.3	166.6	25.0
1988	_92.2_	_4.0_	_93.0_	_100.0_	91.9	5.2	167.0	_56.3_
1989	92.2	4.0	92.9	100.0	91.9	5.2	167.0	58.0
1990	92.2	4.0	92.9	100.0	94.5	5.0	_257.5_	59.0
1995	92.2	4.3	88.2	100.0	94.0	5.0	258.7	64.5

Fig. 6-3. Anomalous reported reserve increases (underlined).

It is to be stressed that no particular conspiracy is implied by the over- or understatement of *reserves* by companies. The management is simply exercising its sensible and normal judgment and prudence on how it should manage its inventory, which in practice is its principal asset. For most purposes and within limits, the size of the reserves and the dating of revisions does not matter much: it only becomes critical when used as a basis to predict future discovery, as in studies like this.

Governments and government companies both overstate and understate, and are much less subject to audit than are private companies. The most blatant case was the huge increases announced by certain OPEC countries in the late 1980s.

Figure 6-3 shows the reserves of these countries as reported to the Oil & Gas Journal. The first apparent anomaly was in Iraq in 1982, when an eleven billion barrel increase was announced, but in fact this was a delayed report of the discovery of the East Baghdad Field in 1979. The next anomaly was by Kuwait in 1984 when a fifty percent increase was announced without any discovery to justify it. Iraq accused Kuwait of exaggerating to secure a higher OPEC quota, which was partly based on *reserves:* apparently with good reason. Then in 1988, Venezuela decided to include about twenty billion barrels of heavy oil, which had been known for many years and which was not in development[9]. It had the not necessarily intended effect of increasing Venezuela's quota, and led Abu Dhabi, Dubai, Iran and Iraq to retaliate by announcing enormous unsubstantiated increases; followed two years later by Saudi Arabia. It is noteworthy that the Neutral Zone announced no such increase: it is owned by Kuwait and Saudi Arabia who had no common position. It is obviously absurd to imagine that Iraq, for example, has increased its reserves four-fold since 1980, when much of the time it was at war or embargoed. Kuwait rather gave the game away when it at first reported that three percent of its reserves had been lost when Iraq fired the wells in the Gulf War. It has recently said that 1.5 to 2.0 Gb had been lost, which implies that the reserves are between 50 and 67 Gb, which sounds reasonable- and less than the 94 Gb it reports.

There can be no doubt that these increases are anomalous[10], but it is less easy to determine if in reality the new numbers were overstated or whether the old numbers were understated, having been inherited from the companies before they were expropriated. In fact it appears that the actual reserves are somewhere in between. The important point, however, is that nothing technical happened at the time to justify the change. I will come back to this in considering the backdating of reserve revision.

In addition to this obvious case, a large and increasing number of countries report unchanged numbers year after year, which is obviously implausible. Production eats into reserves unless matched by new discovery or revision, so it is inconceivable that the reserves should stay exactly the same. In 1996, as many as forty-three countries, including several important producers, reported implausible unchanged numbers.[11]

The reason may be simply that the reserve estimates are not updated, or that those responsible find it politically unpalatable to announce falling reserves. It is prudent, in the absence of other information, to reduce these *reserves* by the *cumulative production* for any period of unchanged reports.

Mexico too has confessed to exaggerating its *reserves*[12], apparently in connection with collateral for debt[13]. The Chicontepec Field is the cause of much of the confusion. It may have as much as 100 Gb in place, but only a small proportion is currently recoverable for geological reasons.

Not all countries overstate their reserves. The most remarkable example of understatement is the United Kingdom, which reports 4.5 Gb for *Proved* when *Median Probability Reserves* are about three times higher. It seems unlikely that there should be such a range of technical uncertainty for a shelf as well known as the UK North Sea[14], but understatement is said to be a British characteristic. Part of the explanation is that reserves are sometimes not reported as "proved" if the field containing them is not yet on production: still another cause of confusion.

These examples demonstrate how difficult it is to come by reliable numbers for reserves. Apart from the obvious case of the countries which had a motive to exaggerate for OPEC quota reasons, no particular conspiracy need be imputed. Much of the problem goes back to definition.

If we speak simply of *Low Case* or *Proved* reserves, they are minimal numbers, which are perforce subject to upward revision during the life of the fields to which they belong. If we speak of *Median Probability* reserves, we endeavour to estimate in advance what the ultimate recovery will be, such that any revisions are as likely to be down as up. If we speak of *High Case* reserves, as does the US Geological Survey, albeit with its own definition terms, we must expect eventual reductions as not all the hopes are likely to be realized. The upward revision of *Proved Reserves* has been widely misunderstood, being attributed to technological progress rather than being a natural consequence of initial understatement or strict definition. This raises the vexed question of the impact of technology and economics on reserves. It much depends on the environment of the field under consideration. Obviously, the scope for the application of advanced technology is greater for small fields in deep offshore waters than for mature giant fields onshore, where well tried established methods are sufficient to extract the oil.

The introduction of the technology of the semi-submersible rig did bring in large quantities of new reserves by opening the offshore to routine drilling. But that was probably the last major technological breakthrough affecting significant global reserves. There have been many innovations since then, including the subsea completion, the horizontal and multi-lateral well, two-phase flow in pipelines, as well as innumerable improvements in drilling and production performance all round. It is probably fair to say, however, that the main consequence has been increased production rate and thereby profit, leading to accelerated depletion rather than the addition of reserves. Such reserves as have been added by technological factors are generally small and in difficult conditions, such as the deep offshore. The impact of economics is even less important. Most of the world's reserves are economic to produce in a price range of less than say $15-$20 a barrel. The huge reserves of the Middle East are producible at less than $5/b[15] (see Figure 6-4). So, not much is added by increasing prices above say $20, remembering always that we are talking about *conventional* oil. The entry of *non-conventional* oil, which *is* much influenced by economics, is another issue to be covered in a later chapter. Gas, for the present, is more susceptible to economic factors, related not so much to production as to transport from remote areas. Higher prices could have a great impact on gas reserves, allowing huge deposits in the Middle East and in places, such as Nigeria and Algeria, to qualify as *reserves,* once they are connected to a market. Again, we have to try to distinguish producible reserves in the ground from those being currently produced. Much could be done to improve the reporting procedures.

It is evident from this discussion that the assessment of *reserves* is an exceedingly difficult issue: not so much in a

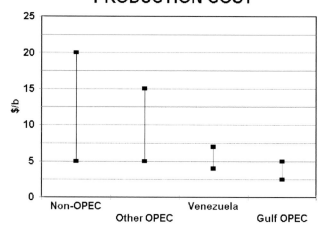

PRODUCTION COST

Fig. 6-4. Production Costs.

technical sense as the procedures are straightforward and well understood, but in relation to definition and reporting procedures. For many purposes, the present laxity does not particularly matter, but to have valid *reserves* is a critical element in assessing the world's endowment and how much is *yet-to-find*. It is by far the most difficult and serious obstacle to be overcome, or at least addressed, in studies of this sort.

Those with access to Petroconsultants' database are able to use *median probability reserves* of every important field outside North America, based on the best information available[16]. But here we will have to be content with published data and endeavour to derive an approximate value for *median probability* by applying a positive or negative factor to the *"Proved Reserves"* as published by the Oil and Gas Journal[17]. While not claiming to be precise, the numbers are probably correct within reasonable limits. They may be compared with those published by the USGS[18], remembering that the latter are *High Case* values, with a probability of occurrence of no more than 5-10%.

RESERVE GROWTH AND TECHNOLOGY

Many analysts, commonly using unreliable reserve data, give emphasis to what they term "reserve growth", which they attribute to advances in technology. Some anticipate that "reserve growth' will continue in the future in response to further assumed technological progress, and solve the indicated coming oil crisis.

It is important therefore to assess the impact of technology in realistic terms. As already discussed, many of the apparent increases in reserves in recent years were simply in the reporting with no technological justification whatsoever. But when we remove these spurious increases, we are still left with some increase that may be genuinely attributed to technological progress.

The introduction of the semi-submersible rig was indeed a breakthrough that opened up the offshore reserves to routine drilling. Offshore production is undoubtedly a high technology business and so routine technological progress undoubtedly has an impact. I think it has two influences. First it can extend plateau production on offshore facilities, which brings economic rewards but in fact accelerates depletion without adding reserves as such. Second it can affect the tail end of depletion delaying abandonment which does indeed add reserves.

I thus accept that advances in technology can add reserves by extending the life of fields, but think that it will have a negligible impact on peak production. We may for our purposes treat tail end reserve additions as effectively *non-conventional* oil and exclude it from the reserve base used for modelling peak production.

The world's total of *conventional oil* reserves at the end of 1996 is here assessed at 836 Gb, which is 142 Gb less than as reported by the Oil and Gas Journal. The discrepancy is largely due to the anomalous increases reported in OPEC countries in the late 1980s. Figure 6-5. lists the ten largest countries and the regions[19]. It will be noted that more than half of the World's reserves lie in just five Middle East countries, designated as the Middle East Gulf Region.

Reserves (1996)	Gb
Saudi Arabia	202
FSU	98
Iraq	71
Abu Dhabi	62
Kuwait	61
Iran	59
Venezuela	35
China	30
Mexico	25
Libya	23
By Region	
ME Gulf	459
Eurasia	129
L. America	77
Africa	63
W. Europe	31
East	22
ME Other	18

Fig. 6-5. P50 Reserves by major country and region.

Glossary of Reserve Terms

With such a plethora of terms,
is it surprising that much confusion surrounds the subject?

Reserves
The amounts of petroleum that are estimated to be recoverable from a known petroleum accumulation as of a stated reference date on certain stated or implied economic and technological assumptions, normally those current or foreseen for the life of the field at the reference date. The estimates may be divided into categories designated as *Proved, Probable* and *Possible* or described in terms of *Probability* (see below).

Cumulative Production
Total produced as of the reference date.

Decline Rate
The percentage reduction in production from one period to the next (month or year).

Demonstrated Reserves
USGS term[20] approximating to Proved + Probable Reserves.

Developed Reserves
Reserves in a field with installed facilities.

Depletion
The process of producing a finite amount of petroleum.

Depletion Rate
Annual production as a percentage of the amount remaining to produce.

Discovered-to-date ("D-t-D") or Discovered or Total Discovered
Total discovered, as of the reference date, namely Cumulative Production plus Reserves.

High Case Reserves (=Proved + Probable + Possible)
Reserves estimated to have a low probability of occurrence (5-10%)[21].

Hypothetical Resources
USGS term meaning Undiscovered oil in a productive basin.

Identified Reserves
USGS term for High Case Reserves.

Indicated Reserves
USGS term approximating to Probable Reserves.

Inferred Reserves
USGS term approximating to Possible Reserves.

Initial Reserves (also Original Reserves)
The reserves as of the commencement of production with any revisions backdated as if known at that time.

Low Case Reserves (= Proved)
Reserves estimated to have a high probability of occurrence (95-90%).

Mean Probability Reserves
The statistical mean of a range of reserve probabilities[22].

Measured Reserves
USGS term approximating to Proved Reserves.

Median Probability (P50) Reserves
Estimates in which the risks that the actual recovery will prove to be higher or lower than the estimate are equally matched.

Oil-in-Place
Estimated amount of petroleum in an accumulation, of which only a percentage is producible.

Possible Reserves
Unsure reserves that fail to qualify as Probable.

Probable Reserves
Less sure reserves that are likely to occur but fail to qualify as Proved.

Proved Reserves (also Proven)
Reserves judged to have a high probability of occurrence: approximating with Low Case reserves.

Proved & Probable Reserves
Reserves of Proved and Probable categories that together are estimated to have a median probability of occurrence.

Recoverable Reserves
Tautologous synonym for Reserves but sometime used to emphasise the distinction with oil-in-place.

Recovery Factor
Percentage of oil-in-place that is producible.

Reserve Growth
Upward revision of Proved Reserve estimates.

Resources
Notional amounts of oil and gas in nature including those lacking reserve status irrespective of technological or economic constraints.

Remaining (Yet-to-Produce or "Y-t-P")
The Ultimate less Cumulative Production or Reserves plus Undiscovered.

Remaining Reserves
Tautologous synonym for Reserves but sometimes used to emphasis that the number applies as the reference date as opposed to the commencement of production.

Speculative Resources
USGS term for Undiscovered oil in a non-producing basin.

Static Reserves (also sleeping or dormant)
Reserves not being produced or developed.

Undeveloped Reserves
Reserves not being currently developed, namely lacking installed facilities.

Undiscovered (or Yet-to-find: "Y-T-F")
Amount of oil estimated to be found and attain reserve status.

Ultimate or Estimated Ultimate Recovery or EUR
Cumulative production when production ends due to depletion; as applied variously to the world, a continent, a country, a basin, a field or individual well. (The term is also used confusingly for the sum of the initial reserves of fields in a basin or country, excluding the Undiscovered).

THE DISCOVERED-TO-DATE

The sum of the *Cumulative Production* and the *Reserves* gives the total discovery. It is a very important starting point from which to determine how much remains to be found and produced. We can study the discovery pattern to see when the discoveries were made, and we can investigate the size distribution of fields.

There is a certain size distribution in nature, which I will

discuss in detail in a later chapter; and there is what might be called the natural environment of exploration, in which the larger and easier prospects tend to be investigated first, reflecting, it could be said, the well known human attributes of greed and laziness.

The pattern of discovery has also been influenced by the impact of evolving knowledge and technology, and of course there have been many political influences. In

The Coming Oil Crisis

general, however, it can be concluded that, although the actors may change, there has always been a strong motivation to find oil over the past Century both by industry and government. Oil is valuable stuff and a source of great wealth. So, in global terms the progress in discovery of oil has not been unduly hampered by artificial constraints: the discovery pattern of even the Soviet Union with its central planning was not markedly different from that of the United States. The pattern of discovery is therefore substantially a natural one, notwithstanding the other influences. It is, accordingly, capable of valid extrapolation.

In earlier chapters, I described the historical evolution of the business, concentrating mainly on the upstream side. As different territories were opened up, the explorers went into action. At first, they had to learn the geology and drill wells to provide the essential information during what can be called the lead-time. Then came the moment-of-truth when the basin either delivered or was likely to remain forever barren, having failed to possess the essential geological characteristics, especially source-rock. Already by 1908, the world's largest petroleum system, the Middle East, with about forty percent of the world's ultimate endowment, had been found[23]. The broad picture became fairly visible early on.

The opening of the offshore in the post-war epoch was another great step forward. The same basic pattern of discovery was repeated but at a greater pace. Whereas the lead-time during the early days onshore could last for a decade; the ability to conduct inexpensive, high-quality and comprehensive seismic surveys offshore reduced it to no more than a couple of years.

When the first well in a basin is drilled nothing is known about the ultimate distribution of its fields, but when the last well has been drilled, everything will be known. It is like watching a photographic print gradually take shape in the developing tray. We have, to-day, reached the point at which the image has appeared, not in fine detail, but in broad outline. This picture shows that now there are virtually no new major provinces left to find[24], and that efforts will have to concentrate both on ever smaller and obscure prospects in established basins, and on trying to increase the recovery from what has already been discovered. The law of diminishing returns applies very much to the discovery of oil.

The pattern of discovery is much affected by the way in which areas are defined. Ideally, a clear-cut geological natural domain with common characteristics should be considered, but in practice the boundaries of geological provinces are often fuzzy. Sometimes, more than one system are superimposed. National frontiers exert an influence in that different countries provide greater or lesser incentives to explore, which are far from constant over time. Furthermore, most of the statistics in the public domain relate to countries, not geological basins.

Discovery is decidedly cyclic. When a new productive trend is found, one discovery follows another with the larger fields coming in first to give a clearly identifiable peak. Some countries or basins have only one such trend,

and their profile has a single peak. In other cases, there may be several cycles, each giving a peak; although taken overall the larger cycles, which tend also to come early, mask the smaller ones.

In terms of distribution, we may consider the number of fields and their size; and plot both parameters over time, noting the inflections, when the rate of rising cumulative discovery begins to decline. To do that requires access to the Petroconsultants' data base, but published information on giant fields (namely those with initial reserves of in excess of 500 Mb) gives a useful indication.

Figure 4-8 showed the discovery of giant fields, and Figure 6-6 shows the World's cumulative discovery illustrating the mid 1960s inflexion as the rate of discovery began to fall off. I will later come to the issue of so-called Swing countries, namely richly endowed Abu Dhabi, Iran, Iraq, Kuwait, Neutral Zone and Saudi Arabia, and the Non-Swing countries. Figure 6-7 shows discovery in the Non-swing countries. As will be discussed in Chapter 8, their aggregate production will have an important impact on oil price. Their discovery inflexion in the 1970s presages their peak in the 1990s, which is a key element in the coming crisis[25].

WORLD DISCOVERY
Conventional Oil

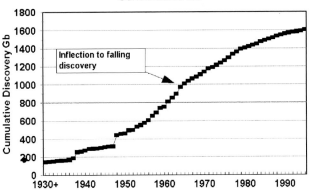

Fig. 6-6. World cumulative discovery.

NON-SWING DISCOVERY
Cumulative Discovery

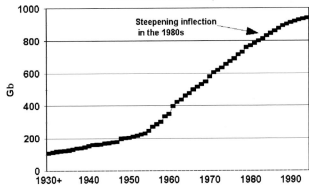

Fig. 6-7. Cumulative discovery in the non-swing countries.

Figure 6-8 lists discovery for the ten largest countries, and the regions. As of the end of 1996, the world as a whole had discovered 1.6 trillion barrels. Peak discovery occurred in the 1960s. Ninety percent of current production comes from fields more than twenty years old; and seventy percent from fields more than thirty years old. These are very telling numbers.

Understanding how much has been discovered so far is a critical first step in any assessment of the world's ultimate endowment. It is not an easy task, given the unreliable nature of much of the data in the public domain. In the next chapter, I will come to discuss how much is yet to find. The *Discovered* plus the *Undiscovered* gives the *Cumulative Production,* when it ends, or in other words, the *Ultimate* recovery. It is a vital concept in depleting a finite resource.

Discovered Reserves	Gb
Saudi Arabia	280
FSU	227
USA	201
Iran	105
Iraq	94
Kuwait	88
Venezuela	83
Abu Dhabi	76
China	51
Mexico	48
By Region	
ME Gulf	646
Eurasia	286
N. America	227
L. America	171
Africa	126
W. Europe	60
East	56
ME Other	37

Fig. 6-8. Discovered-to-date by country and region (1996).

Alain Perrodon
and The Petroleum System

Fig. 6-9. Alain Perrodon, who developed the concept of the Petroleum System.

Q Recently, I have been involved in editing the English versions of several French geological reports, and was impressed by the marked differences between French and Anglo-Saxon geological thinking. It is not just a matter of translation. It seems to me that the Anglo-Saxons give emphasis to disciplined almost sterile description, whereas the French concentrate on trying to understand the broad dynamics of geology, integrating structural and stratigraphic factors. One could say that it is almost a more philosophical approach. I don't know if I put this clearly, but do you see what I mean?

A *Whether or not there is a philosophical reason, certainly we take a more global view of geological phenomena that seeks a coherent integration of the dynamic evolution of the Earth's crust through time.*

Q The contribution of French scientists to petroleum geology has been impressive. Professor Tissot and his colleagues made the critical geochemical breakthrough with hydrous pyrolysis which allowed us for the first time to really understand where the oil came from, and when. This was a critical step in the work that you did to develop the concept of the Petroleum System. Could you explain this historical evolution? And what makes the French contribution so special.

A Petroleum geologists of my generation, especially European geologists, have lived through quite an exceptional adventure. We started our professional lives just after World War II without any petroleum experience, but we were swept ahead by a tremendous current of petroleum history. We engaged in the conquest of vast sedimentary basins, unexplored or sometimes deliberately ignored by eminent experts of those times. In Algeria, and more specifically in the Sahara, for instance, we found our first theatre of operations. We undertook, often with more enthusiasm than technical means, the systematic exploration of territories without knowing in advance if they would contain promising basins. Despite many failures, the global results went far beyond our expectations.

Q Tell us more about how your career unfolded specifically?

A I had a marvellous apprenticeship of field geology, wellsite work and subsurface studies in northern Algeria and the Sahara, lasting for seven years. I then had the opportunity to take part in the first Saharan discoveries: the giant oil field of Hassi Messaoud and the giant gas field of Hassi R'Mel. Three years later, I entered the Bureau of Oil Research, which coordinated the research undertaken by the French national oil companies and the predecessor of the Elf Group, where I spent the rest of my career.

In 1956, having submitted a thesis on the Chelif Basin in Algeria, I started teaching a course of petroleum geology at the School of Geology in Nancy. I became responsible for the Elf Group's geological services and research.

Q What were the milestones in developing the concept of the Petroleum System?

A There is no doubt that the concept of the Petroleum System sprang from the ideas provoked by the theory of Plate Tectonics. It progressively drew attention to a dynamic dimension in geological phenomena that demanded a logical integration. Since hydrocarbons are intimately linked with sediments, it became natural to include them in the broader scheme of things. Geological thinking moved progressively from the analysis of the structural framework to dynamic studies of sediments and their fluids, particularly hydrocarbons. The notion of "petroleum system" replaced that of "petroleum habitat". These achievements are of course in large degree due to the results of technical progress, but they also stem from an extraordinary blossoming of new ideas and from a major revolution in Earth Sciences.

Q It seems to me that you have found a role that is at the same time industrial and academic. You have published widely and received the recognition you amply deserve, including the award by the American Association of Petroleum Geologists. I think you like to see things with a historical perspective. But history does not necessarily only refer to the past. What do you think about the future of the oil industry?

A Increased knowledge means we can better evaluate. We now understand how oil is formed and trapped. We can confidently declare that large tracts are barren of potential. It means that frontier exploration is dying out, even if it is not already dead in many areas. Carried as we are on the powerful wave of technical progress we must not, however, underestimate the fragility of our hypotheses and theories. As J.A. Masters said at the last AAPG Conference, "Technology has its role and I do not mean to diminish it – but the world is changed by dreamers".

Q It seems to me that there are very far reaching consequences for Europe and France itself, if oil production is about to peak as seems inevitable from what we know of the reserves and potential. I sometimes think that we are living on the brink of an abyss, when you realize how dependent we have become on cheap oil-based energy.

A France, more than anywhere, has long realized its dependence on imported energy. It is for that reason that we decided to develop our nuclear capability. Many people fear a repetition of Chernobyl or radiation leaks, but I think that an efficiently managed nuclear industry in state hands can be a solution with acceptable risks. Although oil will become more expensive, it is not about to run out; and there are still possibilities to bring in new gas production. I do not therefore see exactly an energy crisis, but I do think that the energy issue is an increasingly important one. I think it will indeed lead to fundamental changes in the way the world lives. Here in France, we are privileged by having a large rural countryside which will help sustain us, perhaps bringing us back to a more traditional style of life, which personally I would welcome.

NOTES

(For references, see Bibliography)

1. see Stosur, G., 1996.

2. It is important to distinguish "will" from "can": "will" is to be preferred as the more pragmatic case taking into account economic and technological factors.

3. They are sometimes termed *Remaining Reserves* to distinguish them from *Initial Reserves*, although strictly speaking this is a tautologous usage.

4. See the writings of Adelman and Odell, the high priests of this heresy. Also Statoil mistakenly states "Total hydrocarbon resources with the potential to produce liquid fuels are so large that they can be considered infinite for the purposes of this analysis"

5. Laherrère, J.H., 1995, has discussed in several papers the issue of reserve definition, being a firm proponent of the "probability" system. As he points out, there is now standard agreement for equating the deterministic terms of *Proved, Probable* and *Possible* with their probabilistic equivalents, a subject also well covered by Beardall (1996).

6. Sometimes it is said that the reserves have a *better than* 50% chance of occurrence or production. The Probability equivalent of Proved and Possible reserves also differs from one classification to another.

7. see Barry, R.A., 1993, for an excellent objective discussion of the manner in which oil reserves are calculated and reported by the industry.

8. Petrie, P., 1992, describes the Llanos discoveries: see also Cazier, E.C., 1995.

9. Roger, J.V., 1994, reveals this in connection with discussing reserve classification in Venezuela. The impression given is that the Venezuelan state company did not necessarily have OPEC quota in mind when announcing the increases. The reserves in question are producible, although better treated as *non-conventional*.

10. Barkeshli, 1996, an official with the Ministry of Petroleum in Tehran, has now confirmed that these increases were reported for what he calls the "quota wars".

11. Abu Dhabi, Albania, Algeria, Angola, Australia, Bahrein, Bangladesh, Benin, Bulgaria, Cameroon, Chile, China, Congo, Cuba, Czech Republic, Dubai, Ecuador, Equatorial Guinea, Ethiopia, Former Soviet Union, Gabon, Ghana, Ivory Coast, Jordan, Kuwait, Libya, Morocco, Myanmar, Neutral Zone, Oman, Qatar, Peru, Ras al Khaimah, Romania, Serbia, Sharjah, Slovakia, South Africa, Sudan, Surinam, Syria, Yemen and Zaire.

12. Los Angeles Times, 1991, carried a report by an ex-Pemex executive, stating that the reserves had been exaggerated, mainly by inclusion of non-conventional reserves in the Chicontepec Field, which according to Macgregor, 1996, has as much as 100 Gb of oil-in-place. There are serious geological difficulties, meaning that only a small fraction is currently producible.

13. Duncan, R., 1996, explains how the loans granted to Mexico in the wake of the collapse of the peso have oil reserves as collateral.

14. Campbell, C.J., 1996, points out the discrepancy between "proved reserves" of 4.29 Gb reported by the Oil & Gas Journal and Median Probability reserves of 15 Gb, as reported by *World Oil*, although described as "proved". The official "Brown Book" gives for end 1995 605 Mt (4.5 Gb) Proven and 1370 Mt (10.2 Gb) Proven + Probable (defining Proven as having a better than 90% chance of being produced and Probable as having a better than 50% chance of being produced. I don't believe that it is mathematically sound to add probabilities in this way).

15. The issue of cost is however difficult because it involves capital costs, accounting practices and tax issues which differ from country to country.

16. Campbell, C.J., and J.H. Laherrère, 1995, give the most comprehensive analysis of the world's endowment, based on Petroconsultants' material in a report summarized and reviewed by Mabro, R. (1996).

17. Campbell, C.J., 1996, converts published "proved reserves" into "median probability reserves" by use of a factor.

18. Masters, C.D., 1994.

19. See Appendix for a full listing.

20. The USGS terms are not used by the oil industry.

21. Strictly speaking the probabilistic values cannot be related to the deterministic values of *Proved, Probable* and *Possible,* but for practical purposes the correlation shown here is close enough.

22. The Mean or Expected value in a log-normal distribution equates with $((3 \times \text{Proven}) + (2 \times \text{Probable}) + \text{Possible})/3$

23. This is perhaps a generalisation, for several sub-systems can be recognized.

24. Jennings, J.S. (1996), Chairman of Shell, makes this very clear between the lines if not exactly on them.

25. see Campbell, C.J. 1997.

Chapter 7

THE ULTIMATE AND THE UNDISCOVERED

We now come to a semi-philosophical concept: the idea of an *Ultimate* recovery. How can anyone contemplate the ultimate of anything: it is like thinking about infinity. We naturally shy away from any such idea, which is foreign to our way of life, our attitudes and experience. Yet all of us do have knowledge of depleting a finite resource: our own life-span. We are not immortal. Most of us do not however dwell on the thought of our deathbed, or the depletion of our lives: we comfort ourselves with the hope of advancing medical technology and defying the average. The hard-bitten insurance companies approach the subject with more reality when they use actuarial tables to say that I have a life expectancy of only thirteen more years.

So it is with oil. It is a decidedly finite resource, with its own life-span, about which we care not to think. In this chapter, I will discuss how to estimate that life-span. As we shall see, oil is in fact approaching its middle age, or the midpoint of depletion.

"Are we running out of oil?" is a question that is often asked. The short answer is, "yes: inevitably we are, but not for a long time". What matters more is the less often asked question: "When do we reach peak production?" After peak comes shortage, assuming that the demand for oil continues to grow unabated, as it is more than likely to do with the world's expanding population and growing global economy. Shortage means scarcity and higher prices. As I have said, peak production more or less corresponds with the midpoint of depletion, and to calculate that we need to have an idea about the *Ultimate* life-span.

We need to model future supply on these considerations. It is true that we cannot exactly imagine the production of the last barrel of oil: staring down the last wellbore from which nothing emerges. In some regards, it is a case of *reductio ad absurdum*. Nevertheless, it is necessary to develop an idea of the practical limits: even if we cannot know the exact number, nor quite grasp the idea of an *Ultimate*. Already many oilfields have had to be abandoned when their production fell to almost nothing, and it became uneconomic to continue to produce them. To introduce the idea of economic limits, however, brings its own pitfalls. What is the economic limit? Old stripper wells[1], whose production is down to a few barrels a day, are shut-in in the

United States when oil price falls below a certain critical level. It matters to the owners of the wells, but in global terms, it is not really very important because the amount these dying wells deliver is very small in a world context. The difference between very little and nothing is not important in practical terms, whatever the philosophical distinction.

What matters to us is peak production: the discontinuity between rising supply, as we have known so far this Century, and the decline which will inevitably characterize the next Century. It heralds the opening of a gulf between supply and demand with far-reaching consequences.

It is evident, mathematically, that adding ten years to the *Ultimate* life-span, advances the midpoint by only five years. So, even an inexact estimate of the *Ultimate* gives the approximate date of the midpoint. We define the *Ultimate*, as *Cumulative Production* when it ends, but we could as well define it as the *Cumulative Production* at some date in the far future, such as 2075 or 2100. It does not make a great deal of difference as the exponential decline during the tail end of depletion is almost flat at a low level. It is not really relevant to the main issue. Making again an analogy with human life-span, think of our spending pattern: we spend little in the cradle or the coffin but reach a peak in middle age. Looking at the full life-span, it means little if we have an expensive christening or a jolly wake, most of what we spend will still be incurred in middle age.

With these reservations and qualifications, we can therefore try to estimate a practical *Ultimate*. It is the sum of today's *Cumulative Production*, the *Reserves* and the amount *Yet-to-Find*. It should be a rounded fixed number that is subject to only periodic revisions based on a new global assessment. One could consider several cases with higher and lower assumptions to test the sensitivities, but I think it is better to bite the bullet with the best estimate one can make and stick with that until there is good reason to revise it. Revise it, we certainly should, recognizing that it is not cast in stone.

For practical reasons, I include in the *Ultimate* estimate a buffer of some extra oil, described as "unforeseen" so that it does not have to be revised every time an unexpected discovery is made or the reserves are revised.

In the mechanics of the model, it is convenient to derive the *Yet-to-Find* (or *Undiscovered*) by subtracting the *Discovered* from the *Ultimate*. But of course the *Yet-to-Find* is assessed in arriving at the estimate of the *Ultimate*.

METHODS FOR DETERMINING THE ULTIMATE

There are qualitative and quantitative methods for determining the *Ultimate*. Both are built directly or indirectly on extrapolating past discovery in relation to the underlying geology. We can use our heads and deliver old-fashioned judgment, or we can devise statistical methods that generate more abstract numbers. In fact, we need a combination. In a perfect world, the statistical methods would no doubt provide the more accurate estimate, but it is not a perfect world with many uncertainties about the validity of the input and the definition of the domains within which we work. Accordingly, it is well to use common sense and judgment in finding the most reasonable solution. It is not an exact science.

THE INTUITIVE

In Chapter 5, I described making an intuitive assessment of the *Yet-to-Find* of Latin America. I knew the continent well and had an intuitive feel for the characteristics of its basins. I could make reasonable guesses of the potential. Subconsciously, I was asking myself how large the tracts were; how much work had been done; and what the results had been. I said to myself "looks to me as if they could hope to find about half as much again" – or something like that. For many basins, I said "smells wrong: the few wells that they have tried, didn't work. I don't think this place has what it takes". For example, I intuitively wrote off large tracts along the Pacific Coast of Colombia, sensing that they were deficient in source-rock and endowed with poor reservoirs due to the volcanic content in the sediments. Most places with oil exhibit some signs of it in seepages or early wells: and by 1969, when I did the study, most ideas had been tried. The old adage of "where there is oil, there is more" is a good one, but it carries a corollary: "if you find it at all, you find it soon". Of course, I missed the deep offshore Brasil. Our knowledge of the offshore was then limited, and we tended mistakenly to extrapolate onshore conditions in the absence of any other information.

I remember, for example, turning down a prospect around the Natuna Island in the South China Sea because non-prospective basement rocks[2] occurred on the adjoining lands, and I didn't have the imagination to anticipate the unexpected trough in between. It was a bit like that in the early days of the North Sea. The prolific Viking Graben cuts across the older non-prospective Caledonian trend of Norway and Scotland. Now, thirty years on, almost every offshore region has been covered by at least some seismic data, and it is very unlikely that any sizable new offshore basin of interest has been missed.

The intuitive method is not a bad one, if handled objectively by knowledgable geologists with enough global experience to have a good sense of judgment and proportion. Objectivity is however now the difficulty. In corporate bureaucracies, they no longer consult specialists, but hold committees, often of people having no particular insight on the subject. The quest is for consensus, the hallmark of a committee, giving a politically acceptable answer. Often, the management has its own ideas of what it desires from an evaluation, which is then duly massaged into shape. So, it has become very dangerous to rely on studies of this sort made by oil companies. Government agencies, such as the US Geological Survey or the IEA, also have their own politics to worry about, and besides they tend to have a more academic view. In fact, the USGS considers only theoretical geological prospects, saying that it is outside its remit to soil its hands with economic or technical constraints. It explains why they were forced to create their own reserve definitions to hide behind. They have been under threat of being abolished by a cost-saving government: controversy is the last thing they need to survive.

My own intuitive study of 1989, as published in a book the *Golden Century of Oil*, proposed an *Ultimate* of 1.65 trillion barrels, only 150 Gb less than is now indicated with the benefit of vastly improved data and analytical techniques.

THE PUBLISHED RECORD

The record of published estimates of the *Ultimate* is a useful approach, largely reflecting the evolving intuitive judgment of explorers, based on their experience at the time of the estimate. In 1942, when most oil production was onshore, and the United States dominated world production, 600 billion barrels seemed a good number. But the opening of the offshore and the discovery of giant fields, especially in the Middle East, led to progressive upward revision to a peak of 3.5 trillion barrels in 1969. Then, the estimates began to fall with the realisation that giant discovery had peaked, and that much of the offshore was less prospective than had been hoped. The consensus dropped to below two trillion barrels on a falling trend. It is of course important to check carefully into the definitions being used in such evaluations, and to see if the anomalous reserve increases of the late 1980s have been recognized. The latest estimate by the prestigious USGS is 2.3 trillion barrels, but it reduces to about 1.8 once its special reserve definitions have been decoded[3].

CREAMING CURVE

The creaming curve is so named because the better prospects are taken off the top of the list, like taking cream off milk. The curves plot cumulative discovery either against cumulative wildcats or over time. Such curves can only be made where there is accurate data on discovery and drilling, with reserve revisions backdated to the discovery of the fields to which they relate. In practice, it calls for access to Petroconsultants' data base.

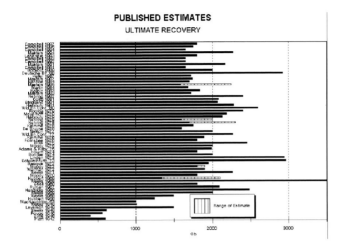

PUBLISHED ESTIMATES

ULTIMATE RECOVERY

Fig. 7-1. Published estimates of Ultimate Recovery (alternate estimates accredited for reasons of space).

DEPLETION TREND
Changing estimates

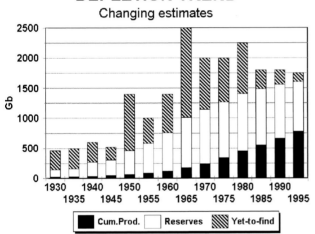

Fig. 7-2. Discovery and potential: changing estimates.

NORTH SEA
Discovery

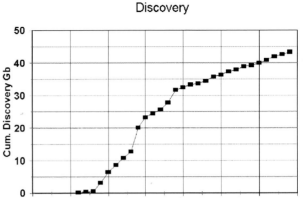

Fig. 7-3. Discovery plot of the North Sea: cumulative discovery over time.

NORTH SEA
Hyperbolic Discovery Pattern

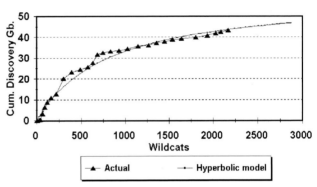

Fig. 7-4. Discovery plot of the North Sea: cumulative discovery against cumulative wildcats.

AFRICA
Past Discovery Pattern

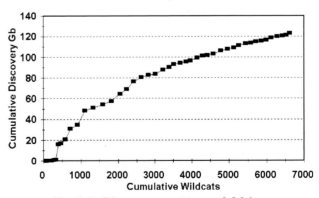

Fig. 7-5. Discovery pattern of Africa.

The plot over time is distorted by political events, such as wars, government licensing rounds or oil price shocks. The plot against wildcats gives a more natural distribution of discovery, but even here there are distortions, such as where wells are drilled to fulfill concession commitments. In the simplest case, in a basin with only one major play, such as the northern North Sea, the plot is hyperbolic, because the larger fields are found first, being too large and obvious to miss. In other cases, where the country or basin contains several different plays, there will be a series of hyperbolas, stacked one upon another, with the overall trend probably also being hyperbolic, insofar as the larger provinces tend to be found first.

The hyperbola can be constructed by taking the relationships between the number of wells drilled at the first discovery; the cumulative discovery at the present number of wells and the discovery when half the present number of wells had been drilled, and extrapolating to asymptote. Alternatively computer programmes with regression analysis can be used. The asymptote corresponds with the

Ultimate, subject to a cut-off to reflect the economics of very small fields, technology and, more generally, the time-span implied, taking into account current and projected drilling rates. Again, we don't need to take an extreme view of what is meant by *Ultimate,* and can be content with notions of how much oil will have been found with, say, double the present number of wells, or after, say, fifty years of exploration drilling on current trends. It comes to much the same within the accuracy of the numbers at our disposal. We must remember that exploration in a province is likely to end when production ends, if not before. So, not all the oil in the theoretical distribution will be found. It is useful to plot creaming curves both against wildcats and over time to compare the results and obtain a hint of any defects or distortions in the input. For example, the listing of wells does not normally identify the objective of the well. In areas where there are both oil and gas systems, some of the wells were aimed at gas and should therefore be excluded from

the oil analysis. There are many points of detail to resolve: should the oil plots be distinguished from the gas plots and how? Should they be combined and considered in terms of petroleum, with the gas being converted to oil equivalence. It is a case of balancing all of the factors, always remembering that the input data are often unreliable.

In addition to the simple curve, we can also plot discovery by size-classes. In many cases, the plot of giant fields has been flat for many years, suggesting that none is left to find, whereas the discovery of the smaller classes of field-size may still be rising. We can plot average size; and the number of discoveries to obtain yet more indications of the pattern.

DISCOVERY BY SIZE CLASS
North Sea

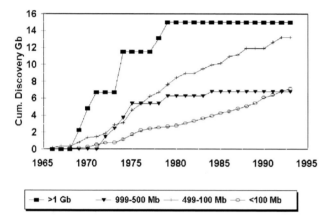

Fig. 7-6. Discovery plot by field-size class: note how discovery of the larger classes has ended.

FIELD SIZE & NUMBER
North Sea

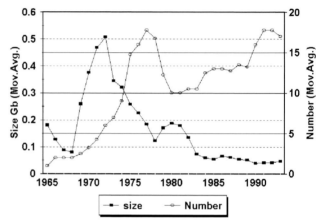

Fig. 7-7. Average size and number of fields: the size is falling but the number is still increasing.

L.F. ("BUZZ") IVANHOE
One of the first to express concern

Q Buzz, I well remember reading one of your articles in the Oil and Gas Journal in the 1980s. Here, I thought, was someone who understands the global situation, and could really explain what is going on. I got in touch and invited you over to speak at a meeting of exploration managers in Norway. Later, you helped organize a study for the Norwegian Petroleum Directorate. I know about your subsequent work, but how did you become interested in the first place?

A *I grew up in Brasil, but had my secondary education in the United States taking a mining degree before going on to do an M.Sc. in geology at Stanford. I had a varied early experience as a mining engineer in Ecuador and running seismic crews in Venezuela and Canada, before joining Chevron.*

Later, I worked for a small company in California which really taught me the oil business at the sharp end. In 1957, I became an international consultant working in Turkey,

Israel, Libya, Peru and Australia for a number of clients including Occidental. The latter retained me as a full time consultant from 1973, following the First Oil Shock.

Q It was a shock, and the scramble was on to find new sources of oil. Everyone was becoming concerned about security of supply. That must have been your baptism on this subject.

A *Yes, there was much concern. My experience gave me an intuitive sense of value, so that I was able to rapidly appraise worldwide exploration opportunities. It was a case of sifting through all the data and trying to find the better areas on which to concentrate. In August 1974, the US government directed the Geological Survey to hold a seminar at Stanford University to review their methodology. It soon became evident that its evaluation was giving much higher numbers than the oil industry experts thought reasonable. The Survey was assuming that the future would be as rosy as the past, whereas we realized that the average size of discovery was falling. When I reported the outcome to the Company, it asked me to make an in-house study of the world's petroleum basins as a basis for its exploration strategy. I believe in visual presentations, and I assembled my data into ten large maps covering the world. I worked on my own, and the Company was very generous in allowing me to publish the general conclusions.*

Q Once you had the basic framework of where the basins were, what was the next step?

A *I set about refining the maps to highlight the more prospective tracts within the basins, and began to quantify their "oiliness". In 1979, I published a paper contesting a Financial Times report "Window on Oil" by Bernardo Grossling who was the USGS expert on global assessment. He had calculated the area of the world's basins and assumed that they would ultimately be as rich as Texas. I planimetered the prospective tracts on my maps, and came up with half the area claimed by Grossling. I pointed out that Texas was not a good analogy and that many basins were much less prolific. In 1983, I was asked to try to put a number on each nation's "oiliness" potential. I did this by resurrecting an earlier method called the Discovery Index, namely the reserves added annually per foot of exploratory drilling. It involved evaluating the validity of the reserves claimed: the reports were often unreliable for political and other reasons. One needed experience and judgment to smoke out reasonable numbers. The Company again cooperated in allowing me to publish the final results in 1984. My estimate of Ultimate recovery, incidentally, was 1.7 trillion barrels.*

Q That is pretty close to my latest estimate of 1.8 trillion. How did it compare with other contemporary estimates?

A *It was less than the 2.0 trillion published by King Hubbert in the June 1974 issue of the National Geographic Society magazine, which concerned me. So I arranged to meet him at his home in Washington. He was then 81 years old, but as sharp as mustard. It did not take long to realize that here was a man of superior intellect. He explained that he was interested in the life-span of oil, recognizing that production had to start at zero, rise to a peak before ending at zero. He predicted in 1956 that production in the US Lower 48 states would peak around 1970 and decline thereafter, which it did. He said that he did not predict global ultimate production as such but simply used the 2.0 trillion Ultimate as a model upon which to build a global depletion curve that the general public could understand.*

Q I think the great contribution of Hubbert was to stress that oil is a finite resource and that peak production will more or less come at the midpoint of depletion. Whether we build the depletion profile on an Ultimate of 2.0 trillion or 1.7 trillion will only shift the date of the peak a few years one way or the other. Given that these numbers are within a realistic range, it means that world oil production will peak in a few years' time and after that will come shortage. It seems an immensely important issue, considering that oil provides about forty percent of the world's traded energy. Do you think that governments are sufficiently aware of the situation?

A *Regrettably no. I do not think that the US public or its politicians have any realization of the constraints to global oil supply. It is like telling a 40-year old jogger that he has cancer and less than ten years to live: he simply does not want to hear the message. Democracies thrive on Good News. One is branded as a crank or doomsday merchant if one points out the reality of the position for which there is ample evidence. Economists, who have never had the practical experience of actually looking for oil, project past trends and misunderstand reserve-to-production ratios to claim that there is lots of oil for the next century.*

I feel strongly that everything possible must be done to publicise the position. There is not much time to prepare. I have accordingly spent the last year organizing the "M. King Hubbert Center for Petroleum Supply Studies". I chose the name to commemorate the remarkable achievements of someone who can rightly be regarded as the "father" of this subject and was ahead of his time. It was inaugurated at the Petroleum Engineering Department of the Colorado School of Mines on October 8th 1996. The plan is to issue a quarterly newsletter to inform newspapers, magazines, politicians and the public at large.

PARABOLIC FRACTAL

There have been many attempts to plot the distribution of oil fields by size. Log-normal plots have been the most commonly used[4]; and the USGS has a thing called the shifted Pareto. Jean Laherrère[5] has investigated all of these approaches, and has come up with a law of distribution, which he terms the *parabolic fractal*. It states that a parabola describes the distribution of objects in a natural domain when size is plotted against rank on a log-log format. It relies on a law of *self-similarity*, whereby a complete segment of the distribution describes the whole. It sounds complicated, but it really is quite simple, although it is not easy to know why this relationship holds true. As an example, we can plot the size of the larger towns in the United States (based on physical city limits rather than administrative units, which are not natural domains) against their rank: New York, Washington, Chicago, San Francisco etc., to determine the parameters of the parabola. It can be extrapolated to the smallest unit, which could be two people in a tent. Such a plot gives the population of the United States to within a few percent of the latest census. Laherrère quotes several other examples, including: spoken languages; the size of species; galactic distances, all of which confirm the validity of the law. Probably, it has something to do with there being no such thing as a straight line in the Universe, everything being eventually parabolic. Throw a stone into the air and it will follow a parabolic trajectory, obeying these same fundamental forces. Entering the log-log domain, giving high numbers, probably brings this into the scale of the Universe. Einstein would have understood but I am afraid I don't, beyond seeing that it works. It is especially useful when applied to oilfields, where the largest are normally found first. It works best in a clear-cut natural domain – a single *Petroleum System* with a common source-rock. The Niger Delta or the Viking Graben in the North Sea are good examples. It can also work to a degree for very large populations that come to form a sort of super domain of its own: such as the continents or the giant fields of the world.

The difference between the parabolic plot and what has been discovered gives the *Yet-to-Find*, subject to a cutoff for small and uneconomic fields.

US POPULATION
Parabolic Fractal

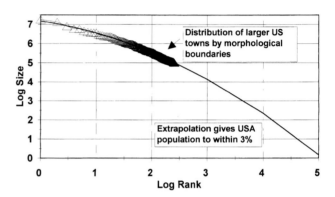

Fig.7-8. Parabolic Fractal of US population.

NORTH SEA
Parabolic Distribution

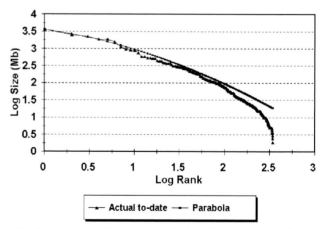

Fig 7-9. Parabolic fractal of the distribution of oil fields in the North Sea.

PRODUCTION EXTRAPOLATION

Since the *Ultimate* is defined as *Cumulative Production* when it ends, we could determine the *Ultimate,* if we could find a way to extrapolate the production curve to zero. M. King Hubbert[6], to whom I have already referred, was distinguished American scientist, who was the first to try this method to predict future oil supply, using a logistic-derivative curve, or simply stated, a bell-shaped curve. It had in fact been used much earlier in population studies. King Hubbert correctly predicted in 1956 when the United States (48 States) would peak, almost twenty years before it did. He was much reviled at the time, but has since been amply vindicated.

His formulae[7] are a bit complicated for non-mathematicians, but in principle, it is quite simple. I suspect that he began his study on very pragmatic reasoning. He realised that the unfettered production of any finite resource starts at zero, rises to a peak, and ends at zero. He made a guess of what the *Ultimate* might be and drew the peak at the midpoint of depletion, namely when half the *Ultimate* had been produced. The simplest result is a triangle, but plotting the actual production to-date showed a curved plot as production increased over time. Hubbert realised that once the rate of increase began to slow, it meant that about one-quarter of the *Ultimate* had been produced. An S-shaped, or sine-curve, was developing. Once this inflection was spotted, the plot could be extrapolated over the top and down the other side, giving the classic bell-shaped curve. So, once about one-quarter of the *Ultimate* had been produced, the inflection in the curve could predict what the *Ultimate* would be: and formulae could be developed to describe it in mathematical terms.

The reason behind it is of course that the production curve mirrors an earlier discovery curve. It is axiomatic that before you can produce oil, you have to find it.

It is a particularly useful tool because it relies only on production data which are generally of good quality and in the public domain. There is no need to worry about the problems of reserve definition and inaccurate reporting. Production could do it all. It works well in a place like the United States where production has been unfettered, save for the period of proration by the Texas Railroad Commission[8]. It works less well for individual countries with a small population of fields or for OPEC countries where production around midpoint has been capped for quota reasons, thus distorting the curve, but even so it can give a hint.

Considered in greater detail, many countries exhibit multiple discovery peaks as different geological trends were developed, each of which being later mirrored in a corresponding production peak. Jean Laherrère[9] has investigated this relationship, and has found that it is possible to correlate discovery and production peaks with a certain time-shift, individual to each country. He then constructs bell-curves for each such peak and sums them to yield an estimate of the *Ultimate*[10]. It is an elegant new approach, which is simple to use and well matches actual production in mature countries.

Duncan[11] has developed a phase diagram, as a variant of the Hubbert method, called the D-Model, that plots annual production against *Cumulative Production,* using translated coordinates. It is based solely on production statistics, and seems to give good results, although labourious to produce.

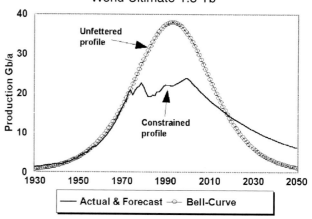

THE BELL CURVE
World Ultimate 1.8 Tb

Fig. 7-10. The bell-shaped curve models the natural unfettered exploitation of the resource. Actual production may be constrained by prorating or other reasons.

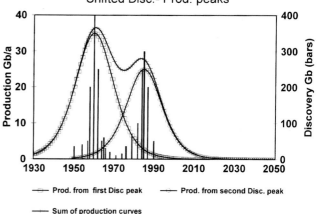

PEAK CORRELATION
Shifted Disc.- Prod. peaks

Fig. 7-11. Correlation of discovery and production peaks, combined with bell-curve modelling is another method.

Mathematical skills, which I lack, are needed to understand it.

These are the methods that can be used. Having been through them all, and having checked the reasonableness against drilling and discovery rate, I conclude that a good number for the *Ultimate* is 1.8 trillion barrels[12]. I stress that this relates only to *conventional oil*[13]. My extrapolation shows that of this less than 200 Gb remain to be produced after 2050. Rounding it off within the accuracy of the model, one could as well say that this 1.8 Gb would be *Cumulative Production* by, say, 2075. It avoids the mental anguish of having to think about the circumstances of depleting the last few barrels and the fuzzy boundary between *conventional* and *non-conventional* oil at that time. It will do well enough to drive the depletion model, which I will come to in the next chapter.

But first, we need to consider the distribution of the *Ultimate,* as depicted in Figure 7-12.

Ultimate				Yet-to-Find			
By Country		By Region		By Country		By Region	
Saudi Arabia	300	ME Gulf	722	FSU	49	ME Gulf	75
FSU	275	Eurasia	339	Iraq	21	Eurasia	53
USA	210	N. America	238	Iran	15	L. America	15
Iran	120	L. America	186	Saudi Arabia	21	N. America	11
Iraq	115	Africa	137	USA	9	Africa	12
Kuwait	95	W. Europe	66	Venezuela	7	W. Europe	6
Venezuela	90	East	60	China	4	East	4
Abu Dhabi	80	ME Other	40	Abu Dhabi	4	ME Other	3
China	55	(Unforeseen)	13	Kuwait	8	(Unforeseen)	8
Mexico	50	WORLD	1800	Libya	2	WORLD	180
Libya	45			Canada	2		
Nigeria	40			Norway	3	Numbers (Gb) may	
UK	30			Mexico	2	not add because	
Canada	28			Algeria	3	of rounding	
Norway	27			Nigeria	4		
Indonesia	25			UK	2		
Algeria	23			Brasil	1		

Fig. 7-12. Distribution of the Ultimate and Yet-to-Find by major country and region (1996).

Only three countries hold more than 200 billion barrels each, and only twenty-two hold more than 10 billions barrels each (less than six months' world demand). About forty percent of the world's endowment is in just one region: the Middle East Gulf.

UNDISCOVERED

The *Yet-to-Find* (or *Undiscovered*) is calculated in the model by subtracting the *Discovered* from the *Ultimate*. It works out to be 189 billion barrels, distributed as shown in Figure 7-12. It could be rounded to 200 billion barrels, but it is better to keep the number as computed, so that things add up properly, recognizing at the same time that it is not as accurate as it sounds.

The Middle East Gulf Region is assessed to have the greatest potential at 75 billion barrels. Again, most of the giant fields have been found, but the rich endowment of source-rocks and the effective seals mean that there is probably much more to find, especially in the complex fold-belts of Iran and Iraq.

Relatively few exploration wells have been drilled in the Middle East, compared with, say, the United States. It is sometimes claimed that this is reason to expect that colossal amounts of still more oil could be found in the Middle East if exploration were stepped up. This may not be so because the Middle East petroleum habitat is a concentrated one for geological reasons. Gentle structures grew over long periods of geological time, so that they were able to effectively drain the catchment areas around them. The fact that seal is as important as source in the Middle East has much to do with it. Half of Saudi Arabia's oil lies in just two fields: Ghawar and Safaniya. What remains to be found is smaller by far. Figure 7-13 shows how closely spaced the major oilfields are.

The Former Soviet Union is held to have the next highest

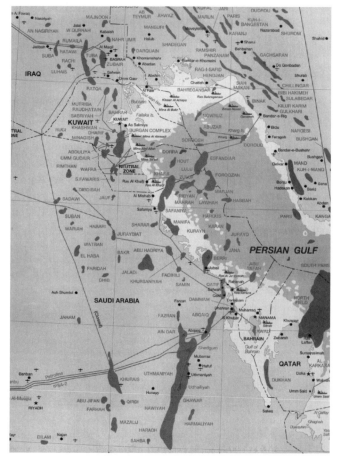

Fig. 7-13. Map of the Persian Gulf area: the wide distribution of oilfields implies that most of it has been thoroughly explored.

potential with 49 Gb. To give a sense of proportion, that is equivalent to about three-quarters of the North Sea, the largest new province found since the War. Although the larger fields and basins have already been found under the systematic exploration of the Communists[14], there may well be much left to find in deep basins beneath a salt seal, and in generally smaller fields. Some of the more remote areas are under-explored, although it begs the question of whether their oil, if any, qualifies as *Conventional*. It is of course difficult to assess the potential on the information available: it may have more than here thought likely, but distant fields are not always greener. The political situation and the difficult fiscal and contractual environment also make it uncertain how much will actually be discovered at least during the next critical decade or so. It is as well to take a cautious number for the purposes of modelling peak production. Even if the estimate made here is eventually due for upward revision, we may be sure that it will not be for a long time to come under any likely politico-economic scenario.

Detailed reviews of country by country potential have been published[15], and my latest assessment is summarized in Appendix 2.

Almost ten billion barrels are attributed to "unforeseen" discoveries in new or currently producing countries. It could occur in stratigraphic traps along the Atlantic margin, or possibly in some rather unlikely areas, such as offshore Namibia or the Falklands. There may be a few surprises in complex fold-belts or remote and difficult areas. But mainly it will come from ever smaller fields in mature basins. Some of the "foreseen" for the Former Soviet Union may not materialise and may have to be reallocated to the "unforeseen" elsewhere.

The offshore is relatively easy to explore. Seismic surveys can be conducted rapidly and cheaply. The results are also of superior technical quality in the absence of topographic distortions, meaning that wells are generally located on firmer evidence. Modern exploration rigs can be towed in without the expense of road building in difficult terrain. It means that most of the major offshore finds have been made, save in the few areas having significant very deep water potential[16]. Attention may therefore return to the onshore, especially to difficult areas or those that are, or have been until recently, closed for political reasons.

According to Petroconsultants, the world is finding less than six billion barrels a year on a falling trend. At that rate, significant exploration will be over within about twenty-five years, even if the tail drags on a little longer. Probably, almost all the important new discoveries will have been made within the next decade or so in areas outside the Former Soviet Union and the Middle East Gulf, where politico-economic constraints may delay and curtail effective exploration. Less than 100 billion barrels are likely to be fully accessible to the international industry.

Some 450 000 wildcats have been drilled out of an estimated ultimate number of about 500 000. But these are misleading numbers because 400 000 of them were drilled in North America under the special circumstances of that region. Many were drilled long ago without the advantages of modern methods. The same amount of oil could be found

with far fewer wells today. Drilling rates are falling almost everywhere, as the number of identifiable viable prospects dries up.

Fig.7-14. Wells have been drilled in all prospective basins and under difficult conditions.

Shell International Photographic Services, London

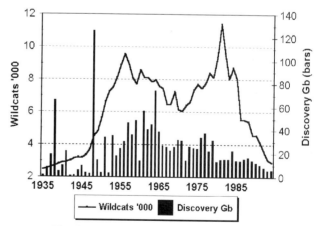

Fig. 7-15. Drilling and discovery rates: effort is not matched by reward.

The high oil prices of the 1980s prompted a surge in drilling, but the results were disappointing. The pending oil price shock may prompt another boom, but the results are not likely to be any better.

It is important to remember that, in many countries,

exploration expense has been deductible against taxable income, so that it already enjoys a considerable hidden subsidy. The subsidies, in effect, meant that dubious, high risk prospects were drilled. Governments, however, may change the tax rules, as has happened in the United Kingdom, when they found that the deduction no longer delivered sufficient revenue producing projects. The subsidy is a form of national investment that has to be justified by results.

On the other hand, much of the potential for exploration lies in countries where State companies have exclusive or dominant positions. They enjoy no tax incentives for exploration, and spend real money for which there are many other competing claims in the national budget.

Profitable niches in exploration will no doubt continue to be available for enterprising independent companies and perhaps contractors, but it is difficult to justify a contribution to world supplies much greater than that indicated, especially when the time and drilling rate constraints are taken into account. It is uncertain what role the major companies will have: they are ceasing to behave like integrated companies, but rather as holding companies with increasingly independent affiliates. I will discuss later the advantages and disadvantages of this development. I am already struck, reading the *World Oil*[17] review of worldwide operations, by how much is in the hands of small and relatively unknown companies.

JEAN LAHERRÈRE
and the Parabolic Fractal

Q Jean: I believe a motor accident in France affected your career causing you to change from geology to geophysics.

A *Yes, I was injured in 1955, and in those days a geologist needed strong legs for field work in remote places, so my boss thought it a good idea to move to geophysics. I had a good grounding in mathematics, and enjoyed the subject, having graduated at the Ecole Polytechnique.*

Q How did your career unfold?

A *I had joined Total, the French oil company, previously known as CFP, Compagnie Française des Pétroles, which had its origins as a founder shareholder in the Iraq Petroleum Company after the First World War. France's strategic needs after the last war called for a more dynamic search for oil as an operator. We concentrated on Algeria, which was then French territory, but was regarded by the US companies as having very limited potential.*

Q What led you to Hassi Massaoud, which was found in 1956, but is still the largest field in Africa?

Fig. 7-16. Jean Laherrère, who discovered the Parabolic Fractal, which helps predict oil discovery.

A *It was found by the now little used refraction seismic technique, which along with aeromagnetics, was the only method to map beneath the thick salt sequence with overlies it. Even modern reflection surveys barely penetrate. My first paper in 1959 was on the anisotropy of seismic waves to calibrate refraction events. The survey*

The Coming Oil Crisis

needed an enormous explosive charge to record refraction arrivals 15 km distant (see Figure 4-2). Incidentally, we used four tons of fuel and fertilizer to make the change: the same as they used in the Oklahoma bombing. My second paper in 1961 was on synthetic seismographs.

Q After the Sahara, you visited and worked in many places around the world. At that time it seemed as if there was still plenty of oil to find. When and how did you begin to question this assumption?

A It dawned on me slowly as my career unfolded and as I began to have responsibilities for exploration throughout the world. I spent time in Australia and Canada, but in 1972 returned to the Head Office in Paris. I had a wide experience, being at different times in charge of negotiations, making Total's first deals in South America, basin studies, and research. I also served as Deputy Exploration Manager. In addition, as President of the French Oil Industry Commission, I supervised the preparation of manuals on exploration techniques. With the passage of time, I began to see relationships such as how the larger fields tended to be found first, as was the case with Hassi Messaoud. During my career I watched most basins become mature exploration areas, and I realised that very few new provinces were being found.

Q You discovered a law of distribution which you call the parabolic fractal. Was this something that resulted from working on the computer? Or did the inspiration come, so to speak, on the back of an envelope?

A I am a visual man: to remember anything, I have to see it. I have always looked for graphic representation, and dream of finding a good and simple representation of the laws of Nature. Being retired, I now have the time to study these things. I am in fact writing a book on natural distribution and inequality. Life is a race, with equality at the starting line but not at the finishing gate.

It has long been a habit of mine to plot relationships and numbers. I plotted oilfields in the order of their size, and then used log paper to bring out the relative sizes. At first, I used a straight line method (as first used by Pareto and Mandelbrot) but it did not fit, especially for the smaller fields. I realized that the plot was curved, and as I thought about it, I realized that many distributions in Nature are curved. The horizon, for example, appears a straight-line but in fact is part of the Earth's circumference. It dawned on me that the simplest curve is the parabola, and I found that the best fit for almost all distributions is parabolic (galaxies, urban agglomerations, languages etc). I then tried to understand more of the theory, delving into fractals, chaos theory, and the law of self-similarity, whereby a complete segment of a distribution describes the whole: one branch mimics the whole tree. Perfect symmetry is linear

from one infinite to the other, but in Nature neither perfection nor infinity exists. Since the self-similarity is not perfect in Nature, the fractal is parabolic. I have written a paper for the Académie des Sciences on the subject.

Q It seems to be a remarkable tool to predict future discovery.

A In certain circumstances, where the data are valid and where there is a clearly defined natural domain, it can give excellent results, but it is not a tool to be used blindly. I believe in approaching each case from as many vantage points as possible realising that in practice the data are often weak and that there are extraneous influences at work. But in general I think that we have reached the point at which we can make a sound assessment of the world's future production of conventional oil, using a combination of these methods.

Q What do you mean by conventional?

A I mean oil that is producible at a price less than about $25/barrel over a certain period of time. A large number of small fields will be left behind, when production ends. The law of diminishing returns explains that it will take an impossible number of wells to find them all. The world has large amounts of heavy oil that are not competitive against ordinary oil such as has been produced so far. Some of it will become viable in the future, but we have to bear in mind the time frame as well as the volume. Heavy oil is slow to produce and will have a very different depletion pattern. There is a big difference between a field with wells flowing at 10 000 b/d and one with wells producing at less than 100 b/d. The amount of work involved is several magnitudes greater. The critical issue is not so much the size of the resource but the number of wells is will take to produce the heavy oil.

Q Do you think that there will be a gradual transition from conventional to non-conventional production.

A No, I anticipate a crisis because we live in a short-term world in which governments are unable to plan ahead. Conventional production is set to peak within a few years on the basis of my studies. About half the world's remaining oil lies in just five Middle East countries whose control of supply must inevitably increase with all that that implies. Even if a crash programme to develop non-conventional oil were put in place immediately, it would be years before it could make a global impact.

Nevertheless, there are few large fields yet to find, and the companies may be able to invest in helping the national companies produce difficult fields.

Q Can you explain the evidence for the coming peak?

A *It is fairly obvious. Before you can produce oil, you have to find it. Countries often have several cycles of discovery as different geological trends are opened up. You can correlate them with production cycles after a time-lag, at least in countries producing at capacity. Peak discovery occurred in the 1960s and is about to be reflected in production, which in world terms will peak around 2000. It is not quite as simple as this because of the uneven distribution, with a few swing countries having an exceptional endowment.*

Q *It is an immensely important message which deserves more attention. What can be done?*

A *We now have the analytical techniques to determine what the situation is. What we need is better information on reserves and stricter definitions. Reporting reserves is a political act ! We like to appear poor in front of the tax inspector; and rich when we meet the banker. The national companies, which own most of the reserves, should be urged to cooperate in providing the essential information. It is not so much a technical problem as a political one, in which the buyers as well as the sellers can exert their influence. They need to know the security of their supply. So long as OPEC quotas are based on reserves, the published numbers will be questionable. More than fifty percent of the countries, whose reserves are listed in the Oil and Gas Journal, report unchanged numbers from one year to the next, which is obviously implausible.*

Q One reads many economic evaluations that say it is just a matter of investment and technology. Indeed, most reputable institutions appear to think that there is not a cloud on the horizon. Are they mistaken?

A *I am sorry to say that they are. One of the reasons is that they have to rely on unreliable published data in which reserve revisions are not properly backdated to the discoveries they relate to giving a totally false impression of the trends. Another is that they misuse Reserve to Production ratios, failing to understand that all oilfields decline after middle age. Economists do not seem to quite grasp the implications of depleting a finite resource. Many people confuse reserves and resources. For example, the Russian classification was not based on realistic recovery factors: they took the maximum theoretical case without regard to economic or technological constraints.*

Q Could you provide an example of one of your plots that demonstrates your concerns well.

A *This is a plot of the remaining reserves over time for the World. The first curve shows reserves with revision taken on a current basis (from the API Basic Petroleum Data Book). The second curve shows the reserves with the revisions backdated as if known at the date of discovery. The first curve shows a plateau after the marked spurious increase in the late 1980s, which was related to OPEC quota. The second curve shows the actual decline, despite all the technology, we hear so much about, and despite unprecedented drilling levels.*

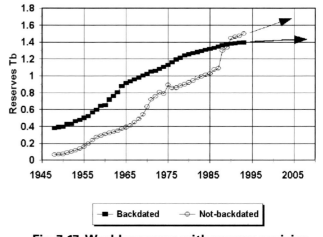

BACK-DATING RESERVES

Fig. 7-17. World reserves with reserve revision taken on a current basis or backdated.

NOTES

(For references, see Bibliography)

1. A term for old wells producing a few barrels a day as are found in several nearly exhausted US fields.

2. A term meaning rocks devoid of prospects, such as the igneous or metamorphic rocks of the ancient shields.

3. See Campbell, C.J., 1995 who explains the USGS definitions and how to decode them.

4. Although the log-normal distribution tends to work well for the parameters used in estimating the reserves in a single field, it tends to under-estimate the reserves of a basin because it gives equal weight to large and small fields, when in nature there are many more small fields yet-to-find than large ones.

5. See Laherrère, J.H., 1996.

6. See Hubbert, M.K., 1956, 1982.

7. Hubbert did not release his formulas until he was 80 years old in his 1982 paper.

8. See also the depletion profile of Germany in Fig. 8-2. It seems to me that the rapid rise in production in the 1950s and 1960s was essentially unfettered. Exploration was pursued vigorously and the larger fields were found. Theoretically this could have gone on had they opened up all the stops to give a natural peak around depletion midpoint, but in fact production was cut back for all sorts of reasons, including good reservoir management, conservation and other practical reasons. It then fell off exponentially, effectively transferring the theoretical peak to the real tail.

9. See Campbell, C.J. and J.H. Laherrère, 1995.

10. In the same way as sound can be defined as the sum of a limited number of harmonics (Fourier analysis), so oil production can be modelled as a small number of symmetrical cycles, each reflecting the practical discovery and production of oil from a separate geological trend or "play".

11. See Duncan, R.C., 1996.

12. My first estimate in 1989 was 1.65 Gb, but I did not then appreciate the industry practice of understating the initial reserves of a discovery. I according increased it to 1.75 Gb to take that into account. This year I have increased it to 1800 Gb and reduced the provision for "unforeseen" on the basis of new information regarding the greater than anticipated potential of the Former Soviet Union and some Middle East countries. I personally remain rather sceptical, believing that even if the higher numbers should be valid, they will not be confirmed for many years to come. The 1.8 Gb is, in my view, an estimate verging on the high side, at least when used as a basis for predicting peak production.

13. In principle, this is conventional crude oil only, but it probably includes relatively small amounts of condensate and non-conventional oil not distinguished in the statistics. It certainly excludes North American NGLs, which are recorded separately.

14. See Tull, S., 1997, for a sanguine assessment of the FSU.

15. See Campbell, C.J. and J.H. Laherrère, 1995, and Campbell, C.J., 1991, provide such studies: the former based on Petroconsultants data, which incorporate Russian official statistics. Masters of the USGS has also evaluated the situation over a period of years, although care must be taken to decode his definitions.

16. Say in water depths of more than 2000 m.

17. See *World Oil*, 51st Annual International Outlook August, 1996.

Chapter 8

PRODUCING WHAT REMAINS

In the last chapter, I described the concept of the *Ultimate:* namely *Cumulative Production* when it ends. Alternatively, if we have a mental hang-up with the idea of actually producing the last barrel, we could call it *Cumulative Production* at some distant future date such as 2075 or 2100. As production declines exponentially, the tail end of depletion is almost flat, so it makes little practical difference.

I have suggested that 1.8 trillion barrels is a good number for the *Ultimate*. If we subtract the 784 billion already produced, to end 1996 we are left with about one trillion barrels to produce in the future (1.016 Tb to be exact). It is made up of 836 billion barrels of *Reserves* and 180 billion barrels *Yet-to-Find* – always remembering that we are talking about *Conventional* oil only. Figure 8-1 shows where it is, both by region and major country. It is immediately obvious that about half of it lies in the Middle East Gulf Region, comprising Abu Dhabi, Iran, Iraq, Kuwait and Saudi Arabia (including the Neutral Zone, owned by Kuwait and Saudi Arabia).

I have already explained that the rate of flow in an oilfield is constrained by both the immutable physics of the reservoir, by what the engineers think is the optimal depletion strategy, and in some cases by market offtake. We cannot have one large champagne party, and blow it all away in a single evening, however much we might try. All fields are subject to decline during the second half of their lives, yielding ever less production. We cannot avoid the issue of rate: depletion rate, decline rate and production rate. It brings a *time* factor to the question of "How Much?": namely "When?".

I have discussed the theoretical and empirical reasons why peak production naturally coincides with the midpoint of depletion. Even where there is artificial interference with the natural pattern, as with the imposition of OPEC quota, peak cannot come more than a few years either side of midpoint, due again both to the physics of the reservoir and the natural pattern of exploration, which finds the larger fields and basins first.

Theory is confirmed empirically in countries at an advanced stage of depletion, such as Germany[1] or Trinidad. It means that we can divide the producing countries of the world into two groups: those that are past midpoint and those that have not yet reached it. The latter group can be

Yet-to-produce (1996) billion barrels (Gb)		Midpoint of Depletion
Saudi Arabia	223	2013
FSU	147	2000
Iraq	92	2017
Iran	74	2007
Abu Dhabi	66	2017
Kuwait	68	2013
USA	37	1973
Venezuela	42	1993
China	34	2001
Mexico	27	1998
Libya	25	2000
Nigeria	22	1999
Norway	18	1999
UK	16	1997
Algeria	13	1999
By Region		
Middle East	534	2013
Eurasia	183	2000
L. America	92	1995
Africa	74	1998
N. America	45	1975
W. Europe	37	1998
East	26	1994
Middle East Other	21	1997
(Unforeseen)	10	2016
WORLD	1016	2001

Fig. 8-1. Distribution of *Remaining (Yet-to-Produce)* oil, and depletion midpoint by country and region.

further sub-divided: first, into those countries which are close to midpoint, such as Norway and the United Kingdom; and, second, the five Middle East major producers (six with the Neutral Zone), which are at a very early stage of depletion. The latter can be termed the Swing Producers because in resource terms they can make up the difference between world demand and what the other countries can produce. The Swing Producers own about half of the *Yet-to-Produce,* and can exercise this swing role for about twenty years until they themselves reach midpoint. I stress *can* because there are many political and economic considerations that will determine if they *will* discharge the role and how they will apportion the contributions amongst themselves.

GERMANY
Ultimate 2.25 Gb

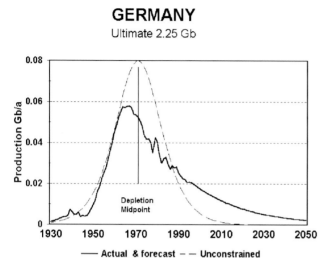

— Actual & forecast − − Unconstrained

Fig. 8-2. Production profile of Germany, a country at a very mature stage of depletion. Peak production came close to the midpoint of depletion. Superimposed is a theoretical depletion curve based on unfettered production.

I do not speak of OPEC as such, because in resource terms it is a disparate and irrelevant grouping[2]. Ecuador and Gabon have already withdrawn. Indonesia will likely be the next to leave as its exports dry up due to resource constraints. Venezuela pays scant attention to the quota obligations of membership. In any event, it is the five Middle East Gulf producers that control the direction of OPEC.

There are several ways by which to model future production. There is nothing particularly new in the basic understandings of depletion[3], but new urgency is given to the subject as we approach the inflection.

BELL CURVES

If world production were homogenous, or still controlled by the major international companies, we could perhaps apply the bell-shaped curve to the world as a whole. With an *Ultimate* of 1.8 trillion barrels, the midpoint peak will be reached when 900 billion barrels had been produced. With 784 billion barrels already produced, it means that there are 116 billion more to produce to reach midpoint, namely in about five years' time, assuming annual production of 24 billion a year.

Figure 7-10 illustrated the bell curve for unfettered world production, built on an *Ultimate* of 1.8 trillion barrels. The profile shows a close match until the oil shocks of the 1970s, when production was constrained. As a consequence the natural peak was not reached, meaning that the future decline will be less steep than would have otherwise been

the case. The production profile for the Non-OPEC countries shows a much closer fit to the natural curve, because they were not significantly constrained, see Figure 8-3.

NON-OPEC
Ultimate 850 Gb

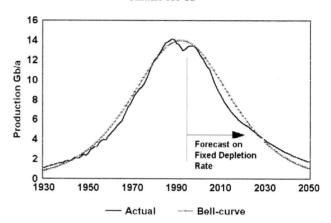

— Actual ---- Bell-curve

Fig. 8-3. Bell curve for Non-OPEC production showing a close fit.

We can refine this approach by applying the same procedure to each country and model each peak and each subsidiary peak for every country or even every geological basin. It would involve a great deal of work and computing effort. The bell-shaped curve is mathematically elegant, but does not necessary match what occurs in practice partly because the theoretical peak production of an oilfield is normally capped for various reasons. The bell-curve model gives the best results when it is applied to countries with a large number of basins and fields, such as the USA or the FSU. I suspect that the model needs a large statistical base and may not mean very much in individual cases. There are simpler alternatives that may give equally valid results.

FIXED RATE DEPLETION

For countries exactly at midpoint, we could calculate a midpoint depletion rate: namely annual production as a percentage of the *Yet-to-Produce*. The model would show production reducing by that amount each year: in other words, it falls at a fixed decline rate[4]. In practice the *Depletion Rate* seems to be fairly constant in countries past depletion midpoint, and the result is similar to that obtained from the multiple bell-curve method of analysis.

For countries past midpoint, we could either plot the midpoint depletion rate, and treat that as the long-term trend, or we could extrapolate future production at the present depletion rate. The differences are usually small.

There are options for countries that have yet to reach midpoint. We can distinguish the *Swing Producers* under alternative scenarios; and for the rest, we can extrapolate the present trend; apply actual forecasts of short-term production, or take an arbitrary increment, such as five percent.

Actually, it does not matter a great deal which model is used, as the differences between the curves reduce towards a crossover, because more produced early means that there is less left for later, given a finite total. Our purpose is to find a valid general model of resource depletion, recognizing that there will always be short-term departures from any model for all sorts of reasons. There is much to be said therefore for keeping it simple and readily comprehensible. Several examples are given in Appendix 3.

SWING CONTROL

The resource constraints I have discussed are important in demonstrating the total amount available, and its distribution, but politics and economics enter into the equation. The world is not homogenous: and the poles of influence change over time. So far, we have in global terms experienced only the left-hand side of the curve[5], which has been characterised by rising supply and demand. But there was interference with the natural pattern as a result of the oil shocks of the 1970s and the changes in ownership of the resources that followed them. I think it was the supply, and its cost, that drove the demand, and not the other way around. It is the same with road construction which increases rather than decreases traffic congestion because it encourages more people to drive[6].

A new environment begins to manifest itself around midpoint, not so much in terms of absolute shortage but in the changing control in the ownership of supply.

NON-SWING
Ultimate 1078 Gb

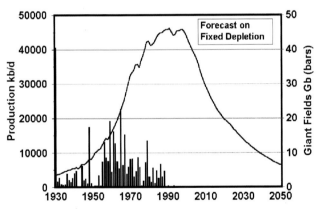

Fig. 8-4. Production profile of the non-swing countries.

I think that the increasing share of world production coming from the *Swing Producers,* which is due to the most uneven distribution of the resource, will be the critical factor determining the scenarios of supply and demand. How these countries exercise that control and what pressures are exerted upon them are the key issues: but it is well to recognize, like it or not, that Nature has endowed them with a disproportionate share of a finite resource. They will not be holding the world to ransom if they point this out in no uncertain terms. Their adverse geography, combined with their bounteous geology, has proved an explosive mix that gives them the character they have, which will no doubt influence their actions. I will later examine the issues more closely, but meanwhile will outline the resource-driven scenarios of supply and demand upon which the model is based.

NON-SWING COUNTRIES

Production in countries past depletion midpoint is assumed in the model to decline at the current (1996) depletion rate. In countries (save the *Swing Producers*) not yet at midpoint, production is assumed to increase either at the present trend or where that seems anomalous at a fixed 5% increase. When cumulative production reaches the depletion midpoint, production is assumed to decline at the midpoint depletion rate.

As already explained, about eight billion barrels have been retained in the model to cover "unforeseen" discoveries or reserve revisions in new or presently producing countries. It is combined with the resources of a number of countries with insignificant production that are lumped together as a group termed "Other". Production from this group is assumed to peak around 2016, and to thereupon decline at the then depletion rate. It is a notional amount, introduced to facilitate the operation of the model, and is of no particular significance.

Examples of the model are given in Appendix 3.

SWING COUNTRIES

As already explained, the Swing Countries comprise Abu Dhabi, Iran, Iraq, Kuwait and Saudi Arabia, together with the much smaller Neutral Zone which is owned by Kuwait and Saudi Arabia. It is assumed in the model that by 2000 each of these countries will contribute, under the identified alternative global scenarios, in proportion to its share of the swing countries' aggregate *Remaining* oil. When each

Mb/d	1995	1996	1997	1998	1999	2000	Percent of Remaining
Abu Dhabi	1.8	2.0	2.2	2.3	2.6	2.7	13
Iran	3.7	3.4	3.2	3.0	3.0	2.9	14
Iraq	0.6	0.8	1.5	2.2	2.8	3.6	17
Kuwait	1.8	2.0	2.2	2.3	2.5	2.6	12
NZ	0.4	0.4	0.4	0.3	0.3	0.3	1
S. Arabia	7.9	7.9	7.7	8.0	8.9	9.2	43

Fig. 8-5. Estimated Swing production prior to 2000 under the base case scenario.

country reaches its depletion midpoint, its production is assumed to decline at its then depletion rate, with the balance being taken up by Saudi Arabia, acting as the swing producer of the Swing group, in view of its large endowment.

Figure 8-5 shows the estimated production of the Swing countries prior to attaining a norm in 2000 under the Base Case scenario. In particular, Iraq's production is expected to grow slowly as the embargo is relaxed or leaks.

In view of the critical importance of these few countries in political as well as resource terms, it may be useful to say a few words about their character which has been forged of their geography, history, religion and politics. A few words it will have to be on a very large, complex, difficult, yet important, subject. Although Turkey has only limited oil resources, it is necessary to include it because it has been very influential and will likely be so again. It seems to be moving away from the West. Turkey has had long historical and cultural links with Islamic countries bordering the Caspian that were incorporated into the Russian Empire of the Tzars and later the Soviet Union. Its ties with these countries, some oil rich, may be due for renewal following the collapse of the Soviet Union. Turkey has occupied a critical position between Russia and Europe throughout history, and its importance may reemerge in the coming oil crisis.

The first point to remember about the whole region is that it is arid, with much of it being desert. A desert is a sort of terrestrial ocean: people move across it from port to port or oasis to oasis. It has not been settled in the same way as has, for example, Western Europe. Accordingly, the concept of land-ownership and frontier is weak and largely artificial, having been in many cases imposed by Britain and France after the First World War out of the ashes of the Ottoman Empire. There is even a tract of territory between Saudi Arabia, Yemen and the Oman which is formally claimed by no one[7].

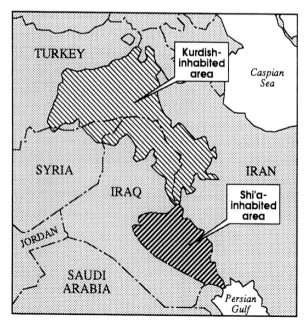

Fig. 8-6. The divisions of Iraq.

That aridity means a shortage of water is axiomatic. The control by Turkey of the headwaters of the two main rivers, the Tigris and the Euphrates, is a critical factor. Water may soon be a cause of more dispute than oil.

In historical terms, we are concerned, put at its simplest, with three main spheres of influence: the Persians who have occupied what is modern Iran since earliest times; the Turks, who were the dominant power through much of history, once with an Empire that stretched from Vienna to Cairo and North Africa; and third the Arabs, a mainly desert people who were previously vassals of the Turkish Empire. A fourth group are the Kurds, who are descended from the ancient Medes and occupy a mountainous belt from eastern Turkey through northern Iraq into Iran. Somehow they were denied national recognition in the carve-up, and have been resentful ever since. The idea of Arab nationalism was fostered by the British in the First World War to foment risings against the Turks, but it may not really be a deeply-based feeling, although it is now again stimulated by the injustices of the Palestinian situation. Most of the Arab nations in the Middle East spring from tribal groupings and the families which controlled them.

Another critical element is the force of Islam. If we look back a few centuries in Western European history, we see that power was divided between Church and Monarchy. The Spanish Inquisition and the Crusades are examples of bigotry and militarism based on religion. This dual power structure has been eclipsed and supplanted by democratic government only this Century. Even the recent experiments of Communism and Fascism had doctrinaire attributes that interrupted the march to democracy. Religious intolerance was widespread in Europe until recently: anti-Semitism was common in many places; and Catholic-Protestant conflicts abounded, with the last echoes still being heard in Northern Ireland.

The Middle East today resembles the Europe of not so long ago: it is still run by Church and Monarch. Democracy has not arrived and may never do so. Whereas we have had Protestants and Catholics disputing the role of the Virgin Mary; Islam has Sunni and Shi'ite factions disputing whether the prophet's son-in-law inherited the mantle. We may think of them are irrelevant issues, but many people die for such causes.

Islam permeates the governments of the Middle East countries, and the issue of holy shrines, including Jerusalem, is a real one. The *koran* is not so much a bible, as a legal and political system. Furthermore, the mullahs are not just religious leaders but secular powers too, leading communities, determining education policy and controlling law and the norms of behaviour. Religious martyrdom is a powerful thread in the national make-up, as demonstrated when the teenaged armies of Iran and Iraq clashed in the First Gulf War with death tolls exceeding those of the killing fields of Flanders.

They are lands of gross inequality with the privileged few and the underprivileged masses, whose lot is made even more difficult by the harsh terrain on which they live. For many of them, Islam offers the only ray of hope. These natural inequalities are exacerbated by oil revenues. Several countries, including Saudi Arabia, Kuwait and Abu Dhabi have huge temporary immigrant workers tenuously and indirectly supported by oil revenues: some now being repatriated.

It seems to me essential to have some glimmering of an understanding of these fundamental characteristics. Western, and particularly American, policy in the area seems often mis-judged. Denying access to the supermarket with embargoes might be an effective penalty for perceived wrong-doing in Detroit, but it is unlikely to work in the environment of the Middle East. Calls for adherence to "international norms of behaviour" can only puzzle the mullahs. They neither know nor care what such norms are: they are leading their lives according to their own traditions, creeds and survival instincts.

Of all of these countries, the most interesting and important is Iraq, under its much vilified leader Saddam Hussein. He did not face an easy job trying to establish a national identify for three very disparate groupings who were thrown together when the British with a red pencil decided to create the artificial state of Iraq. He had to deal with the Shi'ites with Iranian links in the south; the Sunnis around Baghdad, and the Kurds in several factions to the north. Some of his highly repressive measures were forced on him to hold the place together, remembering always that every culture has its own norms of behaviour. The idea of "human rights" is substantially a Western inclination.

Iraq has a large and relatively highly educated population, well capable of technical and scientific achievements, including the manufacture of nuclear power and weapons.

Saddam's two military adventures were disasters, and have left the country embargoed and isolated. In the short-term it is suffering, but out of this suffering may come longer term strengths. Its people have become tough and self-reliant, compared for example with the Kuwaitis, many of whom decamped during the Gulf War to the night clubs of Cairo. They have also had to learn how to live a sustainable life, free of imports and the distortions of world trade and global markets. Vegetables are grown on every available plot of land.

Whereas its neighbours are pumping oil for all they are worth and selling it at depressed prices, Iraq is forced to conserve its resources, and they are very substantial resources. Its *yet-to-produce* oil endowment is the third largest in the world, which very soon will be in high demand.

If Saddam can hold on to power for a few more years – and he seems adept at doing just that – he may come to be revered as the creator of Iraq's *Golden Age*. Menechem Begin, a former terrorist, became the respected and accepted premier of Israel, even being awarded a Nobel Peace prize: memories are short.

I think that Iran is much less relevant in oil terms, since its resources are more depleted[8]. There seems every chance that it will continue to be driven by an Islamic government and policy for a long time to come.

Lastly is the primary issue of the Saudi succession. On balance, it seems that the country will become less Western oriented, especially when higher oil prices restore the lavish revenues of the past and when the notion of market share as opposed to price has been shown to be as flawed as it is, given the resource considerations.

I naturally don't know how these Middle East countries will react to *The Coming Oil Crisis,* but I think that we can be sure that their reactions will be fashioned by a mentality that is very different from Western attitudes. It is claimed that force is the only message they understand. I think that may be a very mistaken view.

THE FORMER SOVIET UNION

Apart from the role of the Swing Producers, a key issue is the extent and rate of production in the former Soviet Union. It is estimated to have almost 150 Gb of *Conventional* oil left to produce, which is 14% of the world's total, second only to Saudi Arabia.

It appears that Russia itself is likely to continue to develop its oil mainly itself, seeing the Western companies essentially as competitors. Given its parlous economic condition, that means that the pace of exploration and development will be slow, and may do little more than reduce, or at most halt, the present decline.

The other former members of the Soviet Union are in a different position. They may at least initially welcome foreign intervention to strengthen their independence from Russia. It is a complex situation because several of these countries contain large ethnic minorities, including Russians, transported to them under Stalin's policy. Some of the leaders are ex-Communists of Russian extraction with an ambivalent view of Moscow.

Particular importance attaches to the Caspian Sea, with its possibly large undiscovered potential, which is the meeting point of Russia itself, Kazakstan, Turkmenistan, Azerbaijan and Iran to the south. Russia's own control is less than secure following the civil war in Grozny, which is seeking independence for the lands bordering the Caspian.

In January 1996, Iran and Russia proposed to the United Nations that the Caspian should be treated as a condominium to be exploited for the common benefit of the contiguous countries. The Caspian, which has no outlet to the ocean, has a very sensitive environment, which could perhaps be better protected by a condominium solution. Azerbaijan and Kazakstan, having the greater oil potential, understandably argue for the establishment of independent territorial rights to the Sea.

In addition to these intractable problems is the issue of pipelines by which to export the oil. At the present time, a 300 kb/d pipeline links Baku with a Russian line to the Black Sea, and another 190 kb/d line connects the north Caspian coast to Russia. Seven pipeline proposals are under consideration:

1. Through Transneft to Russia
2. Through Russia to the Black Sea
3. Through the Caucasus to the Black Sea

4. Through Turkey to the Mediterranean
5. Through Iran to the Persian Gulf
6. Through the Himalayas to Pakistan
7. Through Central Asia to China

All are fraught with serious obstacles and political hazards. Turkey is reluctant to see more oil traffic through Bosphoros because of the environmental risks. Providing Iran or Russia with control of throughput carries its risks, and the physical obstacles of lines to Pakistan or China are immense.

The risks to investment in such undertakings would deter all but the most stalwart financiers, as they are not exactly countries with a long tradition for the respect of contracts.

I, myself, would imagine that sequestration is highly predictable. Nevertheless, I have assumed in the model that FSU production with start rising in 2003 at three percent a year to a late subsidiary peak of about 9 Mb/d by 2010. With much of it having to come from the Caspian area, it is by all means an optimistic scenario, given the political risks and obstacles to agreement by the many parties involved.

GLOBAL SCENARIOS

The global scenarios of future production are here driven by three parameters. The first is the natural increase in world demand due to expanding population and growing energy consumption; the second is Swing Share, which when it reaches a critical threshold is expected to lead to a price escalation that curbs demand; and the third are the resource constraints themselves, whereby production in each country declines after the midpoint of depletion.

The most critical issue is Swing Share, and the oil price consequent upon such control rising above a certain level. The last two shocks were made possible when the Swing Share exceeded thirty percent. It is pointless to try to model every eventuality, and I will be content with three cases. In each of them, it is assumed that price-rises at a certain point curb further increases in demand, giving a plateau that lasts until the Swing Share reaches fifty percent. The "plateau" will in fact be far from flat with many highs and lows as increasing tensions erupt. It is assumed that, by the end of the plateau, the *Swing producers* will be unable or unwilling to meet the demands made upon them under any circumstances. Production will then begin its inevitable decline due to resource constraints. The *Swing producers* will then be collectively very close to the midpoint of their depletion anyway. In fact, several of the individual *Swing Producers* will be past their midpoints, transferring an increasingly insupportable burden onto Saudi Arabia. It will be a very tense situation. Naturally, there is nothing magic about the identified plateau thresholds, and it is easy to eyeball the consequences of delaying or advancing them.

It is tempting to propose that the higher the Swing Share, the higher the oil price, but I think that the strain of transferring such colossal amounts of money to the Swing countries, which in most cases means the families that control them, will limit excessive increases. The world's financial system will be under great stress.

I will outline the three scenarios before discussing them in detail.

High Case
Production increases at three percent a year to a plateau, marked by a Swing Share from 40% to 50%, followed by a decline at the then world depletion rate.

Low Case
Production is flat until the Swing Share reaches 50%, when it declines at the then world depletion rate.

Base Case
Production increases at 2% a year to a plateau, marked by a Swing Share from 30% to 50%, followed by decline as in the other scenarios. A 2% increase is perhaps a conservative assumption, as the International Energy Agency now predicts a 2.5% increase for 1997[9].

Figures 8-7 and 8-8 illustrate the scenarios over respectively the long- and short-term; in the latter case also showing the Swing Share in percent. Figure 8-9 provides the corresponding numbers.

PRODUCTION SCENARIOS
1800 Gb Ultimate

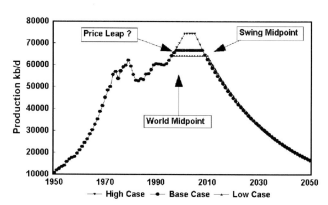

Fig. 8-7. Scenarios of production 1950-2050.

SWING SHARE & PRICE

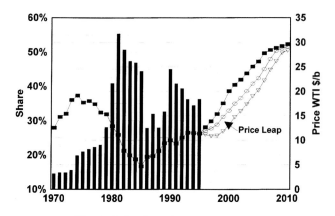

Fig. 8-8. Swing share and price.

The Coming Oil Crisis

	High Case			Base Case			Low Case		
	Plateau 2000-2006			Plateau 1999-2008			Plateau 1995-2009		
	Production Mb/d	Dep. Rate %	Share %	Production Mb/d	Dep. Rate %	Share %	Production Mb/d	Dep. Rate %	Share %
1995	61.4	2.2	26	61.4	2.2	26	61.4	2.2	26
2000	71.2	2.9	39	65.2	2.6	33	61.4	2.5	29
2005	71.2	3.4	47	65.2	3.1	42	61.4	2.8	39
2010	61.7	3.5	52	60.9	3.4	52	59.5	3.2	50
2025	36.1	3.5	51	36.4	3.4	52	36.4	3.2	52
2050	14.7	3.5	60	15.5	3.4	62	16.2	3.2	64

Fig. 8-9. Production, Depletion Rate and Swing Share under alternative scenarios.

No one can predict exactly how the *Swing Producers* will react to their growing control of the market, nor indeed how the consuming governments will react to a radical increase in price charged by the producers. It is stressed that this price is mainly tax, based on the control of the market, as actual production costs are not likely to rise greatly, being already far below alternative supply costs (see Figure 6-4.)

It seems certain that production will commence its inevitable decline by the time the *Swing Producers* control fifty percent of the market, as assumed in all the scenarios. This occurs between 2005 and 2009 depending on how high production rises in the meantime. By 2013, the *Swing Producers* as a group will have reached the midpoint of the depletion of their oil; the world as a whole having done so by 2001. These are the absolute latest dates for decline.

WHAT THE SCENARIOS MAY MEAN

Predicting the future is a very dubious undertaking, especially the near future. Events that seem important today may soon turn out to be irrelevant: and punch lines lose their punch if overtaken by events. Scenarios are not forecasts but logical extrapolations of sets of circumstances and trends. While we can be sure that the actual course of events will not unfold in the manner foreseen in any one scenario, we can reasonably expect that history will unfold within the spectrum of the three scenarios.

Our subject here is oil but there are many other concerns. There are only a few more years left of the present Century, and we face the next one with a mixture of anticipation and foreboding. Will a new age of universal enlightenment and material prosperity dawn? or will the pendulum swing back towards a dark age? Or will it swing in some new direction? and if so, which? These are questions that occupy the thinking mind[10] because there are so many conflicting sign posts.

The brilliant financier, philosopher and philanthropist, George Soros[11], divides the world into Open and Closed Societies. Certainly, the Open Society has been gaining strength as the present Century unfolded. People who live in America, Britain, and France, amongst others, have enjoyed Open Societies for longer than this Century, but they have had to fight for them. The threats of Fascism and its successor Communism have been seen off. Even socialism, with its ideal of a state-run utopia, has been eclipsed. Open Societies are being cemented in Latin America, the East and more tentatively elsewhere. But Closed Societies remain: none more powerful and important than those of Saudi Arabia, Iran, Iraq and Kuwait which are driven by a curious mix of oil revenue and religion as I have discussed. There are relic Communist regimes in Cuba, China, Burma and North Korea, and who knows what may arise from the ashes of the former Soviet Empire. Nationalist and tribal pressures close several African societies, and are a threat everywhere, even in Ireland.

Most people have come to accept that the Open Society is a good way to go. Its freedoms of expression have released Man's creative talents in science and technology which have given rise to a remarkable epoch of material prosperity for many people. Those who don't have it, aspire to it. It has also stimulated a cultural flowering in many areas. But the very nature of an Open Society means that people generally lack a clear direction or a spiritual aim: they just successfully muddle along within a positive environment, which they accept as a normal way of life. They have flourished without being driven by any overriding objective by which to motivate their lives. In a way, this is the other side of the coin and could undermine the foundations of Open Societies: much gets better but much gets worse. Urban degradation and the global increase in crime and violence are obvious concerns. The Belgian paedophiles are not alone[12].

The Open Society's successful economic formula, which is built around the now unchallenged capitalist ethic, has led to the opening of global trade and financial markets. Everything is changing hands, and is doing it so quickly that there is barely time to plant the tree before it is time to pick the apples. It has spawned what could be called the age of consumerism that reaches the threshold of mindless consumerism. It affects everyone. The Borneo native wants to trade in his paddle for an outboard engine, and many have done so. As its name implies, it *consumes* resources, and also creates waste on an enormous scale. People like new things, as children like new toys even if they don't play with them for long.

It is becoming quite evident that this consumeristic

spasm cannot go on indefinitely. Already, it is being asked if the world has enough resources to sustain it, or enough room for the garbage tips. People also begin to ask if it was such a good idea in the first place. They ask how vulnerable they are and what changes can come about: "Even if you win the rat race, you are still a rat".[13,14]

When we look at the resources that are being denuded, our attention is perhaps caught by the rain forests with the mental image of those fine trees being felled along with the nests of the parrots which live in them and the homes of the indigenous people below them. The more scientifically minded may think of the loss of bio-diversity that is eroding the gene bank on which we depend. But when we come down to the most elementary concern, it is for energy. Energy is the lifeblood of the consumer boom. And almost all of what is presently used comes from finite resources of fossil fuels, formed long ago in the geological past and then only under very exceptional conditions. Coal, oil, natural gas and uranium for nuclear energy all have that in common. Of these, the most at risk is oil, which provides forty percent of the world's traded energy. Without it the tractor could not plough the fields nor would there be transport to take either the food or us to market. Manufacturing and trade, as we know it, would grind to a stop without oil. Yet every time we fill up the tank with gasoline, there is that much less left for the future. It is being depleted and is not being replaced. We need to concentrate on the availability of oil as a matter of urgency[15].

The two poles of Open and Closed Societies are not the only ones. The pendulum need not swing between them, but could take off in a different direction. Not only can it, but it of needs must, because the exponential growth of consumerism is clearly unsustainable in a world of finite resources.

The argument for moving into a more sustainable dimension are compelling because it is so inevitable. By definition, Man will sustain himself for his allotted span in the fossil record. He has no other option. So, the issue is not so much about the intrinsic merits of sustainability but about the practical steps, means and above all timetable by which we reach this goal.

Today, no one in the developed countries cares as they shuffle down the shopping malls. They are only vaguely aware that there is, or might be, a cloud on the horizon. Their thoughts are closer to home as they contemplate turning up the heating, or the air conditioning; as they commute to work; worry about insuring their growing possessions against theft; and wait for the next pay cheque, with which to go out and buy some more consumables. In the rest of the world, they are trying somehow to survive in an increasingly artificial economy that tries to mimic the apparent progress of the developed. The Ethiopian subsistence farmer watches his national Boeing fly overhead as he scratches the soil for his next meal, and scratches his head wondering what it all means.

Geriatrics, such as I, have always thought that the world they knew was coming to an end, and speak wistfully of the

Fig. 8-10. The subsistence farmer does n't need the jumbo jet in his country.

Photo: Jørgen Schytte/Still Pictures

good old days. But now, the pace of change has become so great that enormous transformations really are inevitable. There absolutely will be discontinuities as the stresses become uncontainable. To be optimistic, the changes may be to the betterment of Mankind, however difficult the transitions. All the evidence suggests that the oil crisis anticipated in the following scenarios will be a catalyst for radical change. There is certainly a link between oil price and the state of the economy as Figure 8-11 shows. I obviously cannot chart the consequences accurately, but I can at least pose some questions, and invite you to think about them.

Year	Average US oil price ($)	S&P500 Index	Units of Index per 1000 barrels
1970	3.18	92	35
1975	7.67	90	85
1980	21.59	136	159
1985	24.09	211	114
1990	20.03	330	61
1995	14.62	616	2

Fig. 8-11. Relation between oil price and stock prices.

The High Case

The High Case, which envisages a three percent annual increase in oil demand and production, implies an expanding global economy and oil prices not much higher than those of the recent past: namely in the $15-25 range. It presupposes both no major disruptions in the world order, and continuing general economic prosperity. At the time of writing, most of the world's stock markets are close to all time highs. Many investors are concerned, but in this scenario their fears are misplaced: the stock markets continue to soar, and interest rates are held down, helped by increasing unemployment.

World production rises to 74 Mb/d by 2001, of which 31 Mb/d (42%) is supplied by the *Swing Producers*. It implies that King Fahd survives, or is replaced by a like-thinking successor, who is willing to continue to supply the West, and especially the United States, with cheap oil. In return, the US continues, and even steps up, its efforts to protect the throne from dissidents. The Iraq embargo is relaxed or allowed to leak, whether or not Saddam Hussein remains in power. Venezuela, which has already readmitted the international companies, continues to produce at capacity, irrespective of OPEC quota, and Mexico may be forced to admit US companies, having accepted loans with oil reserves as collateral in the 1995 rescue of the peso[16]. Production begins to rise in the FSU, partly from the successful development of Caspian fields by western companies, which allows exports to be maintained in a difficult domestic economic climate devoid of growth. All other countries, including especially the United Kingdom and Norway, produce at capacity, which is now close to peak.

Fig. 8-12. The underclass living in the hostile urban environment.

Photo: Gideon Mendel/Crisis

It is however a sort of fool's paradise, in which serious strains manifest themselves. The global market serves to increase unemployment in the European Union and North America, where urban ethnic ghettoes become ever more desperate. Crime and violence increase everywhere. Anarchist groups, such as the Freemen of Montana, become more active in the United States with random terrorist incidents as the bombing of the Olympic Games in Atlanta. The conflicts of Northern Ireland, Bosnia, the Basques, and Tamil Tigers in Sri Lanka continue to fester. New ones erupt around other minorities in Indonesia, New Zealand and elsewhere. In the Third World, and especially Africa, the elite prosper at the expense of downtrodden and desperate masses, increasingly given to tribal violence. The Tutsis have another go at the Hutus.

Islamic Fundamentalism, while contained, remains a potent but disorganized force, appearing in different guises and circumstances in Algeria, Egypt, the Middle East countries and parts of Asia[17]. It is only indirectly a religious issue, and mainly reflects the last hope of a growing

underclass of people unable to scratch a living from the desert. Little of the oil income, which flows to the Middle East, distorting the natural order in these countries, comes their way.

These undercurrents are there but are contained by authority, especially a US military presence that shores up regimes around the world, whatever their mandate or complexion: many war lords reside in Embassy compounds.

By 2001, however, the Middle East *Swing producers,* now supplying 42% of the world's needs, begin to find themselves no longer able to meet the demands for oil production made upon them, having failed to make available the necessary investment, and prices begin to increase sharply. As they rise, conditions in the Third World deteriorate further, and their often misguided plans for industrialization face set-backs. Exports also begin to slide in the FSU as modest rises in production are matched by equally modest increases in domestic demand.

The world is still a stable place, and the financial system remains intact. World oil production stops growing and plateaus until 2005, with the West turning ever more to gas supplies. Oil prices rise to around $30-35 in 2001 but increase further only gradually thereafter.

By 2009, when the *Swing Producers* supply 50% of the world's supply, demand and production begin to fall in parallel. The earlier economic growth in the industrial West has unexpectedly led to less demand because of structural changes in the way people live. The privileged no longer commute to offices, thanks to electronic communications. They spend more time at home, which leads them to be less consumeristic. Instead, they dedicate their time to gardening, home crafts and serving in vigilante forces to protect their properties. The dispirited underclasses languish in their urban ghettoes. Part of the change is achieved by the aging population: most people by then are old. It has become a less dynamic society.

Fig. 8-13. Old age pensioners abound.

Photo: Help the Aged

Conditions are dreadful in much of the world, but are more or less contained by disease, tribal warfare, and dwindling populations. Sufficient order is maintained to prevent the wholesale migration of peoples. The environment generally recovers as Man's influence on it diminishes.

It is not a good scenario.

Low Case

The Low Case, which implies no increase in demand, is hard to imagine. Perhaps the Brazilian cruzeiro, the Venezuelan peso, and the Indian rupee crash in quick succession, triggering a prolonged recession and a return to high world inflation. Perhaps, there is a general loss of business confidence everywhere, led by isolated and unrelated set-backs in many industrial countries.

Alternatively, perhaps global warming leads to the build-up of a southerly stream of Greenland melt-waters, which reaches a threshold velocity sufficient to deflect the Gulf Stream. The tree-line moves from North Norway to Perigord in South-West France in a matter of a two or three winters, much faster than scientists expected, despite evidence from the Greenland ice-cores pointing to abrupt climate changes[18]. Reindeers are again painted on cave walls in southern Europe by the few survivors. The last time this happened was only 23 000 years ago.

Whatever the reasons, which are hard to picture, the present level of oil production remains on a plateau until 2010, when it begins to fall due to resource constraints. Then, it is only three years before the Middle East *Swing Producers* reach the midpoint of their depletion, and they are no longer in a position to play a swing role anyway. Oil prices rise gradually during the period but are not the driving mechanism for the supply scenario.

It is an even worse scenario than the High Case, and much less plausible.

The Base Case

The High Case and the Low Case scenarios are designed to probe the limits on either side of what seems the most reasonable case: the Base Case. It anticipates that demand will increase at an average two percent a year, a little below the current trend. The Non-Swing countries produce at capacity, but that is insufficient to meet the demand, and the Swing share rises progressively until it reaches 30% in 1998.

This increasing share leads to a gradual strengthening in oil price to about $25-30, the price being undermined on the futures market by fears of a flood of production from Iraq. These fears are fuelled by a confused policy towards Iraq. Steps are taken to partially lift the embargo on one pretext or another, as the West perceives its need for Iraqi oil, and it increasingly leaks through poor enforcement, but there are also reversals as tensions flare for one reason or another. The international companies are invited back to Iraq, but they find it difficult to make an acceptable deal. Production fails to rise radically even as the embargo is relaxed, due to lack of investment, political confusion and disturbances. It is a volatile situation, faithfully reflected by the futures

market which over-reacts to these shifting events and press comment. Astute traders make a lot of money.

Gradually, the *Swing Producers* begin to feel their muscle, noting rising prices, rising share and peaking production in the North Sea, including Norway. They realize that no new oil provinces have been found. In short, they have read this book. In Saudi Arabia, King Fahd's successor moves to reduce his ties with America as a sop to the growing dissident pressures that have erupted there, as in Bahrain, Qatar and Kuwait. There is no major upheaval, but these countries are slow to increase their capacity, comforted by their growing revenues from higher prices. The Iran regime continues in power and does not add capacity either, being hawkish on prices partly for anti-Western political reasons and partly because it fails to make available the needed investment. It may sponsor some acts of terrorism, although it is far from certain that the West's previous claims in this connection were valid. If so, they are far from being the work of religious fanatics, but are designed to exert pressure in a delicate situation. The West now turns an increasingly blind eye, seeking rapprochement.

In parallel with these developments, domestic demand grows in the former Soviet Union, such that exports come under pressure. New investment by the West is slow to arrive due to the difficult politico-economic situation and a resurgence of nationalistic xenophobia, especially in the non-Russian countries.

At an OPEC meeting in 1998, it is decided to cut the quota by five percent. It is found to be a popular political gesture. Demonstrators in the streets shout their endorsement, claiming that it at last represents a rightful reassertion of Arab power. By now, the higher revenue coming to the Arab OPEC countries gives them the ability to manipulate the market. Air Force One makes frequent visits, but it fails to have much impact: the Arab world having been armed to the teeth by unscrupulous Western governments, they don't depend on Western support any more. The surveillance system installed by Erickson of Sweden in each country works well to protect their mutual frontiers. Israel is forced to become more accommodating.

By now, many traders in the West have a better appreciation of the situation and perceive the dangers. Prices have become very sensitive to shifts of opinion and comment, and the OPEC announcement prompts panic buying, as in the Second Oil Shock in 1979. Prices surge above $50 a barrel. One transaction for a Japanese refiner sets a record at $75 a barrel.

The IEA declares a crisis and authorizes the release of strategic stocks, but it is found that some of it has the wrong specifications and is not readily available for all sort of technical reasons. The IEA then reverses its policy and imposes an obligation to increase strategic stocks, having belatedly realised that it was not a temporary shortfall, but the onset of a new epoch of chronic under-supply.

While revellers celebrate the millennium, western governments are hard at work wondering what to do. The dollar comes under pressure, such that the actual cost of oil, quoted in dollars, falls. Kuwait and Saudi Arabia continue to hold large financial reserves in Western banks which cushions the effect of the higher prices.

Western governments belatedly try to recover control of the market by raising taxes on imported oil, reducing at the same time income and social security taxes. To their surprise, the economy and the environment actually benefit[19]. It is a successful formula, and the world price of oil begins to fall. The futures market is subject to rigorous government control with tax penalties for short-term trading.

There is, nevertheless, a general recession due to the higher cost of transport and energy. World production stabilizes at around 67 Mb/d.

The world proves to be remarkably resilient to what is now perceived to be the consequence of the natural depletion of a resource rather than a conspiracy by any interest group.

Energy saving becomes a priority for Western countries, and is enthusiastically embraced by the populace. City centres become free of private motor traffic, save for solar powered vehicles. Everyone rides a bicycle; and finds it an enjoyable experience.

The airline business crashes, but it soon transpires that few people needed to travel anyway: electronic communications having become highly efficient. Business travel ceases when new tax regulations no longer allow it as a deductible expense. Popular overseas holiday travel ends.

By 2008, production begins to fall as the *Swing Producers* approach the midpoint of their depletion. By then, gas pipelines have been constructed to link the Middle East to Europe. Gas production rises throughout the world, and it becomes as valuable as oil. It does not peak until around 2020.

The Third World suffers most from this fundamental discontinuity, but it has a silver lining insofar as movement, migration and military activities become more difficult. The urban conditions are dreadful but rural communities survive, finding new ways to self-sufficiency.

The United States with its huge territory survives well enough, having found that it could easily tighten its belt

from past profligacy in energy usage.

Great advances are made everywhere with the installation of solar energy, and there is a crash programme for nuclear power. Coal mines are redeveloped.

These rather disjointed thoughts try to illustrate some of the features of the discontinuity that the end of cheap oil will, or may, create. There will probably be truly radical changes in life-style that we cannot imagine: yet many may indeed be very positive. It will be, in some ways, analogous to the wartime occupation of France: life went on. It is not a cataclysm. After all, 67 Mb/d is still a lot of oil: about three times as much as was produced in 1960. Perhaps the main impact will be in Man's perception: in 1960, oil production was expected to increase, along with economic prosperity; whereas, by 2008, everyone will understand that it is set to decrease. They will plan their lives to use less.

I find it difficult to imagine these scenarios, which seem extreme in many respects. Can it really come to this? It will certainly not be the first time that discontinuities have struck: think of 1939. On September 2nd, Europe was at peace but on the following day it was beginning a war in which millions died and the whole shape of the world changed. I invite you to figure out your own scenarios for the consequences of a permanent increase in the price of oil for yourselves. The world has never faced anything quite like it before on such a global scale or so quickly. Your guess as to what it will mean should be better than mine, because you have to relate it to your own lives. Difficult as it is to grasp, the evidence presented earlier means that a major discontinuity of some sort is inevitable. If my numbers are wrong, the discontinuity may come a little later, but not much later. Adding 500 Gb to the "Yet-to-Find" delays midpoint by only ten years.

What makes the issue difficult is that it is so imminent. We are not talking about something that may happen a hundred years from now, but something that will be upon us very soon. Our super-efficient short-term economy has galloped blindly to the brink of the abyss. It is not too soon to look over the brink: on the valley below we may spot many solutions, but they will not include raising the production rate of conventional oil.

DR R.C. DUNCAN
on the
Olduvai Theory: Sliding Towards a Post-Industrial Stone Age

Fig. 8-14. Dr R.C.Duncan, who has charted energy usage per capita.

Q Could you say something about your background, early experiences and how you became interested in this subject?

A *My academic background is electrical engineering (practical stuff) - but my hobby is anthropology and archaeology (fun stuff). From 1985 to 1992, I lived and worked in Saudi Arabia, and travelled a lot, visiting some fifty countries. It became obvious that the agenda for global industrialization is a house-of-cards because it is based on rapidly depleting, and environmentally damaging, fossil fuels: oil, gas and coal.*

Q Would you mind if I paraphrase your Olduvai Theory, which I think sums up the predicament facing Mankind in an admirably lucid fashion.

A *Go ahead.*

Q Dr Duncan: I have read several of your papers, and it seems to me that they are very much in tune with the contents of this book. You have established the Institute on Energy and Man. Could you explain more about what it is?

A *It was established in 1992 to study the feedback relationship between energy use per capita and human civilization. It essentially consists of me and my computer. I do all the work from chief scientist to chief clerk. This is the only way I can track over time all the annual energy data and control the mathematical analysis. A "feeling" for our energy situation is what's really important. If I handed it over to "graduate assistants" I'd lose that.*

THE OLDUVAI THEORY

In 1989, I concluded that the life-expectancy of Industrial Civilization is horridly short. I think that it can be based on a measurable index: world energy-use per person. I have since managed to collect data on world energy and population, and improve the model. I call it the Olduvai Theory implying an impending return to the Stone Age.

My interest in the subject was prompted by a lecture I went to long ago, entitled Of Men and Galaxies by the famous Cosmologist, Fred Hoyle. He made the point that Mankind had one shot to make a go of it. If Man used up his resources of minerals and energy, which are the prerequisites of high-level technology, he will fail, and with him this planet's system. My travels around the world have led me to realize that industrialization is not evolving towards sustainability: just the opposite. Fred. Hoyle is most likely right: it is a one-shot affair. The question therefore becomes: How long will it last ?

I began to search for references on estimates of the life expectancy of Industrial Civilization, and can summarise the findings as follows:

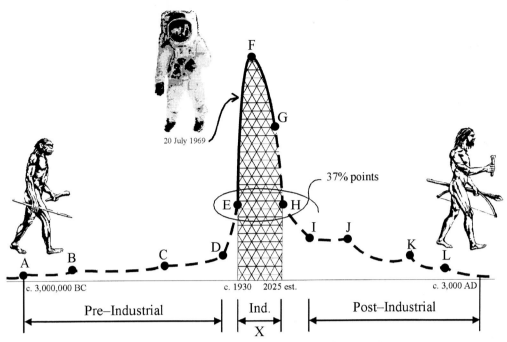

Fig. 8-15. The short span of Industrial Civilization.

Haldane	1927	"39 million years"
Russell	1949	"it cannot long continue"
Drake	1961	one million years
Watson	1969	potentially "millions of years"
Forrester	1971	natural response, about 200 years
Meadows, *et al.*	1972	natural response, 100-200 years
O'Neill	1976	"even our success becomes failure"
Leakey	1977	about 100 years
Harris	1977	"a bubble-like nature"
Crick	1981	short to 10 000 years or more
Laszlo	1987	"extremely short" to very long

In 1989, I divided human history into three phases:

The *pre-Industrial Phase* was a very long period of equilibrium, when economic growth was limited by simple tools and weak machines;

The *Industrial Phase* was a very short period of non-equilibrium that ignited with explosive force when powerful new machines temporarily lifted all limits to growth;

The *de-Industrial Phase* lies immediately ahead during which time industrial economies will decline toward a new period of equilibrium, limited by the exhaustion of non-renewable resources and continuing deterioration of the natural environment.

As my studies continued, I began to realize that Industrial Society can be described as a single-pulse waveform of fixed duration as measured by average energy-use per person per year. In other words, it had the classic depletion pattern of a finite resource: zero-peak-zero. Access to data on population and energy-use came in, showing that peak per capita energy usage was in about 1977.

Figure 8-15 shows the trend graphically.

1. Pre Industrial Phase (c. 3 000 000 BC to 1765)
   A – Tool making (c. 3 000 000 BC)
   B – Fire used (c. 1 000 000 BC)
   C – Noelithic agricultural revolution (c. 8 000 BC)
   D – Watts steam engine of 1765
2. Industrial Phase (1930-2025)
   E – Per capita energy-use 37% of peak value
   F – Peak energy-use
   G – Present energy-use
   H – Per capita energy-use 37% of peak value
3. Post Industrial Phase (c. 2100 and beyond)

The database is naturally less than perfect, and there is more than one way to look at it. The average of four different studies gives a peak in 1977 and a decline of 0.9% a year. There are other scenarios of course: we may not make it even back to the Stone Age.

It is an important issue. Potentially there is much that Mankind could do to manage the transition to a virtually sustainable economy. But it's not happening; and the Olduvai Theory says it won't happen. I hope I am wrong.

NOTES

(For references, see Bibliography)

1. Hiller, K., 1996, gives a valuable demonstration of the depletion pattern of Germany, based on very reliable data.

2. OPEC, the Organization of Petroleum Exporting Countries was formed in 1960. The following countries belong: Algeria, (previously Ecuador and Gabon), Indonesia, Iran, Iraq, Kuwait, Libya, Neutral Zone, Nigeria, Qatar, Saudi Arabia, UAE (esp. Abu Dhabi) Venezuela. There is also a group, known as the Organization of Arab Petroleum Exporting Countries, OAPEC.

3. See Foley and van Buen, 1978, who include a depletion model for 1978 by Mortimer of Shell, which is little different from that presented here. It is built on an Ultimate of 2 trillion barrels and peaks in 2000 with a 3% growth rate. It is as valid now as then, but we are now much closer to the inflection than they were in 1978.

4. Depletion Rate is annual production as a percentage of the *Yet-to-Produce,* whereas Decline Rate is the percentage difference in production from one year to the next. The two rates may be the same.

5. This is true in global terms, but some countries, such as the United States are already well down the right hand side of the curve, providing incidentally a convincing demonstration of its fundamental validity.

6. Fleay, 1995, discusses this irony at length.

7. See Roberts, J., 1995, for an excellent description of this and other Middle East issues.

8. See Takin, 1996, for a review of the situation in Iran. Between the lines he questions the official reserves, and explains that the country is at an advanced stage of depletion.

9. International Energy Agency Oil Market Report, September 1996.

10. McRae, H., 1994 and Kennedy, P., 1995, give thoughtful analyses of the future.

11. Soros, G., 1995, has written a brilliant book on his career in the financial world where he achieved great success by recognizing the concept of boom and bust and the discontinuities between trends. I think he would have much sympathy for what is written here.

12. A reference to a particularly gruesome case of kidnap and murder that shook Europe in 1996.

13. A telling aphorism by Duane Elgin, quoted by Howgego, 1996

14. See Duncan, R.C., 1996, for a particularly cogent discussion of the inevitable slide back to the stone age as non-renewable energy is depleted.

15. It is very encouraging that Australia is coming to understand the situation, see Fleay, B.J., and J.H. Laherrère, 1997.

16. Duncan, R., 1996.

17. George, A., 1996, has given a particularly penetrating analysis of the Middle East, pointing out that fundamentalism is a convenient blanket term that covers many different pressures, by no means all motivated by religion.

18. Tickell, C., 1996, discusses climate changes in history, some rapid. See also Egg, M., 1995 on the Ice Man, a 5000 year old corpse recovered from an Alpine glacier.

19. See Barker, T., 1995, who explains such a model for the United Kingdom aimed at cutting pollution.

THE RISE AND FALL OF HYDROCARBON MAN IN PICTURES

by Ann Slettebø

1. IN THE BEGINNING

Some theories say that the Universe came into existence 15 billion years ago with a "big bang". Planet Earth goes back to 4.6 billion years when it was thrown off the Sun. Life is estimated to have commenced 3.3 billion years ago, although the fossil record does not go back that far.

2. THE FOSSIL RECORD

3.3 billion to 5 million years ago

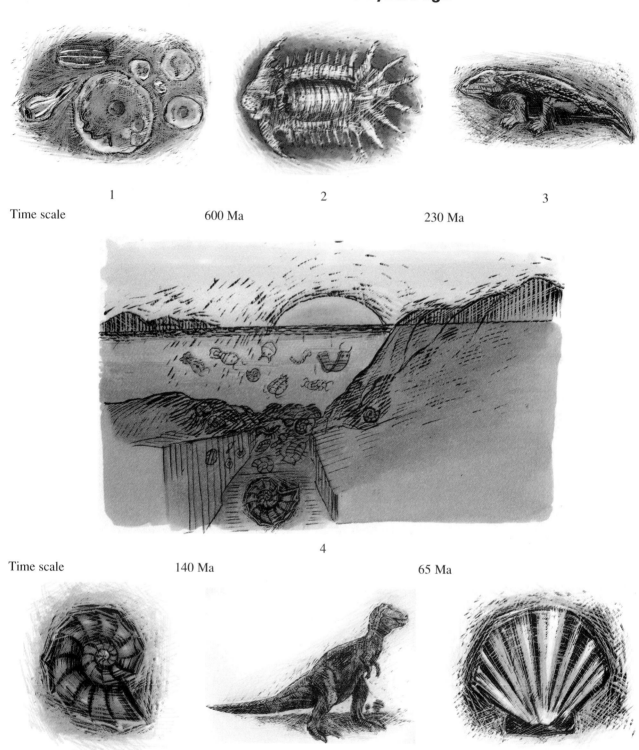

1

2

3

Time scale   600 Ma   230 Ma

4

Time scale   140 Ma   65 Ma

5

6

7

1. Very early life.   2. Terapsis, a trilobite.   3. Seymouria, an amphibian.

4. The debris of early life fell to the bottom of stagnant seas and lakes to become the raw material of petroleum but the conditions for preservation were exceedingly rare in time and place.

5. Lytoceras, a Jurassic ammonite.   6. A Cretaceous dinasaur.   7. Pecten, a Tertiary clam.

3. EARLY AND MODERN MAN

Five million years BC to Christ

1	2	3

Time scale 1M years BC 100,000 years BC

4

Time scale 3000 years BC 200 years BC

5	6	7

1. Pithecanthropus, pre-Man. 2. Neanderthal Man. 3. Modern Man.

4. Early Man followed the retreating Ice front in search of game. Living in the difficult conditions of the Ice Age sharpened his wits.

5. Horse-power for hunting. 6. Power for agriculture. 7. Wind power for transport.

4. SUSTAINABLE MAN

Christ to 1850

1

2

3

Time scale   500 AD   1000 AD

4

Time scale   1600 AD   1800 AD

5

6

7

1. He began to worry about the future.   2. Wind power for conquest.   3. Mechanical power.

4. Wind power for exploring the world, finding the new continents, and beginning world trade.

5. Water power for industry.   6. More travel.   7. Steam power from fossil fuels.

5. INDUSTRIAL MAN

1850-1950

1   2   3

Time scale   1900   1910

4

Time scale   1930   1940

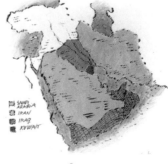

5   6   7

1. The first oilwell in the USA.   2. Oil powers steamships.   3. Oil begins to power War.

4. Coal, concentrated solar energy from the geological past, fuels modern industry and in turn capitalism, followed by the socialist reaction to the unequal benefits from cheap energy.

5. Oil for popular motoring.   6. Middle East oil power.   7. Japan goes to war for oil.

6. HYDROCARBON MAN

1950-2050

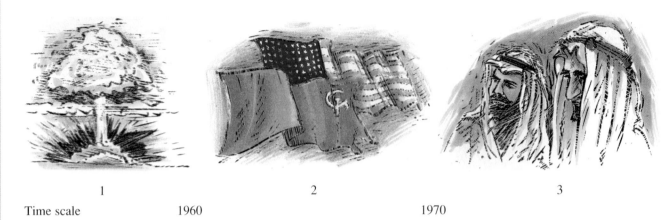

| 1 | 2 | 3 |

Time scale   1960   1970

4

Time scale   2000   2025

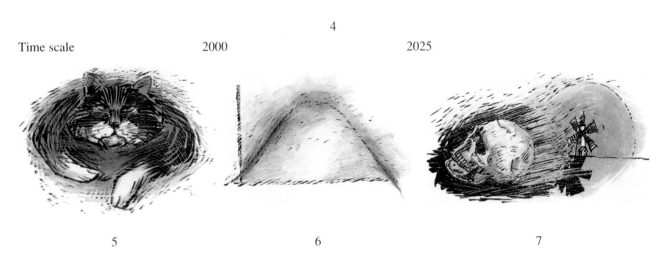

| 5 | 6 | 7 |

1. Nuclear weapons.   2. The Cold War.   3. The oil shocks.

4. Man becomes almost totally dependent on oil which permeates every aspect of his life. He spends much of his time in an automobile, which begins to pollute the atmosphere and change the climate.

5. The affluent society.   6. Oil production peaks.   7. The end of consumerism.

7. NEW SUSTAINABLE MAN

2050-2500

For most of his time on Earth man has lived a sustainable life. It was based on what Nature provided without consuming the accumulated mineral and fossil fuel resources of the Earth, which were formed over eons of geological time. He will have to adapt to doing so again when these resources are exhausted.

8. AFTER HOMO SAPIENS

Man can expect to become extinct like all other species in the fossil record. The simple limpet Lingula has managed to survive since the Cambrian, but more highly specialized and adapted forms died out after relatively brief spans of geological time. Man has become so highly specialised that his future must be at risk. But when he has gone, other forms of life will survive. At least Man better make the most of it while it lasts.

Chapter 9

NATURAL GAS, GAS LIQUIDS & NON-CONVENTIONAL OIL

The last chapter ended with a sort of body-blow: it was the punch line. Conventional oil production is about to peak, and there is nothing we can do about it. The scenarios considering the consequences of this discontinuity in the way the world lives were difficult to write, and may be hard to accept: they sound extreme and alarmist. Surely, we ask, things are not as bad as they seem; surely some happy solution will be found. Perhaps so, but anyway, why not sit up and think about it?

I won't belabour the point further. It is time to step back from the main theme, and fill in some of the other supporting data and ideas, before returning to a synthesis in the final chapter.

As discussed in Chapter 2, petroleum is a family of hydrocarbons in gaseous, liquid and solid states. Taken together, and ignoring the size of individual accumulations, the total resource is enormous. Economists, who fail to understand the distinctions and differing depletion patterns of these substances, are misled into stating:

"Total hydrocarbon resources with the potential to produce liquid fuels are so large that they can be considered infinite for the purposes of analysis".[1]

For them, the resource is treated as near infinite, and production is just a matter of supply, demand, profit, investment and technology[2]. In fact, these several species of hydrocarbons have very different characteristics and depletion patterns, which need to be taken into account when forecasting supply.

In the previous chapters, I have considered *conventional oil*, the substance which has provided most of our hydrocarbon fuel to date, and which is likely to continue to do so over the next critical twenty years. There are other hydrocarbons which will become important when conventional production falls. They are subject to very different depletion patterns that control their production rates. There cannot accordingly be a seamless transition from one to another, least of all if no plans are made to adjust to the decline and eventual end of *conventional oil*. In this Chapter, I will describe them. It is not a particularly inspiring chapter which could easily be skipped, but I have

to cover it to close off this escape route for those searching for an easy way out of the predicament the world faces.

CONVENTIONAL NATURAL GAS

Natural gases fall into two categories: the combustible gases, comprising methane (CH_4), ethane (C_2H_6), propane (C_3H_8), butane (C_4H_{10}) and hydrogen (H_2); and the non-combustible gases, including nitrogen, carbon dioxide and hydrogen sulphide. Methane is by far the most abundant, making up more than eighty percent. Typically, seventy-five percent of a gas accumulation is made up of combustible gases.

Conventional natural gas comes from two sources: specific source-rocks which are rich in organic material derived from plants; and normal oils which are cracked to gas on being heated excessively on deep burial. Many basins contain both oil and gas fields, reflecting the admixture of source-rock types and variations in the depth of burial. But some basins, such as for example the Southern North Sea, contain only gas – in this case because the gas comes from deeply buried coal deposits, which have been subject to natural coking. Many fields produce both oil and gas, the gas commonly having separated as a gas-cap above the oil in the reservoir.

There are two main types of combustible gas, termed respectively *dry gas* and *wet gas*. *Dry gas* normally consists mainly of methane, although in some fields it may be mixed with nitrogen, carbon dioxide, sulphur dioxide (sour gas), as well as, rarely, helium. The composition of the gas reflects particular source-rock characteristics, as well as the thermal and bacterial conditions to which it has been exposed. *Wet gas* contains higher hydrocarbons, including butane and pentane, and is generally associated with oil accumulations.

In general, gas prone source-rocks are much more widespread in Nature than are oil source-rocks, and they are especially abundant in young geological sequences. But on the other hand, gas is more mobile than oil, due to its smaller molecular size. As a result, gas requires a much stronger seal than does oil. Salt is the most effective seal, and many gasfields depend upon a salt cover. Permafrost conditions in Arctic environments are also effective. These two factors – of wider source but the need for better seal –

counter-balance each another, so that the total endowment of conventional gas is in fact less than oil. Some gas reservoirs at shallow depth are still being charged from the source-rocks, and are accordingly productive despite weak seals.

Figure 9-1 gives some tentative estimates of the *Cumulative Production, Reserves* and the *Undiscovered* potential of conventional natural gas by major country and region[3].

Figure 9-2 illustrates the trend of published estimates of the *Ultimate* gas recovery.

Gas is at a much earlier stage of depletion than is oil, mainly because of the high cost of transporting it from remote areas. Much was flared from oilfields in earlier years, and some is reinjected into the reservoir to enhance the production of oil. Gas is still being extensively flared in Nigeria, despite the penalties imposed by the government. For these reasons, the statistics on reserves and undiscovered potential are even less reliable than is the case for oil. Furthermore, economic factors have a much greater impact on defining the boundary between *conventional* and *non-conventional* gas than is the case for oil, making it harder to distinguish the two categories.

Gas flows more easily through the reservoir than does oil, which affects both the recovery factor and the depletion pattern. As much as eighty percent of the gas is normally recoverable, compared with less than about fifty percent in most oil fields. Peak production is usually capped to optimise the facilities, or meet the terms of the supply contract, which generally prefers a long constant supply to a high peak. The production profile is therefore characterized by a rapid rise to a long plateau, followed by an abrupt fall. The terminal decline is commonly controlled by compressor capacity: where compressors have been installed when the natural flow drops. Seasonal fluctuations in demand determine the detailed production schedule.

It is clear therefore that the production profile is primarily driven by economic factors although the resource constraints do eventually bite.

The production of gas associated with oilfields is however closely linked with the oil. Gas is commonly reinjected to maintain pressure, enhancing the oil production, until that falls to a low level. At a certain point, late in the field's life, the operator may decide to draw down the gas cap, effectively converting the field from an oilfield

Tcf	Cumulative Production	Reserves	Yet-to-Find
FSU	514	1977	747
USA	858	164	102
Iran	42	742	235
Qatar	9	250	5
Saudi Arabia	48	186	47
Abu Dhabi	15	205	4
Venezuela	41	140	18
Algeria	37	128	33
Nigeria	19	110	26
Iraq	10	110	24
WORLD	2205	4933	2862

Fig. 9-1. *Cumulative Production, Reserves* and *Yet-to-Find* of conventional Natural Gas.

PUBLISHED ESTIMATES
Ultimate conventional gas

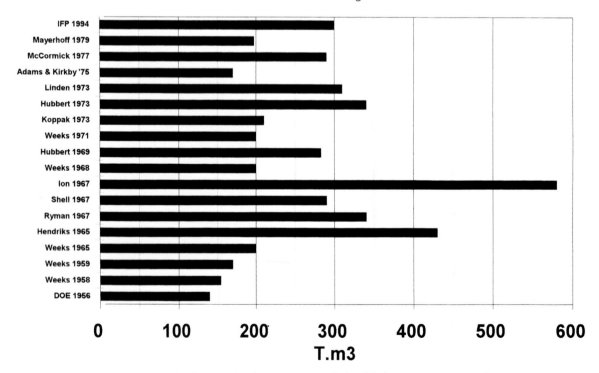

Fig. 9-2. Trend of published estimates of the *Ultimate* recovery of gas.

to a gasfield. The Brent Field in the North Sea is a well known example[4].

It raises the issue of how to define oil and gasfields, which in turn touches on the conversion factors by which to equate gas with oil[5] (here 1 boe = 10 000 cf, based on value not calorific equivalence). This relationship will naturally change as gas becomes more valuable. Fields with more ultimate gas than oil should be classed as gasfields.

Whereas it was possible, in the case of oil, to model the production profile around fairly well understood resource considerations on a country by country basis, it is not possible to do so for gas because a large part of the supply is driven by confidential supply contracts.

That said, we may still suppose that the midpoint of the plateau of production will more or less equate with the midpoint of depletion. We may also identify a certain Middle East Swing role, which will go into effect when that region is linked by gas-lines to Europe and the Indian sub-continent.

Figure 9-3 depicts a very generalized global production and depletion profile for conventional gas. It is assumed that production rises at about two percent per year until 2005, by which time the gas-lines will have been built to tie in the Middle East supply. Production will then rapidly rise to a plateau of about 120 Tcf/a until 2030, which is effectively constrained by pipeline capacity around the world, before a resource based decline of about three percent a year sets in. It is not possible to be too specific in the estimates, but it is assumed that non-conventional gas, especially coalbed methane, is included.

By 2050, gas production in terms of oil equivalence will slightly exceed conventional oil production.

Figure 9-3 shows that gas production is likely to plateau for a number of years either side of the midpoint of depletion around 2020. Gas will clearly enjoy a boom during the first few decades of the next Century, as *conventional* oil supply dwindles and becomes much more expensive. Gas prices, too, are likely to rise when it begins to compete more effectively with *conventional* oil.

Natural gas is normally transported by pipeline, but it may also be liquefied by lowering its temperature below −250°C and transported in special insulated and refrigerated tankers. Accidents have been few but the dangers inherent in transporting large quantities of gas in this way can be imagined.

NON-CONVENTIONAL GAS

As discussed, the boundary between *conventional* and *non-conventional* gas is even less easily drawn than in the case of oil. By and large, however, it can be said that the production of the categories here classed as *non-conventional* gas will follow the *conventional* category, and accordingly will not become important until far into the future.

The resources of *non-conventional* gas may be very large. Some are already in production, where they are located close to market, as in the United States or in the Po Valley of Italy. Exactly how large they are is hard to estimate. I identify the following categories:

Biogenic Gas

On deposition, organic material is subject to microbial action at more or less surface temperatures. The larger molecules are broken down with the release of methane and carbon dioxide, giving a reducing environment, which is incidentally important to the subsequent generation of oil.

The methane, so released, fills the unconsolidated sediments with which it is in communication. Much is lost to the surface. Permafrost can act as a seal for such gas at shallow depth, and some of the Siberian gas fields contain biogenic gas. Permafrost may itself have released large quantities of methane to the atmosphere during periods of global warming. In other cases, as in some of the gas fields in the Po Valley of Italy, shallow reservoirs can be continuously charged from the source, compensating for the loss of gas through weak seals.

Many basins, such as the North Sea, the Gulf of Mexico or the Niger Delta, contain disseminated deposits of biogenic gas at shallow depth, but individual accumulations are generally small, and would be rapidly depleted if exploited. Such deposits form a hazard in drilling: the release of large quantities of gas can so reduce the density

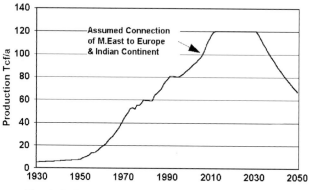

GAS
Ultimate 10 000 Tcf

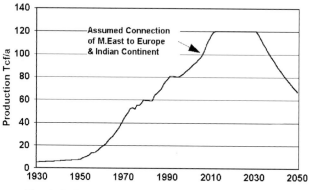

Fig. 9-3. Conventional gas production profile.

Fig. 9-4. The 12 000 cu .m LPG carrier "Vestri" that operated in the 1970s for the Smedvig Company of Norway.

of sea-water that the drilling rig above may lack buoyancy and sink, a situation made even worse if the gas ignites.

It is perhaps arguable that biogenic gas should be treated as *conventional* because it is locally in production, but in general I think it should be treated as *non-conventional:* the potential resource is huge and almost impossible to

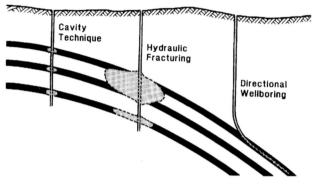

Fig. 9-5. Producing coalbed methane.

quantify. Most of it will not be of interest for a very long time, if ever, and so those few accumulations that are exploited had better be treated as anomalies.

Dissolved Gas

Gas is soluble in water under geopressures, and deep aquifers may contain large amounts of dissolved methane, depending on the salinity of the water, and the pressure. A pilot production project was undertaken in the United States but proved uneconomic because of the difficulties in disposing of the brines.

Coal-bed Methane

Coal deposits contain large amounts of methane. When combined with air, it forms an explosive mixture, which has been the cause of many coal mine accidents. The methane is physically absorbed on the internal surfaces of the coal, with only about ten percent occupying the pore-space. Initially, coal-bed methane was extracted for safety reasons, but it is now viable in its own right.

The production technique is fairly labourious. The pressure in the coal-bed is reduced by pumping out the formation water in the joints, which causes the methane to desorb from the coal matrix into the pore-space. Three alternative stimulation techniques are then applied: hollowing out a cavity; hydraulic fracturing or directional drilling.

Production is most advanced in the United States where it is likely to increase to 1.3 Tcf per year by 2000, contributing about six percent of the country's anticipated gas demand. Reserves stand at about 60 Tcf[6]. The removal of tax incentives in the USA in 1992 has prompted several American companies to try to develop coal-bed methane overseas, especially in Europe, but so far progress has been delayed by a number of legal and environmental issues. The long term prospects seem, however, excellent.

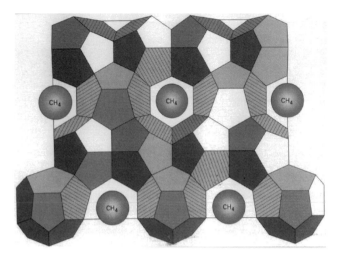

Fig. 9-6. Methane bound among water molecules in hydrates.

Gas in Tight Reservoirs

The smaller molecular size of gas means that it can accumulate in reservoirs lacking sufficient porosity and permeability to hold oil. There are, for example, large amounts of gas at depth in such circumstances in the Alberta Basin of Canada. If, however, the combined pressure and temperature conditions are too high, most such gas will be in solution, and the amounts are then constrained by the volume of the aquifer. Over-pressured shales also contain substantial amounts of gas, but it is not exploitable.

Hydrates

Methane occurs in hydrates, which are ice-like solids found in Arctic regions and in deep water. Hopes of exploiting such deposits appear to be doomed because, being a solid, the gas is unable to migrate and accumulate in commercial volumes. Reports[7] that the Messoyakha Field in Siberia produced gas from hydrates are apparently erroneous.

NATURAL GAS LIQUID (NGL)

Liquid hydrocarbons dissolved in gas may condense at surface conditions of temperature and pressure. The process of condensation gives rise to cooling, and it is curious to see an oil well in the tropics coated in ice. The product is a light clear liquid, similar in appearance to a refined product. Indeed, it can be used as an inferior motor fuel. It is known as *Condensate*.

In some instances when a gas reservoir is uplifted to shallower depths by earth movements, condensate may separate in the reservoir itself. In this case, it is known as *retrograde condensate*.

Gas-condensate fields occur in the deeper parts of a basin, where oil has been cracked to gas, leaving a large liquid fraction dissolved in the gas.

Liquid hydrocarbons, such as butane and pentane, may also be extracted from gas by processing.

Condensate is often combined with oil production, somewhat confusing the statistics. It is likely to become an increasingly important commodity in the future as more gasfields are brought into production. It is assumed that each trillion cubic feet of gas generally yields about forty million barrels of NGL.

NON-CONVENTIONAL OIL

I use the term *non-conventional* as a basket for oil that does not qualify as being *conventional* oil for any one of a variety of reasons. It is perhaps a better term than *unconventional* which is also used. It is a broader usage than, for example, that employed by the USGS, which restricts it to deposits having no clear water contact, being termed *continuous-type* resources. As discussed earlier, the boundary between *conventional* and *non-conventional* oil is fuzzy, although the general distinction is valid enough, and indeed critical.

Included in the non-conventional category are:

– oil from oil shale
– oil from tar sands
– heavy oil
– oil from enhanced recovery
– oil from infill drilling
– oil in very hostile environments
– oil in very small accumulations.

Together they comprise a substantial resource, and have two essential characteristics: they generally become viable only in a high price environment; and they have a different depletion pattern, rising only slowly to a long, low plateau before eventually declining. The production of some categories are closely linked to the production of the *conventional* oil with which they are associated. They will not make much of a contribution to world production until long after *Convention Oil* production has peaked.

Whereas *conventional* oil is primarily resource constrained, *non-conventional* oil depends more on economic factors, and its production is more analogous to mining. The distinction between *reserves* and *resources* has more meaning when applied to *non-conventional* oil than is the case with *conventional* oil. The sheer scale of the operations needed to extract significant amounts of *non-conventional* oil is a major constraint.

Shale Oil

Oil can be distilled from certain shales rich in organic material, in the same way as oil can be distilled from coal, as was done in Germany during the War. The organic material consists of kerogen that has not been converted to oil. Accordingly, the liquid so produced is strictly speaking not a natural oil at all. Shale oil should in fact better be classed as part of the coal domain rather as a hydrocarbon source, but it is mentioned here because it is often considered in connection with oil.

The most advanced exploitation of shale oil is in the Piceance Basin of Colorado. It is reported that deposits yielding more than 10 gallons per ton are potentially commercially exploitable[8]. The process carries a high environmental cost in terms of the disposal of waste, some toxic, and the large amount of water consumed. The waste material has a very fine particle size and occupies more space than it did before it was processed. It is unstable when heaped up in tips.

Although considerable investments were made in US shale oil projects in the aftermath of the Oil Shocks of the 1970s, most projects have now withered away. There are similar deposits in many other countries, including Australia, Brazil, FSU, Zaire, and China. The resources are very large, but actual world production is unlikely to exceed 500 000 b/d for a long time to come.

Australia, which has limited oil resources, has recently started exploiting shale oil near Gladstone on the coast of Queensland[9]. The project has been encouraged by government tax relief. Production is expected to start two years from commencement, rising to 14 800 b/d in the eighth year. If all goes well, further developments may be undertaken, eventually yielding 250 000 b/d. Production costs are anticipated to be about $11.50/b, possibly falling to US$6.50 when the facility is in full operation. While it is evidently a very promising and valuable project, doubling Australia's present Bass Strait production, the long time-frame of such operations is well demonstrated. The slow rise in production to a long low plateau is very characteristic of *non-conventional* oil.

Oil from Bitumen, Tar Sands and Heavy Oil Deposits

When oil migrates to shallow depths on the margins of basins, it is attacked by bacteria, which remove the light ends, leaving behind sticky viscous materials known variously as bitumen, asphalt and tar. These substances grade into heavy oils, and there have been difficulties in knowing how to classify them precisely. Their characteristics vary, depending on the composition of the oil from which they were derived and the subsequent alteration processes. The boundary between bitumen and Heavy Oil is drawn at 10°API gravity and a viscosity of 10 Pa·s. Asphalt is a type of bitumen with a gravity around zero degrees API[10].

The two largest deposits are the tar-sand deposits of Athabasca in Canada and the heavy oil deposits of the Orinoco area of eastern Venezuela, which are each estimated to have over one trillion barrels in place.

Of the two, the Canadian deposit is at the more advanced stage of exploitation. The deposit is mined in mammoth open pits, as well as being exploited by in-situ steam stimulation. A new development has been the use of horizontal wells for steam injection as well as production. The mining operation involves stripping off the overburden; separating the bitumen with steam, hot water and caustic soda, and then diluting it with naphtha. After centrifuging, liquid bitumen at 80°C is produced, which is then upgraded in a coking process and subjected to other treatments, eventually yielding a light gravity, low sulphur, synthetic oil. The process is economically viable, and it is estimated that costs can optimally be reduced from \$12-13/b to about \$9/b by 1998, depending on the cost of capital etc.[11] About 400 000 b/d are being produced to-day, and there is clearly scope for expansion[12]. A huge work force is engaged in the operation.

In Venezuela, the heavy oil, which has an average gravity of 9.5° API, is extracted with steam stimulation and chemical dilutants from reservoirs at depths of 150 to 1200 m. It is estimated that about 270 Gb could be recoverable.[13] Typically, patterns of five wells are drilled on a regular grid. Steam is injected through the peripheral wells. It drives the oil to the central well, which can produce initially at up to 600 b/d, for a period of a few months until the catchment is drained. There are new proposals to extract it directly with the help of horizontal wells and submersible pumps, expanding the catchment area to increase flow rates to 1400 b/d, even without steam injection. In the 1980s, Venezuela commenced marketing a product made from bitumen, known as Orimulsion, which is used as a commercial boiler fuel for electricity generation. It consists of an emulsion of 70% bitumen, processed to a particle size of 20 microns, mixed with water and 2000 ppm surfactant. It has a relatively high sulphur content, which can, however, be largely removed by conventional scrubbers in power stations. Production is expected to rise to about 400 000 b/d by 2000 and 600 000 b/d by 2005. Almost half the investment is in precessing facilities. Exports to Europe have dwindled, however, partly because of the emissions, but new markets may be found.

It is obvious that the resources of tar-sand oil are enormous. The largest are the Canadian and Venezuelan deposits, but there are many others around the world, including two large ones, known as Aldan and Siliger in the Former Soviet Union[14]. Although they may be economic to produce, at least to a certain scale, they use a large amount of oil in steam generation and are very environmentally unfriendly both for producer and consumer. Undoubtedly, production will rise in the future, but probably to a low ceiling, constrained by the sheer scale of the operation, and only when *conventional* oil is much scarcer than now.

Other Heavy Oil Deposits

The classification of what constitutes heavy oil is somewhat arbitrary, the boundary being variously drawn at 10, 15 or 20° API – here we prefer the upper number. There is a very large number of heavy oil fields, which are found in virtually all producing basins, generally at shallow depth. Heavy oil generally has a high viscosity, and production rates are low, commonly requiring the pump. The production profile consequently rises slowly to a long low plateau before declining gradually. Many heavy oil deposits have been neglected, or produced slowly, in the past, because their economics compared unfavourably. A larger proportion of what will be produced in the future will be heavier oil: one estimate suggests as much as 37% of the *undiscovered* will be heavy[15]. Deepwater finds tend to hold heavy oil because of low geothermal gradients from the thick water cover[16].

Enhanced Recovery

The proceeds of secondary recovery methods, such as waterflood and gas injection, are included in *conventional* oil. Additional tertiary methods can be employed, including the injection of steam, polymers, carbon dioxide, nitrogen and miscible gas to sweep the oil through the reservoir. Steam injection is by far the most commonly used procedure. Heavy oil deposits are the primary candidates for enhanced recovery, largely because the initial recovery is low. Furthermore, progress is being made in improving the efficiency of such techniques, increasing recovery and production rate.

Fig. 9-7. Shooting a well with nitroglycerine was a dangerous activity. Here the explosive is being poured into the "torpedo" which was lowered into the reservoir to shatter it and thus improve production.
(E.S. Blakey, © 1988, reprinted by permission of the American Association of Petroleum Geologists, from the Cities Service Co.)

A new approach involves actual mining. Several shafts are being drilled through the shallow Corsicana field in Texas. They are connected by underground galleries which collect oil drained from the overlying reservoir[17]. The first oil in Texas was found near Corsicana in 1893, and this mining development a hundred years later is a commentary on the full cycle of depletion. The gushers are over, and the Texans are reduced to almost digging out the last few drops with shovels. It says a great deal.

Infill Drilling

A field with homogenous reservoirs can be efficiently drained by fairly widely spaced wells: the norm in the United States being one well per 40 acres, described as a 40-acre spacing. Fields with inhomogeneous reservoirs can however require a closer well spacing. Successful infill drilling campaigns on 20-, 10- and eventually 5- acre spacings can be conducted on appropriate fields, after the initial 40-acre development has been completed, especially in cases where that was undertaken long ago without the benefit of modern technology. Such activities are common, for example, in parts of the Permian Basin of Texas, where much oil was by-passed in the initial development due to

The Coming Oil Crisis

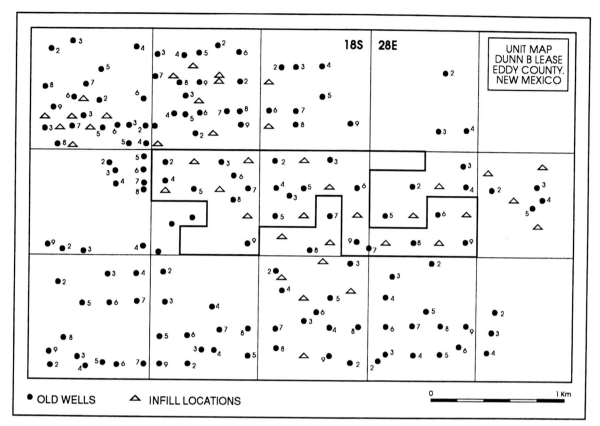

Fig 9-8. Infill drilling in the Permian Basin of Texas.

● OLD WELLS   △ INFILL LOCATIONS

18S 28E

UNIT MAP
DUNN B LEASE
EDDY COUNTY,
NEW MEXICO

0   1 Km

the particular nature of the reservoirs and the primitive stimulation, which involved "shooting" the wells with nitroglycerine. There is nothing magic about 40-acres, although it was a normal spacing in the United States.

The recent introduction of so-called 4D seismic facilitates early infill drilling. Seismic monitors are placed permanently over a field and can track the movement of oil, so that infill wells can be optimally placed. As this approach becomes more routine, infill drilling becomes a standard procedure as part of the initial development, and the proceeds can be treated together with conventional production.

Oil in Hostile Environments

The frontiers have been pushed back over time. The North Slope of Alaska was considered out of reach in earlier years, as were the deep and stormy waters, off northern Norway, but both are now fully accessible.

Even so, it is probably as well to recognize that some areas are still effectively out of reach in the price environment of the next twenty years. The Sea of Okhotsk, the Siberian offshore, the Greenland icecap or Antarctica are candidates. Exploration can be conducted in very deepwater, but there are some production constraints.

There is no reason to think that such areas are particularly oil-bearing, but a notional amount of *non-conventional* oil is attributable.

Small and Very Small Accumulations

In theory, there is a huge number of ever smaller accumulations of oil still to be found, down eventually to a thimble-full in the chamber of an ammonite somewhere. To-date, most of the smaller fields were found in the course of looking for larger ones. Once found, such small fields were in many cases developed, although they would not have formed viable exploration targets in their own right. Advances in technology have improved seismic resolution to the point at which it becomes increasingly difficult to misjudge the size of a prospect. Once the stock of larger prospects has been consumed, there may be insufficient

ALL HYDROCARBONS

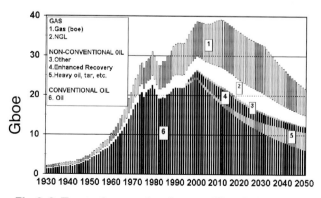

GAS
1. Gas (boe)
2. NGL

NON-CONVENTIONAL OIL
3. Other
4. Enhanced Recovery
5. Heavy oil, tar, etc.

CONVENTIONAL OIL
6. Oil

Fig 9-9. Tentative production profile of all hydro-carbons including gas and non-conventional oil.

incentive to explore deliberately for small and very small prospects, at any rate until higher prices give a new incentive. The threshold of what is viable shifts over time and with changing oil prices. It is reasonable therefore to attribute some notional non-conventional oil to this category. Furthermore, there is the important practical constraint of timing. Exploration is likely to end in a country when production ends, if not before, leaving behind some untapped resources.

It is not easy to quantify the amounts of *non-conventional* oil that may occur in the several categories described above. By far the greatest potential is for heavy oil and synthetic crude made from tar sands. Enhanced recovery applied to heavy oil may also make a useful contribution. Macgregor, 1996, provides data on oil-in-place, with: Canada 1.7 Tb; Russia 1.6 Tb; Venezuela 1.5 Tb, compared with Saudi Arabia at 500 Gb at the head of the list. Much of the endowment is *non-conventional:* an important but not easily accessible resource for the future.

Probably about two million barrels a day of what should properly be called *non-conventional* oil are already in production, but cannot be distinguished in the database. Figure 9-9. is a tentative production profile of all hydrocarbons including gas, NGL and *non-conventional* oil. Production of the latter is expected to rise to about 6 Gb/a by 2050, by which time some 270 billion barrels will have been produced. Of this, heavy oil and tar sand production provides about 5 Gb/a, which can probably be sustained at that level through the 21st Century.

NOTES

(For references, see Bibliography)

1. Statoil, 1996, makes this very misleading statement in a report.

2. Adelman, and Odell, in several papers listed in the Bibliography, make the underlying assumption that the resources are near infinite and that production is driven by economics and politics. Odell in the 1970s accused the companies of deliberately understating the North Sea potential as a conspiracy to raise prices. He was correct about the understatement, although probably not for the motive. But that was at a very early stage of exploration and on the basis of much less data and knowledge than are now available. Things have changed, but he sticks to the same idea.

3. See Laherrère, J.H., A. Perrodon and C.J. Campbell, 1996, for a comprehensive analysis of gas based on the authoritative Petroconsultants database.

4. See Moody-Stuart, M., 1996.

5. Gas is equivalent to oil as follows
   calorific 1 barrel of oil = 6 000 cu ft
   value 1 barrel of oil = 10 000 cu ft
   (at $1.8/ Mcf & $18/b)
   10 tcf gas = 1 Gboe (rounded)

6. A good account of coalbed methane is given by Preusse, A., 1996.

7. An account of gas is given in McCabe *et al.,* 1993 in USGS Circular 1115.

8. Ion, D., 1980, published a useful account of the availability of world energy in 1975, in the aftermath of the First Oil shock, which makes interesting reading as to the attitudes of the time.

9. See Petroleum Review, 1966, for a discussion of the Australian shale oil development.

10. Cornelius, C., 1987, discusses the classification of heavy oil, bitumen and asphalt.

11. Hobbs, G.W., 1995 gives a useful review.

12. See Will, 1996, for a discussion of the situation which is attracting more interest and tax inducements. Even so production is unlikely to exceed 1 Mb/d for at least a decade.

13. See Abraham, K., 1997, for a useful description.

14. Ulmishek, G.F, 1993, speaks of huge remote Siberian deposits; and D.S. Macgregor, 1996, lists them.

15. Masters, 1994

16. See Imbert, P., 1996.

17. See Hart's *Petroleum Engineer International,* September, 1996 p. 15. for a description of this novel approach and also See M., 1996).

Chapter 10

ECONOMISTS NEVER GET IT RIGHT

I am not an economist. So perhaps I am being unfair to write a chapter with such a title. A man in a greenhouse should not throw stones. All the same, I am not alone in expressing scepticism of the economist's skill in predicting events, as the economic forecasts of most countries confirm. They do not seem any better in dealing with the oil business. I have witnessed many false economic evaluations of prospects, and have seen companies fail in their endeavours by allowing themselves to fall too much under the influence of economists. Certainly, the great international oil companies were built without the intervention of economists. I read economic reports that seem incapable of grasping resource constraints, and I am therefore less than convinced that they are working from the right premises in their work.

Where I live in southwest France are to be found many castles built during the hundred years war between the English and the French in the 11th Century. Looked at even by today's engineering standards, they are amazing structures, but when one pictures the conditions under which they were built, they are truly remarkable. Each stone had to be quarried, and transported to the site by horse. It had to be labouriously faced with a hammer and chisel, which was at least a day's work. Lime had to be slaked and cement made. It took forty years to build a castle, and even if the workers were not paid much, they certainly had to be fed. An unfinished castle was no use to anyone, so it represented an investment with a pay-out of something like fifty years. It was a project that would never fly under modern discount economics, yet the castles were undeniably valuable investments, in many cases still standing centuries later.

In the past, people certainly operated under longer term economic theory than is practised today, whether building castles or planting orchards for their grandchildren. In fact, the idea of discount economics is quite modern, being credited to Donaldson Brown[1], the finance director of General Motors in the 1920s. He wanted to measure the relative performance of different factories, and developed a system of discounting future value, so that an asset ten years hence was deemed to have no current value. He was living in a rapidly expanding economy, when time was money. The World's experience over the past Century has been one

Fig. 10-1. A medieval castle could not be built under modern discount economics.

Fig. 10-2. Donaldson Brown, who invented Internal Rate of Return.

of unprecedented growth, made indeed possible by the abundance of cheap oil-based energy. So an economic theory that gives weight to the time-cost of money was perhaps the best one for the epoch.

If I understand it right, the famous Keynes proposed a theory whereby governments could pump up sagging economies by printing money, which fed investment and in turn created growth. It seemed to work for a while, and was embraced by the socialists, who found that it was a quick means to redistribute wealth indirectly by the consequent inflation. It now seems to have fallen into disrepute, having been found to be a quick fix rather than a long-term solution.

Brian Fleay in his brilliant book[2] describes this short-termism as Neoclassical Economics, being the science of human relationships in the market place, which ignores the role of energy and natural resources.

At the heart of economic thinking are the well known ideas that supply and demand determine everything. If wheat prices rise, farmers plant more in the next sowing, and the system readjusts. Inventories are treated as burdens to the system, to be held as low as possible. In essence, economic theory is built around human agency, which does indeed cover most aspects of life. The cost of coal is deemed to be nothing more than the cost of the miners and the capital investment: the resource itself being there for free. If the reserves were infinitely large, perhaps there is no need to consider them other than as a gift from Nature, but there are some danger signals in this proposition.

Hotelling[3] in 1930 wrote a classic paper on the economics of depleting a resource. He concluded that there was no real charge attributable to the resource itself, suggesting that it should be essentially priced at the discount rate. He recognized that it was a finite resource, but thought that it would be subject to a natural substitution if it began to be in short supply. Thus, firewood was naturally superseded by coal, and coal by oil, in a well ordered progression under understood economic principles. It was not really a finite resource, but simply finite at a certain price.

This thinking still infests the economic community. They write[4], as I have already quoted:

> "Total hydrocarbon resources with the potential to produce fuels are so large that they can be considered infinite for the purpose of this analysis".

In one sense they are right, because if you include infinitesimally small accumulations, down to a molecule lurking somewhere out of reach, I suppose that the resource is effectively infinite. But they don't really mean that. To their way of thinking, if you want more oil, all you have to do is drill more wells; or increase the price; or invent some technology; or improve the terms or tax. Thus, Odell[5], one of the high priests of this heresy, claims that the North Sea will not have been fully explored until every formation down to granitic basement has been exhaustively tested.

With this false notion of an infinite resource, the economists spare themselves having to think about the concept of *Ultimate* recovery. That denial itself excludes the very idea of depletion: you cannot deplete an infinite amount. It explains how they can extrapolate the growing production of the past into the future. They show graphs with production rising to a certain future date at the end of their forecast period, oblivious of the implication that it must fall like a stone on the day after if it is to respect the resource constraints. Further, they can glibly refer to the Reserve to Production ratio, saying that the reserves support present production rates for a given number of years, without recognizing that production must fall as the reserves are depleted. It really is very surprising that they can sustain such specious arguments, when the simple observation of an individual well or field provides ample evidence of decline and depletion. In the United States, the tax man even grants a depletion allowance in recognition that it is a dwindling asset. Yet reputable economists and reputable organizations still persist in this idea of a semi-infinite resource[6]. Worse than that, they influence governments into pursuing very short term energy policies[7].

While on the subject, it also seems to me that economic measurements are far too restrictive because they ignore the substantial and critical economic input of non-traded activities, such as housework, raising children, home carpentry or reaching self-sufficiency in the vegetable garden. The readings from the false measurements may influence policy: for example, encouraging women to go out to work in the active economy when they were happier and more useful in the passive economy. Whether or not we can do it in practice, all the economic books have to balance. Many people live in the most gruesome architectural surroundings because the system does not attribute value to intangible things like happiness or beauty. All this may have to change in a post-consumeristic world brought about by the coming oil crisis.

RISK

Much of the practical work of economists in the upstream sector of the oil industry is concerned with the management of risk. It is thought that there are economic trends and tools that can improve the judgment of oil men in making their decisions, but it is well to realize that the flawed underlying perceptions of the economist in relation to resource constraints permeate also their short-term calculations.

The industry likes to depict itself as having to face exceptionally high risks, listing:

Exploration risk – they may be looking in the wrong place;
Geological risk – the geological interpretation may be wrong;
Contract risk – the lawyers did a bad job;
Government risk – the terms may be changed;
Political risk – war, sequestration;
Development risk – the engineers get it wrong;
Natural risk – the hundred year wave;
Terrorist risk – somebody blows it up;
Tax risk – the rules change, even retroactively;
Environmental risk – they spill some oil and have to clean it up;
Corporate risk – their stock suffers, or they are subject to a takeover bid;
Commercial risk – prices fall or costs rise
Labour risk – the workers strike.

Personally, I think that all these risks are greatly exaggerated, both absolutely and especially relatively. Most business works with lower margins and higher risks: an arbitrary change in government policy cutting subsidies can bankrupt the farmer after years of work; the arrival of a supermarket puts long established and successful small traders in the town out of business; the lifting of trade barriers may destroy efficient local enterprise.

What distinguishes the oil industry are the huge sums involved and the very high marginal tax rates. The latter amply cushions most of the risk to which they are exposed.

CASH FLOW

The primary economic challenge in exploration is to model actual or anticipated cash flow. Figure 10-3 shows a typical study either of an actual development project or of a hypothetical one, undertaken to see if an exploration drilling would be viable if successful.

The parameters are quite simple:

- Profit is revenue less investment, operating expense and tax.
- Revenue is production times oil price.
- Investment is the cost of the facilities, including the drilling; and operating costs are the running costs of labour, insurance, tariffs on pipelines, and contracted services.
- Tax is tax, including royalty.

The calculation may be made in terms of *money of the day* in which estimated actual future costs are taken, or *real money,* in which the effects of inflation are discounted.

The next step is to calculate what is called *discounted cash flow* to determine the present value of future earnings. Thus $1 000 000 in five years time at a 10% discount rate is worth today

$$\$1M / (1 + 0.1)^5 = \$620\ 921$$

The sum of each future year's discounted cash flow over the life of the field gives the *Present Value* (PV), from which may be calculated the *Rate of Return*. There is also the concept of *Payout:* how long to wait til the investment is recouped, and the project moves into profit.

This describes the simplest outline of the procedure. There is great scope to make it ever more complex, by addressing multiple scenarios and risking each element using statistical probability theory, and so forth[8].

Companies normally have what is called a hurdle rate of return, namely the minimum return that they can accept under their investment policy.

This is on the face of it all good stuff. The geologist provides his estimate of reserves; the engineers feed in more about the numbers of wells and producing rates; the construction people estimate what it will cost to build the thing; and a committee of economists is dragged out to pronounce on future oil prices. The calculator whirrs, and out comes the answer: the project flies or it does not.

If it does not, well the geologists can invent some more reserves, or the construction man can have second thoughts about the costs. So, if those involved want it to fly, they can usually massage it into shape. They are often under pressure to make it work, whatever their personal judgment, because they may be bidding in a competitive situation where there is much more at stake than the specific project. Failure to participate may create a bad impression with the host government, which would have wider significance. The stock-market too encourages companies to explore, naturally being ignorant of the real geological risks.

If it is made to fly, the proposal is now blessed with a notional number showing it to be sufficiently profitable, and it passes up the management hierarchy, each level having less and less knowledge of the actual situation.

I remember a golden case when, having discussed a project in Sumatra for some time, the chairman of the committee, looking at his agenda, said "that is all very interesting, but when are we coming to the Indonesian proposal". He was not well travelled. In reality, all that really sinks in at that stage is what the magic rate of return number is, and what is left in the budget.

The management desires a notional level playing field and excludes local tax situations so that they can pretend to fairly compare the rate of return from investing in a refinery extension in Texas versus an exploration well in Norway.

Year	Prod. Mb	Oil Price	Gross Revenue	OPEX	Op. Income	CAPEX	TAX	Net Cash flow	Cum. Cash flow
0	0	20	0	0	0	65	0	−65.0	−65.0
1	4.6	21	96.6	40	56.6	0	0	56.6	−8.4
2	4.2	19.5	81.9	40	41.9	0	10.4	31.5	23.1
3	3.9	18	70.2	40	30.2	0	9.4	20.8	43.9
4	3.5	16	56.0	40	16.0	3	4.0	9.0	52.9
5	3.1	16	49.6	35	14.6	0	4.5	10.1	63.0
6	2.6	15	39.0	33	6.0	0	1.9	4.1	67.1
7	1.9	16	30.4	31	−0.6	0	−0.2	−0.4	66.7
8	0	17	0	0	0	5	−1.6	−3.4	63.3
TOTAL	23.5		423.7	259	164.7	73	28.4	63.3	

Fig. 10-3. An economic evaluation of a field.

They thus fail to notice that 85% of the risk of the well in Norway is borne by the Norwegian taxpayer, who is willing to accept that the cost of putting it in the wrong place is deductible from taxable income. Companies without the taxable income in Norway soon withdrew from the game.

To be fair, the economic analysis does force those involved to think about all aspects of the project, which they might not have done otherwise.

Even so, it is a dangerous tool. It tends to influence companies to search for oil where the best deal lies rather than where there is the best chance of finding oil. It has sustained so-called frontier exploration for many years longer than justified by the geology. New areas, on which knowledge is limited, commonly have good concession terms, and pass the system much more easily than sound prospects where the conditions are onerous.

The one factor that really affects the economics, however they are conducted, is oil price. On that, the economists have little to contribute, because it has been largely politically contrived[9] and because they do not accept the finite nature of the resource. They are not therefore in a good position to assess its distribution, and accordingly cannot take into account the growing control of the resource by a few critical producers, which must surely influence the price more than ordinary economic factors.

Companies tend to have committees to assess future oil prices, mainly comprised of economists. They read the Wall Street Journal, and consult Salomon Brothers, thinking in terms of supply and demand trends. Consequently, they normally come up with one or more bland scenarios, whereby oil price is above or below inflation by so many points. There is talk of the gentle ramp. Their record in forecasting has been abysmal.

But if all this seems rather negative and dismissive of the economist, in fairness I do admit that it is difficult to see how else centrally controlled global companies could run their affairs. They clearly have near limitless opportunities to explore for oil, which is not, however, by any means the same thing as finding it. They can invest in many different things upstream or downstream, buying reserves or other companies. So, they do need some yardstick by which to choose, and perhaps the economic analysis, in a very general way, does provide a comparison, even if far from reality itself. Their bland oil price assumptions are also understandable as it is difficult to plan for a crisis, even if crises are a normal fact of life. The system more or less

helps the management avoid serious mistakes, even at the expense of not getting much right either. Above all, it shields them from responsibility by allowing them to justify their actions. That is often their motivation as they approach pensionable age.

It appears that this system is coming to self-adjust as the companies downsize themselves. The new business unit or the asset team as it may be called, may be sufficiently close to the actual business to have fewer real options. They no longer need to be much guided by hypothetical economic analysis, as local circumstances, many either beyond their control or offering an obvious preferred choice, will dictate most of their actions.

I have the suspicion that I have been too harsh and sceptical, and have possibly overstated the argument. Perhaps the economists understand perfectly well, but are not free to express their views[10].

You would think that the depletion of oil-based energy, which has driven economic experience for most of this century, would also be fundamental to economic theory. But that does not seem to be the case.

It seems to me that the premise behind discounting is being eroded, and that value, if not enhanced value, does somehow attach to future assets, especially critical reserves of oil. Oil supplies forty percent of all traded energy and will continue to do so, so long as it is available and competitive. That, one would have thought, should transcend the short-term profit and loss account, at least until some viable substitute for oil is in place.

I think that the economists need to go back to the drawing board and figure out some new principles. In fact, some have. For example, Odum[11] and others propose that all value should be perceived to come from Nature in a thermodynamically open system embedded in the environment. For them, the human economy is like a living organism. Its metabolic processes require a continual one-way flow of complex organized energy and matter from nature to replenish and maintain the economic structures inherited from the past; to build new ones; and to provide the goods and services that humans need. Degraded energy and matter are returned to the environment. Above all, the new principles should grasp the finite nature of oil, which has become the main provider of energy in the modern world. These are hopeful directions[12], but meanwhile, we should accept the pronouncements of neo-classical economists on oil supply and price with grave misgivings.

B.W. PACE
An explorer who teaches economics

Fig. 10-4. B.W. Pace.

Q Bill, you started your career in 1953 as a geophysicist with a seismic contractor working in international exploration. Later, you joined an oil company during the early days of the North Sea exploration and began to face the dilemma of how to allocate budgetary resources between exploration costs and technical excellence.

A Yes, I was lucky to have the opportunity to work in the Far East, South America and the USA as well as in several European countries. I saw the development of modern digital techniques from the earlier analogue methods. It was a fascinating progression from the days when a geophysicist functioned as a jack-of-all-trades responsible for the acquisition and interpretation of the data to the current state where he is frequently a specialist within one area of the science. However, the requirement of technical integrity has remained unchanged throughout this time. In my experience, most companies view the situation pragmatically and are prepared to reduce their interest if budgets are too tight or the risks are too high. Nowadays, when much of the exploration is carried out offshore where data acquisition is relatively cheap compared to drilling costs, this is less of a problem. Many companies recognise the economic merit of 3D exploration surveys, particularly in areas of proven hydrocarbon potential. Not only does it optimise the mapping of reservoir geometry and the first well location but it facilitates an immediate appraisal plan and a rapid field development in the event of success.

Q Later, you became General Manager of Unocal in Norway and the United Kingdom, but it seems to me that at heart your interest was always in the technical field. It raises the interesting issue of whether management can be made a technical business based on scientific and mathematical principles. Personally, I never thought so. To me, it was more a case of being entrepreneurial to smoke out opportunities by the seat of the pants, and to become skilled in whatever form of "corruption", if that is the word, applied in the country concerned. It was the job of trying to extract preference for the company from the government. Even internally, it was more corporate politics than reality.

A Management is not susceptible to universal mathematical solutions since many of the problems cannot be expressed in such terms. Nevertheless analytical techniques should be applied in those areas where they are applicable. The role of politics becomes increasingly more apparent as one moves from the purely technical area into management, and one of the responsibilities of a manager is to ensure that the Company is viewed favourably by host-country agencies responsible for the conduct of petroleum development. This may be achieved by various means, no doubt, but in my opinion the demonstration of high technical ability and honesty are the most important. However, the relationship with the government, that is the politicians, is often more complex, and corporate influence can become an important factor. I believe that it was Nikita Kruschev who said that a politician is someone who will promise a bridge even where there is no river to cross. I suppose that the perfect lobbyist must adopt similar attitudes. As far as internal politics are concerned, they are practised in all companies, and the bigger the company, the more extensively they exist. It is a very destructive factor, dissipating the energy of the Company by internal power-struggles, petty jealousies, interdepartmental skirmishes and so forth; but it seems to be inevitable. It is one of the reasons why a small group of experienced, motivated professionals working in an Independent company can out-think and out-run the large companies, given that they are not initially shut-out by the strong political influence of the latter.

Q Presumably, you used economic analysis and probability theory to guide your actions. Did you believe in the results? Or was it rather a satisfying mental exercise which almost incidentally led you to better decisions?

A *I know that you have a low opinion of economists in general so I suspect that your question is loaded! But business opportunities must be valued, and the relatively simple techniques used by most companies provide a way of achieving this: the effectiveness of the process is generally a function of how carefully the opportunity has been researched. For example, in licence round reviews, we always assessed minimum reserve cases for the different areas for use as filters, and prepared estimates of the risked economics for the selected blocks. Given that the engineering solutions and costs were conducted with care and that the exploration work was of high integrity, the results had value in a relative sense. The discipline imposed by the analyses was undoubtedly valuable to the exploration team. It was always more effective to involve the petroleum engineers within the exploration team so that a sense of mutual trust and understanding could be developed. When the disciplines operate in relative isolation as was often the case in the old days, the reverse can often occur ... they don't trust us and we don't understand them! The starting point, however, is always technical excellence in the exploration phase: all the most sophisticated economic analyses are to no avail if the prospect is half-baked ... unless you stumble on a juicy strat-trap in the Paleocene when drilling for an ill-conceived Jurassic play, an outcome which is not without precedent, as I am sure you will agree!*

Q *I notice that you use the word "relative" in your comments. I often think that economists are historians rather than prophets in that they are skilled in documenting the past but inept in their predictions of the future trends. Does your use of the word indicate your recognition of this inherent trait?*

A *Economic analyses of petroleum ventures can never be an absolute science due to all the uncertainties of world politics as they affect stability and the accuracy of the future oil price predications, and we all know how inadequate this has been over the last twenty years. But when the larger part of the World's total reserves exist in an area so riven by complex macro and micro politics, it is like trying to predict which horse will win the Derby in five years' time! So all economic analyses are relative to the current perception of the future political and price stability, in addition to the company's cost of capital borrowing at the time that the analyses are made. Another important factor is that frequently it is obvious that the short-term, net-profit driven objectives of the institutional investors in Wall Street conflict with the need for long-term planning which is the basis for sustained corporate growth. There is a fundamental conflict between the financiers and the economists with the management squeezed in the middle; the outcome tends to lie in favour of the former, particularly in the case of US companies. One could have a*

long discussion on that issue : suffice it to say that every system contains within its structure the seeds of its own self-destruction.

Q How is the political and economic risk normally assessed?

A *When a company forms a new venture in a foreign country, it will have gauged the local political risk as best it can, based on perceptions of stability within that country and its relations with the adjacent states. The financial risk is usually handled by using high, median and low predictive functions both for the oil price and the exchange rate, and if the project is marginally viable for the less favourable outcome, then it is generally acceptable. Ascribing a value for geologic risk is always problematical, perhaps not so much for the estimates of the team working in the area but in the standardisation of risk assessment across a number of groups or divisions within a large company. Unless a means of controlling this variable in a lateral sense is achieved, the figures for risked economics recorded in Head Office will probably contain relative distortions. Conversely, the risked economics produced by a team assessing the blocks offered in a licence round should be reasonably consistent and provide a credible means of ranking the opportunities. So in answer to your question: economic analysis is an important tool in the decision-making process; it will seldom produce the right answer in an absolute sense but it acts as an economic filter for individual projects and as a ranking tool in portfolio selection. It is also an essential exercise in that it imposes discipline on an evaluation and creates an understanding of the entire process throughout the team.*

Q How useful was the evaluation in the internal politics of the company between the affiliate for which you were responsible and the head office? Did the cold economics overly influence the higher management, causing them to miss out in the local environment? For example, I was surprised when Unocal went through the agony of developing Vesslefrikk in Norway, which was fraught with difficulty and conflict, yet when it had been successfully accomplished, the Company sold out. I wondered if this was an economic decision or an emotional one. Perhaps the company said "If that is what success is like, we sure don't want to take risks to get there".

A *So far as Vesslefrikk was concerned, I was not involved in that specific case, having returned to the UK somewhat earlier, but with the high operating costs associated with the Norwegian theatre and the perceived oil price profile at the time, the decision was probably economically controlled, particularly since Unocal was trying to reduce its debt-asset ratio at the time. With regard to your first question, I am sure that the explorationists in every affiliate office feel that Head Office has occasional blind spots when it comes to ranking their prospects within the total portfolio opportunities available to the company but there are generally financial and political policy requirements which are inevitably superimposed on the assessment. One may disagree with the policy, of course! But with all the political as well as geological uncertainties associated with international exploration, it is seldom easy to make a decision and every successful manager will admit to having missed some significant opportunities during his career, and will also recognise that luck has played some part in his successes!*

Q To what extent do you consider that intuitive judgement is an essential element in the process of making decisions under uncertainty? Is there a relationship between luck and intuition?

A It depends how you define intuition. To me, intuition which has value stems from the ability of an enquiring mind to recall, relate and weigh a wealth of acquired knowledge in an unbiassed manner and to compare the elements of the current problem with that data-base. It does this in a free-cycling, subconscious mode rather than by use of conscious logic. But the essential component of acquired knowledge can only be gained by dedicated research of the subject both in the library and by systematic analysis throughout an individual's career. Intuition of this calibre must always be respected and is often at the root of all good decisions, although to a casual observer it may sometimes appear to be merely good luck. But it is not to be confused with simplistic solutions based on limited experience and sold with the authority of a silver tongue; I am very suspicious of charismatic people who wave their arms and make rash judgements.

Q Now you teach petroleum economics. I suspect that preparing your course material led you to investigate issues that you had not actually used before. They say that the best way to learn is to teach.

A Much of the short seminar I teach to MSc students in Petroleum Geoscience deals with the fundamental tools used in the valuation of a project, combining risk analysis with economic appraisal. Such techniques were commonplace when I was working. But with the advent of cheap, desk-top computers, these techniques have become more sophisticated, and the modern explorationist

has much more power available to tackle the problems. For example, for most of my career, Monte Carlo simulation was something that was run in a Head Office and other than providing the basic data, the explorationist had very little to do with the process. A schematic diagram of a comprehensive Monte Carlo simulation is shown below: it is taken from a book by Ikoko and was based on the earlier work of Paul Newendorp in adapting simulation techniques to petroleum economics. Nowadays, everyone can churn one out on a PC which makes it important that they should appreciate what the programme is doing; how the dependent variables are handled; how misleading the process can be if it is used in a cavalier manner and so forth. So it is inevitable that some of the material I teach relates to techniques which I never used in my career but the underlying principles are generally based on concepts established in earlier days. All the computer modelling methods describing the geological and geochemical history of an entire basin and its potential economic viability merely supply an answer in a few minutes which could have been achieved previously by a costly team of scientists in months of labourious work. But all such techniques are merely tools, and pre-suppose that the input data are the outcome of exploration analysis of the highest quality which recognises the issues at stake: without that, the results are meaningless.

Q How do you view the future of the oil business? Do you agree with Robert Anderson, the former head of Arco when he said "It is a sunset industry, and the sun is low in the sky"?

A Ever since I joined the industry in 1953, people have been questioning how much longer it could all last, yet major discoveries continued to be made as new basins were developed. It would be nice to envisage this going on indefinitely, but there must obviously be an end to it, if only in economic terms. World oil demand is increasing as more countries are becoming industrialised, and has outrun the discovery rate so we have to recognise that we are already on the World depletion curve. The larger fields in most basins have been found, and there are few new provinces to go after, with the exception of the very deep water areas where the development costs will be very high. The Former Soviet Union countries present opportunities, of course, particularly since their technology is twenty or thirty years behind modern methods; in their case, geological risk appears to have been largely replaced by political risk. In addition, I am assured by many explorationists that the reserves in the prolific Middle East countries will be considerably upgraded, so there is some hope there. As the price of oil increases when supplies gradually dwindle, perhaps the economics of shale-oil and tar-sands will become feasible, and there are some considerable reserves already mapped of such resources.

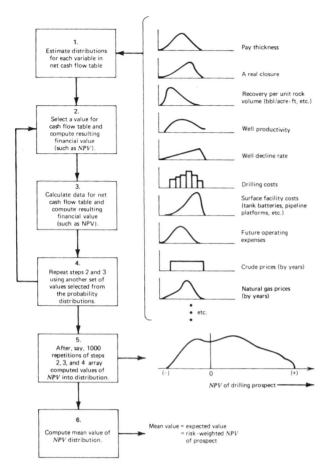

Fortunately, there appears to be plenty of gas to be exploited and probably a lot more to be discovered, although the transportation logistics in the more remote areas invoke economic problems. The LNG option will become more commonly used. Avoiding the temptation to dwell philosophically on the future energy scene, the inevitability of nuclear power if western living standards are to be maintained and so forth, it must be recognised that the traditional industry's sun appears to be past high-noon but perhaps is not yet too low on the western horizon! It might be possible to slow the process down by improving recovery factors but one faces the pressures of the market-place if this is to be achieved on a significant scale. The next half-century should prove to be very interesting with the complex hazards of the politics and religious fundamentalism in the Middle East, the shift of economic power to the East, a probable move towards nuclear energy, the amalgamation or disappearance of companies with household names and so on: there is a rich fabric of business risk for future managements to contemplate! I'm rather pleased to be able to observe it from the sidelines!

Fig. 10-5. Economic analysis procedure.

Ikoko (John Wiley)

Fig. 10-6. Richard Hardman: explorer, executive and President of the Geological Society.
Photo: Amerada Hess

RICHARD HARDMAN
As I See It

Q Our paths first crossed in Colombia in the early 1960s when we were both working for BP. You had arrived from Libya. What had been your earlier career? And what were your impressions as a young petroleum geologist, fresh from Oxford?

A Yes, I had been working in Libya doing wellsite work and subsurface studies, namely understanding the geology as revealed by borehole data. Before that, I had spent some time in Kuwait, working on one of the world's largest oilfields. BP in those days was much influenced by its Iranian experience. Its geological interpretations always had a certain Iranian imprint, with the shallow geology somewhat detached from the deeper zones, as was the case in Iran where the salts of the Lower Fars Formation provided a glide plane. As a young geologist, I naturally knew no different, but with hindsight it may be that BP was then somewhat less technically oriented than were the French or American companies. I am not saying that this was necessarily a disadvantage, because gifted and dedicated amateurs can often perform miracles.

Q When you came to Colombia, you embarked on a project of field mapping in the Eastern Andes, which was to contribute greatly to the understanding of the Llanos, which has recently yielded some giant fields, including Cusiana and Cañon Limon. It was a new experience. What did you make of it?

A As I look back, I realise how privileged I was to have had this experience. Very few modern petroleum geologists have worked in the field. It was a special experience that taught self-reliance, self-discipline, and indeed management. I was on my own, and each day had to decide what to do: which outcrops to investigate; how long to spend on them; and how to deal with the logistics and labour force. It was physically demanding, climbing the mountains and forcing your way through rough country to search out the critical sections. But there was more time: one grew to know the country and get an intuitive feeling for the rocks and fossils: to see patterns and relationships. As the survey progressed, my knowledge of the geology and structure evolved: I began to know what I needed to know and where to look for it. I also learned to match effort against reward: I could not hope to cover everything, so I had to try to prize out the key information by being selective.

Q So Colombia was a formative and critical chapter in your career. But you left BP and went to work for Amoco in London when the North Sea was opening. Why did you do that?

A There was a mixture of motives. We had been living like gypsies, moving from country to country on short-term assignments. My family needed more stability, especially with schooling becoming more important as the children grew older. Also, I did not exactly have a positive response from BP about my career prospects: the local manager was not encouraging. I thought it would be fun to see how the Americans did it, and the North Sea was just opening. It seemed a good idea to be in at the start.

Q Were your expectations met?

A I suddenly found myself earning double, so that wasn't bad. But I encountered for the first time the hierarchial nature of a major American company, which was very different from the club-like atmosphere of BP.

Q Can you explain more about this? I certainly understand what you say about the Club-like environment of BP: they still have reunions for old Libya hands, and retired BP people like to reminisce about the fate of their erstwhile colleagues.

A Amoco had what I call a long-chain structure. The explorer at the front-end made his proposals through a long chain of command. In principle, each higher level should have added expertise and experience. But the reality was otherwise: these people were effectively filters, who had more to gain in the corporate politics by taking the low-risk easy option than by grasping the real opportunities. As a result they missed the main chance. It was a frustrating bureaucracy. But at the same time, I liked the pragmatic approach of the Americans I worked with: they would say "if that's what the Boss wants, that's what we better give him". Technically it was an invaluable experience, as I had a chance to be part of the pioneering exploration of the North Sea. It was an exciting epoch.

Q Then in 1976, you were transferred to Norway as Exploration Manager, and played a key role in the discovery of the Valhall Field.

A It was indeed a challenge. I recruited a small team of young Norwegian geologists, and we began to solve the intractable problems of Chalk geology. Valhall looked like a dough-nut on the seismic records. What was the hole in the middle? Some people thought it was a salt intrusion, but our studies eventually led us to conclude that the escape of gas had affected the seismic velocities in the overlying strata, so what looked like a hole was in fact the higher part of the structure. We were vindicated when Valhall came in as a major field. I became immersed in the subtleties of chalk geology, which remains a particular interest to this day. I was recently able to turn this knowledge to good effect, discovering a chalk field in Denmark which had been rejected by another company that had forgotten its chalk lessons.

Q I am interested in the idea of there being certain insights that companies have, lose and sometimes recapture. One searches for new ideas, only to find that people have been there before. Only the other day, I found a depletion model by Shell in a book[13] published in the 1970s which is almost identical to that I now propose: little is new.

A Yes, there are basic geological insights that tend to have lives of their own. They may form in the minds of explorers as no more than tentative ideas which are at first rejected, but they should not be allowed to die.

It is one of the reasons why it is of such cardinal importance to document studies in written reports. Later generations can often learn from early work, and use it as a foundation for new hypotheses as new information comes in. The world has now been very thoroughly explored: some work has been done almost everywhere. In the rush of modern business, the ideas of the past are often ignored, but there may be gold-mines in the technical files. BP's Colombian discoveries are a case in point. We proposed the Llanos prospects in the 1960s, but they were then deemed economically hopeless, being located on the wrong side of the Andes for export. Indeed the Company pulled out of Colombia altogether. But then later, they were searching for new frontiers and turned up our old reports, which led them to try again. This time they succeeded. It has taught me one thing. Geologists must use relentless logic to support their ideas: but they must not give way easily to all the pressures that stand in the way of testing them. The economist can always find compelling arguments for not exploring or for not making the highest bid for the best acreage. There is a polarity in all of this: ventures either succeed well, or they fail absolutely. It has to do with nature: the average, which tends to underlie economic analysis, does not bring rewards. It takes a kind of intuitive judgment to spot the winners, and to have the courage to go after them: the mealy-mouthed average does not deliver.

Q From Amoco you moved to Amerada Hess in London, becoming the Director of Exploration. Now you were able to implement your ideas about how things should be run. What did you do?

A Amerada was blessed with a dynamic leader in the form of Leon Hess, its founder. He knew his own mind and did not need to surround himself by self-serving committees. We did not exactly design a management structure, but circumstances and shared attitudes created it. The decision process was inverted in what I call a short-chain structure. The senior management determined the budget and the broader policy, and then instructed the local office to go out and do it. In effect, it delegated authority down the line, providing enormous challenges and a sense of satisfaction to those on the ground. We became a dynamic group and one of the most successful North Sea explorers. We were small compared with the major companies, who in many cases acted as our operators. It meant that much of our effort had to be dedicated to persuading the operator to follow our wishes: in many cases to save them from their own ineptitude. We often found ourselves expressing views held by the operator's own technical staff, but which had been frustrated in the long-chain command structure. It was a sort of diplomatic inductive process: to somehow get the ideas tested, despite all the back-sliding.

Q I do understand. I remember when I became your successor in Norway, I often found myself powerless to implement what I wanted to do. I sometimes found it a great help to discuss proposals first with partners, including Amerada, so that they could help push it through the management filters. Long-chain companies often prefer to be voted into action, as it shields the chain from responsibility. You have had a very successful career, and I suppose that the crowning glory is to have been elected President of the Geological Society.

A Well, I don't know that it is a crowning glory, but it is certainly a privilege to tread in the footsteps of some of Britain's greatest geologists. I would like to leave a mark on the Society by bringing it more into the public arena, and giving it the standing it deserves.

Q What sort of public issues do you have in mind?

A At first, you would think of geology as a rather academic subject, but it isn't. It has a much wider and philosophical role. After all, we live on this planet, and geology provides us with a means to know it better. The whole issue of resources and their depletion is a critical one for Mankind, and geology is at the heart of the debate. There is the growing environmental awareness and concern on which geology has much to say. Lastly, there is the issue of professional standards, qualification and training, which has to be modernized to meet today's needs. The publishing division of the Society sets an example of a highly efficient operation and what can be achieved from what was once a scientific journal. Now we have to think about manning levels in the profession: the provision of appropriate training and facilities. One thing I would like to do is to encourage field work to be again the core of geological training not only for professional geologists but for those who take it as a subsidiary subject before moving on to other walks of life. The hammer can often tell you more than the electron microscope, and it costs less. I cannot imagine a better background than a sound grounding in field work as we discussed earlier.

Q Good Luck ! I am sure you will succeed in changing things for the better, bringing together your industrial background and the political insights that come from it, with your basic love of geology. No doubt some of the old guard would like the short chain to become a noose, but if you succeed in this direction your achievements will have the recognition they deserve.

NOTES

(For references, see Bibliography)

1. Brown, D., 1977, describes his experiences as an industrialist in America's great age of growth.

2. Brian Fleay's book "The Decline of the age of oil" gives a brilliant exposé of the economics of growth and the consequences of the coming oil shortage for Australia in particular.

3. Hotelling, H., 1931, wrote the seminal work on the economics of resources.

4. Statoil, 1996.

5. Odell, P.R., 1996.

6. Mitchell, 1996, Statoil, 1996, World Energy Council, Institute of Energy, Odell, Adelman.

7. European Commission Green Paper.

8. There are several articles on this subject in Doré, A.G. (Editor) 1996. Quantification and prediction of hydrocarbon resources. Publ. Elsevier ISBN 0-444-82496-0.

9. By "political" I mean the whole role of the Middle East, and OPEC, including the sequestration of the companies' rights, to which may be added the creation of State oil companies with their own agenda. The special US/Saudi relationship is an example of the political factors that fundamentally affect the price of oil.

10. See an excellent paper by Haldorsen (1996). He demonstrates his understanding of the short span of the age of oil and humorously shows a traffic stop-sign overlying the notion that oil prices are set to rise radically. He comments:" Today very few companies maintain the 'hockey-stick' [radical increase in price] approach to oil price forecasting. The industry has become very cautious: too cautious, some will say, particularly so far as long term oil price is concerned. Some scientists note that the world's hydrocarbons may run out in 50 years as the world's population doubles and fission/fusion still cannot power cars". He is absolutely on the right track, but then demonstrates how to ignore his own words of wisdom.

11. Odum, H and E. Odum, 1981, Energy basis for man and nature; McGraw Hill, New York, explained more fully by Fleay, 1995.

12. Having written this chapter I now come upon a reference to a new economic theory by Prof. Laughton which comments: "Because of oil companies' use of the highly flawed DCF/IRR approach, the world is littered with under-utilized petroleum production facilities ..." . Evidently I am not alone in questioning neo-classical economics. (see Adam (1997).

13. Foley, G. and A. van Buren, 1978.

Chapter 11

OIL TRADING AND OIL PRICE

As I have said several times already, the oil industry has been plagued throughout its history by an environment of cycles, or in other words "boom or bust". It reflects the very nature of oil production and the fundamental pattern of depleting a reservoir. The cycle starts, figuratively speaking, when pressure is high, and the resulting flow floods the market. The flood kills further efforts, but leads eventually to a natural decline, shortage and high prices, which prompts a repetition of the cycle. The cost of the marginal barrel[1] has been generally very low. That, anyway, is how it was, so long as there were plenty of new finds to be made to fit the economic cycles. Different patterns will almost certainly emerge in the future.

But it is not as simple as that because of the very uneven distribution of the discoveries and the changing reserves. As each new major territory was opened up, it held sway for a few years: first, the United States and Russia; then Mexico and Venezuela; North and West Africa; and finally the North Sea. The Soviet Union too had its turn, although not quite as much part of the global market as were the others. Overshadowing them all was of course the Middle East with its huge endowment that gave it a lasting disproportionate influence.

From Rockefeller's day onwards, there has been a need to somehow regulate production to stabilize prices, and provide an environment for the long term investments required. Over the past fifty years, it has boiled down to finding a way to control Middle East production because it was the largest, and even had the potential to be larger yet.

TEXAS RAILROAD COMMISSION

It is now hardly more than of historical interest to look back at the operations of the Texas Railroad Commission that prorated Texas production. It was established in 1891 to supervise the railroads, and since much oil was moved by rail, it had the ability to regulate oil supply. Rockefeller, after all, had used rebates on the railroads as a means of monopolizing the market. When the oil price collapsed in the Depression of 1931, the Governor of neighbouring Oklahoma, none other than the colourfully-named "Alfalfa Bill" Murray, declared an emergency and sent in the State Militia to shut down the oilfields until the price improved. His action was followed by the Governor of Texas, who closed the East Texas fields. When the medicine worked, and prices did begin to recover in the following year, the

Railroad Commission was given stronger powers to curb production by allowing operators to produce for only so many days a month. There was widespread cheating and smuggling.

In 1933, President Roosevelt appointed Harold Ickes to be Secretary of the Interior, with a mandate to bring order to the oil industry as his priority. He resolved to reduce the country's production by three hundred thousand barrels a day, and to secure a workable allocation between the producing regions. Within four years, he had substantially succeeded in making a national strategic policy work, even in a place as capitalist as the United States. It sets an example the World could now do well to emulate.

Fig.11-1. The first train carrying oil from Oklahoma.

Photo courtesy of Exxon Corporation

Proration continued until about 1971, by which time US production had peaked, and they needed all they could produce: there was no more surplus capacity to control. It had little to do with investment, technology, supply or demand, and everything to do with the depletion of a finite resource. It is a cautionary tale about to be re-enacted on the World stage.

THE POST-WAR WORLD

Since the United States was the dominant producer in the inter-war years, the activities of the Texas Railroad Commission had a global impact, controlling the marginal barrel. But in the post-war epoch, things changed with the colossal growth of overseas production from Venezuela,

North Africa and the Middle East, the latter rapidly emerging as the dominant force.

Now the mantle of regulator fell upon the seven major international oil companies, the "Seven Sisters". Five were American; one was British; and one was Anglo-Dutch. These huge integrated enterprises controlled everything from the wellhead to the forecourt, and despite their competitive relationships were sufficiently monolithic to exert a practical unified control.

In the immediate post-war epoch, they had a pricing system based on the marginal barrel in the Gulf of Mexico, from which differing freight costs were discounted. Although the marginal barrel was in the Gulf of Mexico, it was not the cheapest barrel, as production costs in the Middle East were very low.

Discounts were allowed against the nominal Gulf of Mexico price. The major oil companies nevertheless accepted this marker and commonly exchanged oil with each other around the world to meet their individual marketing needs, paying a cash adjustment to cover only the freight charge. Other companies were at a disadvantage because they had to pay full list price plus freight.

The companies were at the same time under pressure from the producing governments who wanted to maximize their revenues. Their national affiliates in the producing countries paid a royalty, normally equivalent to one-eighth of their production, as well as tax on profits. Since most oil was traded between the affiliates of the same company, the parent was able to rig the pricing structure to its best global advantage, so it was difficult to determine exactly what profit was attributable to the production in any particular place. It led to the concept of a "posted price", at which the companies declared themselves willing to sell oil to any comer on an arm's length basis. It was used as the basis for tax paid to the producing government. The posted price was a tax reference but did not represent the real price of oil, which was known only to the companies as they traded it amongst their affiliates around the world. It is well to remember this when considering the historical oil price trends: actual prices were mainly less than reported.

As product prices fell with the growth of production in the post-war years, the companies sought to reduce the posted price for crude to better bring it into line with the product market. Naturally enough, the producing governments objected to such a move that would reduce their revenues.

OPEC

Venezuela's oil minister, Perez Alfonso, was particularly strong in his view that the companies were denying his country a fair share. He resolved to try to form a "global Texas Railroad Commission" to prorate production, and set about enlisting support from other producing countries. He succeeded, and on September 14th 1960, his efforts resulted in the formation of a new organization by which Saudi Arabia, Venezuela, Kuwait, Iraq and Iran agreed to cooperate to preserve the price of oil. It was known as the Organisation of Petroleum Exporting Countries or OPEC.

But even in its early stages it proved rather ineffectual because of differences of policy between its members. It set up a secretariat in Vienna in 1965 to strengthen its executive function.

Middle East crude was being lifted on a posted price that purported to yield a fifty per cent take to the government: in reality it was often closer to eighty percent, since the posted price was above the market price. But it was not as simple as that because of the inevitable distortions of tax. Under double taxation agreements, the companies were able to charge the tax paid to the producer as an expense against their taxable income in the consuming country[2]. In effect, the consuming taxpayer was paying the tax levied by the producer. Furthermore, the parent company charged an unrealistically high price for the oil it sold to its affiliate in the consuming country, which therefore operated at a notional break-even or loss-making level, paying little, if any, tax. The scope for manipulating inter-affiliate profits and losses, and freight charges, was huge, and provided the parent with much of its profits. While the companies did not make inordinate apparent profits for their shareholders, they were able to build up huge feather-bedded establishments, all treated as a deductible expense against tax somewhere. The German government finally moved to stop the abuses of the apparently loss-making marketing affiliates in its country, and began to levy tax on profit based on its own assessment of a deemed fair cost of imports, irrespective of what the affiliate paid its parent.

With the establishment of OPEC, the companies lost the ability to set "posted price", which was now imposed by the producing governments. But nothing much else changed, as most oil was still traded by the affiliates of the same major companies as before.

This arrangement lasted until the early 1970s when the national affiliates of the major companies were expropriated in the main production countries, or their rights sequestered. The expropriation itself was, naturally enough, motivated by a desire of the host governments to secure a higher revenue from their oil, although there were also nationalistic influences. They realized that most had been found, and were not overly concerned to lose the foreign investments for further exploration and development: they rightly figured that *the past was worth more than the future*.

Trade on the surface continued much as before. First, the companies and, later, the governments, looked to long-term supply contracts as a form of security. But in fact much did change beneath the surface, and especially in the minds of the oilmen: they no longer felt in control of their main sources of production. It led to two reactions: to pump like hell; and to do whatever was possible to find alternative sources that they did control, even if such was much more expensive oil.

The search for alternative supply during the 1960s and 1970s was remarkably successful, as there was still much left to find, and the eternal pattern of "boom and bust" was repeated. Flush production from the new giant fields, found early in the new provinces such as Alaska and the North

Sea, began to flood the world. Furthermore, the Soviet Union had brought in major new oil provinces during the 1950s, and was now beginning to export its flush production.

I remember sailing in Dublin Bay in the early 1970s. We would often tack under the stern of a rusty tanker at anchor flying the hammer-and-sickle, which had been discharging oil for the Ringsend Electricity Power station. To witness this ship from behind the mysterious Iron Curtain was somehow impressionable at the time. It was helping to flood the world with cheap oil.

The Yom-Kippur War and the decision on October 17th, 1973, by several Arab countries to apply an oil export embargo was another turning point in the evolution of the market. The organization announced that its members would begin cutting production by five percent a month, and absolutely embargo shipments to the United States and the Netherlands, who had declared particular support for Israel.

For four brief years, OPEC was in command, and the World hung on Sheik Yamani's every word as he jetted in his burnouse from Riyadh to the Intercontinental Hotel in Geneva, making oblique comments and veiled threats. His power did not last, however, for OPEC's revenue surplus of 1974 had been transformed into a deficit by 1978. By then, the founder members had been joined by Algeria, Libya, Nigeria, Gabon, Indonesia, Ecuador as well as other smaller Gulf producers. Ecuador, Gabon and Indonesia probably joined as a matter of prestige, since their relatively small exports were not very critical. Ecuador and Gabon have now left, and Indonesia can't be far behind. They have no advantage in paying membership fees and incurring obligations to restrict their exports. Indonesia will soon be an importer, and will have nothing to gain from an exporters' club.

It was a short reign because the demand for oil began to fall as a result of improved efficiency in the industrial countries, as exemplified by the jet engine and other factors. Demand for OPEC oil fell further in the face of competition from the new flush production. Many of these finds were offshore, where the high investments prompted a rapid build up of production to secure a quick pay out.

In March 1982, OPEC discovered that setting arbitrary posted prices did not work, and concluded that there was no recourse but to cut production. Complex negotiations were held to resolve how to apportion the quota amongst the members: two key factors being population and oil reserves. Both were subject to exaggeration in efforts to secure higher quotas[3].

But in the meantime, new companies – especially those with Libyan production to sell – were entering the European market. Unlike the established "seven sisters" they had no ready-made refining and marketing organization to absorb their oil. They started to trade product on the Rotterdam Spot market. This in turn put pressure on the local affiliates of the "sisters" who found themselves being undercut. Before long, they too began to behave like independents, reducing their ties with their parents and using their wits to buy and sell on the spot market.

During one baleful epoch of my life in the late 1980s I was obliged to attend the Board meetings of a marketing company in Oslo. I well remember arriving to glum faces as the autumn cold began to grip. One said "last year it was October 15th"; another said "but the leaves turned early this year". What they were debating was when to lace their diesel with paraffin. They were no longer buying premier product from their own refinery, but making spot purchases of Polish oil, much of which was contaminated with water and solids. If it sat long enough in the tanks, the junk would settle out, but if not, there was a sort of freezing emulsion that clogged everything. At next month's meeting, the faces were even glummer. They had got it wrong, and an entire bus line in north Norway had been immobilized as the diesel froze. I was always worried when I saw one of their tank trucks approaching an airliner I was travelling on.

Before long, the spot market for product was extended to crude oil[4]. For the first time, prices became transparent, although the major OPEC producers tried to resist the spot market, which they felt undermined their sovereignty. Not only did they not sell on the spot market, but they prohibited resale of their oil bought under contract.

But the spot market controlled the price, whether the OPEC countries traded on it or not. Even term contracts were commonly linked to spot prices.

In 1985, Sheik Yamani tried a new formula: the net-back, in which the producer was paid in relation to the price at which the refiner sold his product. It was a boon to the refiner who was assured of a satisfactory margin, no doubt further enhanced by imaginative accounting. Refinery output increased, and crude prices and tax revenues to the producer fell further. The system did not survive for long.

As North Sea production grew rapidly in the 1980s, prices again sagged: ultimately to a low of about $6/b in 1986. The Middle East share was down to sixteen percent. The market was not, however, the only factor, for one of the causes of this collapse was related to President Reagan and Mrs Thatcher's resolve to bring down the Soviet Empire by economic warfare. They persuaded King Fahd to open the tap[5], as already discussed in Chapter 4.

Many of the US independents and several contractors faced bankruptcy as a consequence. Stripper wells in the United States were shut in.

It was another watershed: security of supply, which had caused much concern in the 1970s, ceased to be an issue. What mattered now were prices and price volatility. Price volatility is what makes money for the stockbroker, and in 1986 the market saw the increasing role of financial traders, familiar with the well known principles of paper transactions, futures and derivatives. They were called the Wall Street refiners, who traded in paper more than the physical delivery of oil. Oil became traded as any other commodity, like pork-belly futures, and Exchanges developed in New York and London to trade in it. The brokers needed a benchmark price against which to trade the various oils around the world. West Texas Intermediate[6] became the benchmark for North America, and Brent from the North Sea for most other transactions. In addition, Dubai crude had a role for Eastern Hemisphere trade.

A convoluted market evolved in which physical trade was supplemented by forward markets and futures. There was a wide variety of crudes and products with complex inter-relationships. The forward market provided a form of low-cost hedging, whereby the companies could minimize the risks of volatility. Brent Crude from the North Sea was the main marker, and the other crudes and products were linked to it via differentials based on quality and freight charges. It was itself, however, a relatively small market, controlled by a few privileged members, and its days must now be numbered as Brent physical production declines. Modern communication made it a highly efficient market, so that virtually any transaction could be hedged by another.

For the first time, prices became truly transparent, as they moved up and down with the benchmark reflecting a huge pyramid of hedged future transactions. In large measure, the hedging became a mechanical procedure run by sophisticated computer programmes aimed at minimizing the exposure by buyers and sellers to perceived or computed market risk.

The industry will always have some physical stocks: much oil is tied up in transport in pipelines and tankers which is a form of storage; and refiners cannot exactly match the changing seasonal demands for different products so they have to maintain some product in storage.

When stocks are low, prices tend to be in what is termed *backwardation*, meaning that the price of physical delivery is above the price of future delivery, because a greater value attaches to oil available for immediate needs. The opposite situation, known as *contango*, arises when stocks are high[7]. This relationship naturally assumes no fundamental supply shortfall and is no more than a short-term response to stocks and market. The large traders are laying off price risk without really speculating about what the future price will be. It could be reversed when a future shortfall in actual supply is perceived. A combination of high physical stocks and *contango* might be the market's warning signal to watch for, meaning that judgment for the future rather than mechanical hedging was beginning to affect prices.

The reporting of price also has become more efficient, with Argus and Platt's providing a credible service, used in all sorts of transactions, and as measures of performance.

Prices continued to languish for the next few years until the Gulf War successfully supported them by the removal of about two million barrels a day, an outcome that poorly enforced OPEC quotas failed to achieve. The Gulf War itself was a great boost to the new trading system. It was a precise crisis, to which a foreseeable positive outcome soon became apparent as Desert Storm got under way. At the height of the crisis, prices reached about $40/b but with an end in sight, it was tailor-made for hedging on the futures market. The market gained credibility that it could enforce the performance of the contracts for which it was responsible.

There can be no doubt that it is an efficient and transparent market representing a great improvement over the murky world of inter-affiliate transactions of the past. But I am not sure that efficiency is the same as wisdom. I think that mechanical hedging effectively is an extension of today's environment, and leaves little room for prudent physical storage against the unexpected crisis. Could the whole pack of cards implode? One can lend to the future, but one cannot borrow from the future if the tank is empty. The future market says little about the actual future price of oil but it gives an indication of the relationship of present stocks and needs.

At the time of writing in the autumn of 1996, physical stocks are at close to a four-year low[8], and prices are firming despite moves to partially relax the Iraq export embargo. I wonder if it is the lull before the storm, for all the reasons expounded in these pages.

Last summer, the United States launched an essentially unprovoked missile attack against Iraq. It could easily have galvanized Arab and Iranian attitudes against the West which could have triggered an oil shock, for which the fundamentals set the scene. There was no spare capacity anywhere else to meet a curtailment of supply from this region. The futures market is an efficient one to deal with normal conditions, but I think I would feel happier to have some physical oil in my tank. There have been runs on respectable banks, when their customers lost faith: there could be a run on the futures market too.

By the time you read this, you will know what happened. Charting the short-term future is an unenviable task.

Crude Oil Price 1860-1996

West Texas Intermediate and Average US Wellhead in US dollars

	WTI		WTI		WTI	Av		WTI	Av
1860	9.6	1900	1.2	1940	1		1980	38.0	21.6
1861	0.5	1901	1	1941	1.1		1981	36.1	31.8
1862	1.1	1902	0.8	1942	1.2		1982	33.7	28.5
1863	3.2	1903	0.9	1943	1.2		1983	30.3	26.2
1864	8.1	1904	0.9	1944	1.2		1984	29.3	25.9
1865	6.6	1905	0.6	1945	1.2		1985	28.0	24.1
1866	3.7	1906	0.7	1946	1.4		1986	15.1	12.51
1867	2.4	1907	0.7	1947	1.9		1987	19.2	15.4
1868	3.6	1908	0.7	1948	2.6		1988	16.0	12.6
1869	5.6	1909	0.7	1949	2.5		1989	16.7	15.9
1870	3.9	1910	0.6	1950	2.5		1990	24.5	20.0
1871	4.3	1911	0.6	1951	2.5		1991	21.5	16.5
1872	3.6	1912	0.7	1952	2.5		1992	20.6	16.0
1873	1.8	1913	1	1953	2.7	2.72	1993	18.4	14.3
1874	1.2	1914	0.8	1954	2.8		1994	17.2	13.2
1875	1.4	1915	0.6	1955	2.8	2.82	1995	18.4	
1876	2.5	1916	1.1	1956	2.8	2.82	1996	22.2	
1877	2.4	1917	1.6	1957	3.1				
1878	1.2	1918	2	1958	3				
1879	0.9	1919	2	1959	2.9	2.98			
1880	0.9	1920	3.1	1960	2.9	2.97			
1881	0.9	1921	1.7	1961	2.9	2.97			
1882	0.8	1922	1.6	1962	2.9	2.97			
1883	1.1	1923	1.3	1963	2.9	2.97			
1884	0.9	1924	1.4	1964	2.9	2.95			
1885	0.9	1925	1.7	1965	2.9	2.92			
1886	0.7	1926	1.9	1966	2.9	2.94			
1887	0.7	1927	1.3	1967	2.9	3.03			
1888	0.7	1928	1.2	1968	2.9	3.07			
1889	0.8	1929	1.3	1969	3.1	3.3			
1890	0.8	1930	1.2	1970	3.2	3.2			
1891	0.6	1931	0.7	1971	3.4	3.4			
1892	0.5	1932	0.9	1972	3.4	3.4	Note		
1893	0.6	1933	0.7	1973	3.9	3.9	1976 Posted WT1		
1894	0.7	1934	1	1974	6.9	6.9	1984-96 Spot WT1		
1895	1.1	1935	1	1975	7.7	7.7			
1896	1	1936	1.1	1976	12.23	8.2			
1897	0.7	1937	1.2	1977	14.22	8.6			
1898	0.8	1938	1.1	1978	14.6	9.0			
1899	1.1	1939	1	1979	25.08	12.6			

Fig. 11-2. Crude Oil prices.

BP
After Two Lives, The Old Lady May Contemplate Marriage

British Petroleum, or BP, began its life as the Anglo-Persian Oil Company which was formed in 1909 out of a syndicate created by William Knox D'Arcy (see Fig. 3-17), who had signed a concession for 480 000 square miles of the Zagros Foothills in Iran on May 28th, 1901. After initial disappointments, he struck oil at Masjid-i-Sulaiman in 1908. Under the influence of Winston Churchill, the British government took a 51% interest in the Company just prior to the First World War, both to secure oil for the British Navy and to establish a stronger British presence in the Middle East.

For fifty years, Iran was to be central to BP's enterprise. Discovery followed discovery, and the Company used this sure supply to build up markets around the world, especially in Europe and the British Empire. Much has been written and conjectured about its political role in Iran. The Iranians are almost united in seeing it as a dark force behind the scenes conspiring to deny them their proper rights and aspirations, being widely thought of as a front for the British Government. The truth is probably that it was doing no more than trying to hold on to its rights in a difficult and uncertain country. As the provider of the bulk of Iran's revenue, it was a hot seat to occupy.

The Company strengthened its position in the Middle East by taking up a share of the Iraq Petroleum Company when that was formed in 1928, followed five years later by securing a fifty percent interest in a concession to Kuwait. In both of these steps, it was much assisted by the British Government.

Whereas the major American companies grew up in a highly competitive environment and achieved their size by acquisition and merger in a country where there was a huge number of independent wildcatters that needed support to bring in their discoveries, BP operated under long-term concessions where there was no competition. The Iranian and Kuwait operations were staffed by BP technical personnel, and the Iraq Petroleum Company had its own staff, although using BP personnel on secondment. These were very large scale, single-minded operations, which did not in fact call for a huge exploration establishment. The geology of Iran was itself on a grand scale, with huge clearly identifiable anticlines offering prospects, obvious to the most casual observer. As a consequence, BP's

exploration staff was made up of quite a small group of people, who knew each other intimately. It was more like a club than a corporate hierarchy.

At the head of it was G.M. Lees and later Norman Falcon, who died only recently at the age of 92. From Cambridge he had joined Anglo-Persian as a field geologist in 1927, and together with J.V. Harrison had made the pioneering geological map of Iran, which remains the bedrock of modern studies. He was a highly respected man who was elected a Fellow of the Royal Society in recognition of his scientific achievements. He was Chief Geologist from 1957 to his retirement in 1965. In those days, the position of Chief Geologist was a highly respected one with ready access to the Chairman. Today, if the position exists at all, it is probably concerned with training and vacation schedules: such is progress. Being spared the pressures of wheeling and dealing, so central to exploration

Fig. 11-3. Norman Falcon, legendary Chief Geologist of BP.

Photo courtesy of British Petroleum

in the American environment, BP's pioneering geologists were able to dedicate themselves to the science and practice of exploration. They had the rather 19th Century style of the dedicated and gifted amateur, but were very effective.

Although BP relied during this epoch wholly on its Middle East production, it did maintain a modest interest in other exploration possibilities. J.V. Harrison went on roving commissions to Latin America, investigating Argentina, Colombia and Mexico. The Company tried to establish itself in Colombia in the 1920s, but withdrew without drilling. It mounted another abortive endeavour in Papua-New Guinea. It had little incentive to find more oil as its own production greatly exceeded its market, making it a substantial seller of crude oil. The other prospects it looked at around the world must have compared very unfavourably with what it already had. It sometimes had local marketing incentives to make a gesture to exploration, not to mention local income against which to expense it.

This chapter came to an abrupt end in 1951 when the Company was ousted from in Iran, the foundation stone of its enterprise. Its directors were cast in an imperial mould, and probably assumed that at the end of the day they could rely on the British Government to protect their rights. Surely, Britain, which had only recently emerged victorious from a world war would not allow a major asset to slip from her grasp. She did, and probably not to the dismay of her ally then seeking a greater stake in Middle East oil.

Now began the second successful life of BP. It embarked on a vigorous campaign of exploration throughout the world to replace what it had lost in Iran. Its supply situation was not critical as it still had Iraq and Kuwait, although they were to be lost too within twenty years. Norman Falcon led his team to find new fields, and he was remarkably successful.

One of the first new efforts was in Canada through the acquisition of Triad, a small company with a large land-holding. BP's team was led by Dan Ion, who incidentally had strong interest in world energy resources and their depletion[9], and A.N. Thomas. It was a new experience for them to work in the highly entrepreneurial environment of North America, making deals and trading rights, which probably sharpened their commercial wits. It also brought them to investigate the Arctic, which thirteen years later was to lead to the remarkable Prudhoe Bay discovery in front of the Brooks Range of Alaska. With reserves of over 13 billion barrels, it is by far the largest field in North America.

The story of its discovery makes interesting reading. BP, again with its Iranian background, at first concentrated on the foothills of the Brooks Range, and having teamed up with Sinclair as an American partner drilled a number of dry holes in this province in the mid 1960s. It then turned towards the less disturbed platform to the north finding a huge structure at Colville, but the results of drilling were discouraging. The structure was an early uplift and lacked a reservoir development, a feature nicely called a "bald high". Sinclair decided to pull out, but a combine of Arco and Esso turned attention to the area, approaching BP to see if it

would sell, which it nearly did. Interest was focused on a structure at Prudhoe Bay which was smaller than Colville but with better prospects. Arco and Esso made a successful high bid for the crest in a Lease Sale that was being held in 1967, whereas BP submitted lower bids for the entire structure, ending up with the flank. Arco realized that it was a critical test of the Alaska province and persuaded all the companies to contribute to the cost of the well in proportion to their holdings. The well was drilled on the crest and tapped the gas cap of a large structure, but most of the oil lay on the flank in BP's territory. BP's development of the field was another remarkable pioneering endeavour, reminiscent of the Company's early days in Iran. Credit for the discovery goes largely to A.N. Thomas, Harry Warman and Peter Kent. It was a remarkable coup for an outsider to find the largest field in North America, the home ground of the major US companies.

Another venture was launched in Libya in the 1950s, where one of the obstacles were the minefields left over from the War. Whereas many companies concentrated on the Sirte Basin near the coast, BP headed into the interior, where it brought in the giant Sarir Field in 1961. With reserves of 4.5 billion barrels, it is the second largest in Africa.

The Company also joined forces with Shell to explore respectively East and West Africa. The latter soon yielded major discoveries in Nigeria. Other exploration ventures were conducted in Papua-New Guinea, Trinidad, Abu Dhabi, Australia, Algeria, Senegal, Colombia, Sicily, Switzerland and the United Kingdom itself, of which some were successful. The Company continued however to be a victim of the imperial past of its mother country, being driven out of Libya and Nigeria for political reasons unconnected with the Company itself.

North Sea exploration commenced in the 1960s to follow up the trend that had given the giant Groningen gas field in Holland in 1957. BP naturally took up a key role, making the first offshore discovery in British waters with the West Sole gasfield. That was in turn followed by other major discoveries including the giant Forties field. The Company was also active in Norway, but rather as a late starter having been put off by the outrageous terms of the Fourth Round of 1979. It farmed into the licence that yielded the Ula and Gyda Fields, and it was a partner with Shell in the Draugen Field off mid-Norway. The most recent and somewhat unexpected discoveries West of the Shetlands cap the list of a very successful exploration campaign.

In charge of exploration during much of this period was Sir Peter Kent, and the achievements were made by what was still quite a small closely-knit team, although eventually a more American style emerged with faceless committees, economists and bureaucracy.

Mrs Thatcher's policy of privatization in the 1970s allowed BP to buy out the former state company, BNOC, further strengthening its position. It also resulted in the Government disposing of its fifty-one percent shareholding. There was a fear that this might have been acquired by Kuwait in an ironic twist of fate, but that attempt was

Fig. 11-4. Sir Peter Kent, whose early work led to Britain's onshore fields.

Photo courtesy of British Petroleum

determined man with a dramatic voice who risked borrowing to pay the dividend in the conviction that oil prices were about to rise. They didn't, and after a generous handshake, he went on, surprisingly, to run British Rail.

Having flirted with Statoil in a global "alliance" for exploration cooperation, an interesting third life for BP may be in the offing with a possible marriage with the American giant, Mobil. They have already agreed a $5 billion deal to merge their European downstream operations. Mobil has never been a strong explorer, but has a huge global market. A recent article[10], entitled "oil buccaneer", describes how Mobil is buying high risk upstream oil and gas interests in remote places as what sounds like the last throw of the dice in an attempt to replace its declining production in Indonesia which provides half of its upstream profits. It is a telling insight. For reasons already discussed in earlier chapters, the major international companies of the past are becoming eclipsed. They have lost their principal sources of supply, and there is now no hope of replacing them, as this book explains. To merge two major oil companies into one may buy a few more years of life. The "Seven Sisters" became six when Chevron absorbed Gulf; and if this marriage should happen they would be down to five[11,12]. It is a sign of the times, and delivers a message about oil generally.

If such a merger should happen, it could have very serious strategic consequences for Britain in the *Coming Oil Crisis,* when the country might very much want to have a national company of this size to meet the challenges of the times. Think too of poor Ireland where the valiant efforts of its national company Aran were sold down the river to Statoil. The global market may soon have much to answer for.

thwarted, and the shares were widely placed, many being taken up by American investors.

One colourful incident was the unheard of dismissal in the late 1980s of the Company's Chairman, Bob Horton, a

NOTES
(For references, see Bibliography)

1. The price of oil is heavily influenced by the marginal barrel, that is to say the often relatively small amount by which the supply is above or below the demand.

2. Nasmyth, J., 1996, gives an excellent review of the period 1945-1985.

3. Barkeshli, F., 1996, an Iranian official, confirms that reserves were exaggerated in the "quota wars".

4. See excellent articles by J. Rieber for a fuller discussion of the evolution on markets.

5. Schweizer, P., 1994, and R.V. Allen, 1996, describe the CIA conspiracy with Saudi Arabia.

6. West Texas Intermediate is a posted price which did not match realised prices, as listed in Figure 11-2.

7. See Long, D., 1996, Verleger, P., 1996 for useful discussions of stocks and the futures market.

8. See Simmons, 1995, regarding low stocks.

9. He published a valuable book in 1978 on the subject.

10. Economist, 30th November 1996.

11. I have already mentioned the proposed merger of Texaco, Shell and Saudi Arabia in US marketing and refining.

12. Mortished writing in the Times in February 1997 also suggests that such a merger is in the wind.

Chapter 12

THE FINGER POINTS AT NORWAY

The millions of viewers around the world, who watched trolls emerge from clouds of freezing mist at the ceremonies marking the 1994 winter Olympic games at Lillehammer in Norway, must have thought that this is a fairy-tale country. Indeed, anyone who has visited it, or still more lived there, will agree that here is indeed an earthly paradise. That is exactly what I thought after living in Norway for ten years. It is a marvellous place in almost every regard. Its natural beauty is truly outstanding: from its mountains to the fir-clad islands and channels of its rocky, protected coast. Its towns too with their neat, white clapboard houses and tidy gardens, with the immaculate Volvo parked outside, are pictures of quiet unostentatious dignity. But most of all, it is the fine people with their striking pure features, reserved manner, stalwart and simple characters. They are, or were, almost universally honest: if you dropped a gold watch on a train and advertised its loss, the finder would telephone to tell you at which police station it was to be found. I am an enthusiastic admirer of Norway. Why then do I say that the finger points at this marvellous country?

Every country faces a dilemma over its policy towards developing natural resources, especially oil: Who will benefit and when? This generation or the next? So far, most interest attaches to the issue of *who?*, but at the end of the day it is *when?* that matters more. Apart from *who?* and *when?* is also the small matter of *how much?* Let us therefore zero in on Norway to see how it dealt with its oil industry, whose arrival came as a great surprise to everyone. They started right but eventually were misled by short-term money, and that is why the finger points at these intrinsically splendid people. Norway is now exporting its birthright at a rate only surpassed by Saudi Arabia. That is particularly why the finger points at them. How were they misled? It is instructive to investigate this important oil country. No criticism is implied: most other countries are much worse when it comes to protecting their resources.

Fig. 12-2. The fine featured Norwegians became victims of their success.

Fig. 12-1. The beautiful coastline of Norway.

HISTORY

Norway is a mountainous country on the western seaboard of Scandinavia, populated by some four million people. Life has been difficult on this rocky coastline that extends for 2000 km into the Arctic Circle. From southern Norway to North Cape is farther than to Rome. The natural conditions have left their mark on the people.

It is an ancient land, although the modern State came into existence only in 1905 after more than two centuries of domination by Sweden and Denmark. It is a sort of Ireland of the North. Independence sprang from a cultural and political flowering in the last Century which mobilized a sense of fierce nationalism and common purpose, still manifested by an abundance of flagpoles in the country.

Fig. 12-3. The Norwegian flag flies everywhere.

Photo by Egil Eriksson

The sea and the mountains were its resources: the sea for fisheries and shipping; the mountains for hydroelectric power. Norsk Hydro, the largest private company, was formed with French capital in the early years of the Century to produce artificial fertilizer, fixing nitrogen from the air by hydro-electrical power. That itself was a major economic discontinuity.

The shipping business had humble beginnings during the last Century when much of the fleet was run by owner-captains, who traded on their own initiative, sharing the proceeds with their crews. It was one of the foundations for the spirit of cooperation which is a national trait.

From these beginnings grew a number of shipping dynasties: the Olsens, Bergesens, Smedvigs, Wilhelmsens, Uglands, to name a few. They competed with entrepreneurial drive abroad, carving out for themselves profitable niches in world trade, yet at home they lived modestly, being very conscious that they were the breadwinners of their communities. The capitalist of the high seas changed his coat on entering home waters.

Fisheries, too, provided opportunities and hazards. Fortunes were made in good years, but unpredictable changes in currents and water temperature sometimes meant that the herrings did not arrive in the fishing grounds on the appointed date. Fishing for cod in an open boat off North Norway in winter was no fun, but they did it through a blend of independence and inter-dependence that is such a national characteristic.

Norway was plunged into the trauma of War by an unprovoked German invasion in 1940, but not before the veteran curator of an ancient fort on the Oslo Fjord managed to sink a battleship of the grand fleet by firing an old cannon. Much of North Norway was fired by the retreating Germans at the end of the War under their "scorched earth" policy. The disgrace of Mr Quisling's puppet government was redeemed by an active resistance movement.

The experience of War left an indelible mark on the country's subsequent history. Post-war reconstruction had to be closely controlled by the government. Control once grasped was never surrendered, and Norway remains one of the most intensely regulated countries on Earth. While regulated, there remains however an intense individualism: to reconcile these opposing threads is one of the conundrums of Norway.

Like most of Europe, Norway's politics moved to the left in the post-war epoch, as a natural extension of the then essential government control of everything. It was not however doctrinaire socialism built on envy and class-warfare as in Britain, but a much more positive cooperative version. Taxes became very high, but most people were happy to pay them. In fact, the actual burden was not as high as it seemed because of the facility to deduct interest on loans. It was an extension of control: everyone was in debt to the banks which also controlled their lives. There is a gruesome office, called the *Ligningskontor*, which supervises tax collection under the logo of an unblinking eye.

They also have had a thing about alcohol. It is a state monopoly to which people sheepishly went with unmarked bags. There was however no shortage of drunks, one of whom, Oivind, a sailor back from the sea, lived next to us and had a party, paid for by social security which lasted for several years. It was not perhaps the best form of national cooperation, but he was a nice chap all the same.

The nation's wealth grew in the twenty years following the War. Poverty was eradicated, and high standards of living were attained. The average Norwegian found himself living in housing that would be the envy of the elite in most countries, and many had country cottages or pleasure boats as well. The banks probably owned them, but the people used them, and led happy simple lives, heading into the mountains with leather backpacks at weekends. They did not have much ready cash for a pint of beer, but their surroundings and lives were superb.

It was a cooperative society in which everyone was supposed to have a useful role, which once recognized would be supported through thick and thin by community and government action.

How did oil come to this model land? It came from Holland where, in 1936, Shell exhibited an operating drilling rig at a congress in the Hague. To everyone's intense surprise, it struck shows of oil, which stimulated exploration that resulted in several finds. Another chance discovery was made in 1957, when a weekend communications failure led to an unintentional deepening of a well at Groningen. It fell into a giant gasfield in Permian desert sandstones, which no one had previously thought of as a possible objective for oil or gas. It did not take long to extend this productive trend offshore and into British waters, where more large gasfields were found.

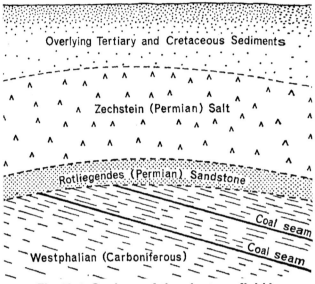

Fig. 12-4. Geology of the giant gasfield in Groningen, where gas from deep coal is sealed beneath salt.

By the mid 1960s, the industry was turning northwards, wondering if the same trend, which relied on the natural coking of deeply buried coal, could extend into Norwegian waters. This interest was greeted with incredulity in the country, where no one could imagine oil or gas beneath the stormy waters of the North Sea. But in 1963, Norway proclaimed sovereignty to its continental shelf with the agreement of the neighbouring countries[1], and on April 13th 1965 offered licences to explore and produce petroleum to the industry. The business was managed by the Ministry of Industry and Handicraft, whose very name illustrates how far Norway then perceived itself to be from the world of oil.

It was a successful offering in which seventy-eight blocks[2] were let under twenty-two licences. There were six major applicants: Shell, Esso; the Phillips-Fina-Agip Group; the Amoco-Noco Group; Texaco and Chevron (then Gulf), and the French companies, Elf and Total, in partnership with Norsk Hydro.

The first discovery was made by the Phillips Group in 1968 with a modest find, called the Cod Field, near the UK median line. Later in the same year, the Amoco Group drilled Well 2/11-1, which, although aimed at a deeper objective, unexpectedly found oil shows in the Cretaceous Chalk. This is the same rock as that of which the White Cliffs of Dover are built. It had not hitherto been considered an oil objective. A few months later, the Phillips Group brought in the giant Ekofisk discovery on a Chalk prospect in the neighbouring Block 2/4.

The remarkable combination of geological circumstances responsible for Ekofisk makes it almost one of the wonders of the world[3]. The Upper Jurassic had generated oil in the strata below it, which, having nowhere else to go, had migrated upwards along self-made fractures. The Chalk, which normally lacks adequate permeability and porosity to hold oil, just here had been deposited as submarine slumps, analogous to avalanches, in which the porosity had been exceptionally preserved. Lastly, early faulting and deep-seated salt movements gave closed structural traps to hold it. This remarkable find, along with its satellites found later, hold about three billion barrels of recoverable oil; one-eighth of Norway's total discovery.

Fig. 12-5. The miracle of Ekofisk.

Photo by Egil Eriksson

The Government had no particular policy towards oil in 1965, other than to find it. The terms were normal, with a one-eighth royalty and standard corporate tax on profits. Companies with adequate technical and financial qualifications were welcome, working either singly, as in the case of Shell and Esso, or as groups.

Another round of licensing was held in 1969, but did not result in anything particularly interesting.

Meanwhile in Britain, across the median line, important finds were being made by the industry which had negotiated comparable rights from the government on similar terms. My friend, Myles Bowen, whom I had first met in Borneo in 1956, led his company, Shell, to test a structure close to the Norwegian median line, finding another giant field in Jurassic deltaic sandstones[4], which was named the Brent Field.

Britain was then a socialist country bent on bringing everything under public ownership. Once the oil had been found, it was soon regarded as a national birthright being exploited by undesirable capitalists and foreigners. A state oil company, BNOC, was established. In a disgraceful chapter of bad faith, the rights previously freely granted to the companies were clawed back under duress to give to this new State entity.

Already socialist Norway saw what was happening across the border, and decided in 1973 to set up its own State company, Statoil, with the idea of protecting the national interest in the Norwegian extensions of this new trend which had already been found in Britain. To its great credit, however, honest Norway did not renege on its previous contracts.

Fig. 12-6. Myles Bowen, who opened up the rifted Jurassic prospects of the North Sea.

Fig. 12-7. Arve Johnson, behind a smiling face was a forceful man with global ambitions.

Photo courtesy of Statoil

A large highly prospective structure offsetting the Brent Field had been identified on the seismic data. Although the fledgling State company was now in place, the government felt that it was not qualified to take on a project of this scale, and accordingly granted rights to a large group of companies, run by Mobil. Statoil was given a fifty percent interest. It was a great achievement for Mobil to be Operator[5] of what was to become the North Sea's largest field, Statfjord, but it probably got its rights by agreeing to transfer its operatorship to Statoil at a later date, little thinking that the option would be exercised: it was.

This development coincided with the First Oil Shock when oil prices increased five-fold. Norway realized that it had become an oil nation in no uncertain manner. It moved to improve its administration of the business by setting up the Norwegian Petroleum Directorate in Stavanger and the Department of Energy in Oslo. *Directorate* was the operative word: it decided to do just that – direct and control. In many other countries, government supervision of the oil business is often entrusted to retired Shell oilmen, with a mandate to follow developments in a substantially passive sense. But in Norway, where virtually every aspect of life, apart from the offshore shipping business, was managed by the government, it was no great step to decide to run the oil business too. The Norwegian bureaucrat is used to being in charge.

Farouk al-Kasim, an Iraqi, who had trained in petroleum engineering at Imperial College in London and had married a Norwegian, was recruited to take charge. He was extremely effective. Working with him was Egil Bergsager,

an imaginative and optimistic geologist from Haugesund. These two, under Frederik Hagemann, the Director, cooperated closely with the Ministry in Oslo, which was more concerned with the legal and political aspects. Together, they set about planning Norway's oil policy, aided by a staff of mainly young men and women, straight from university, who found the growing control of this enormous and expensive business a heady start to their careers of power.

At first, the momentum of the past continued with the third round of licensing, when various tracts were infilled without any spectacular results. The newly established Statoil began to recruit its staff, and to intervene forcefully under the abrasive leadership of Arve Johnson, who was the protégé of the Labour Party.

In 1979, came the Second Oil Shock when oil prices rose to almost $50/b. Norway reeled under the promise of new riches of unprecedented proportions. Meanwhile new seismic surveys had been shot over the northern North Sea, and it became obvious that Norway had some superlative prospects with every chance of yielding giant fields.

THE GLORIOUS FOURTH

Against this backdrop, Norway moved to organize the Fourth Round[6] of licensing under new terms. And new terms they were.

The first move was to license a large prospect, close to the Statfjord Field, applying a national solution: 85% went to Statoil, and the balance going to the two other national companies, Norsk Hydro, of fertilizer fame, with nine percent, and Saga, with six percent. Saga, representing ninety-six shipowners and industrial companies, was the flagship of private enterprise in Norway, which while in eclipse was far from dead. The Directorate, however, was not delegating any of its powers even to nationals, and imposed as a condition of the licence that the group had to drill eighteen appraisal wells to test different compartments of what was a much faulted structure. Each well probably cost around $15 million: so the commitment cost over $200 million. It was worth it, for the giant Gullfaks Field was the outcome. It has reserves of almost two billion barrels which at the then oil price of around $30 meant that a gross revenue of around fifty billion dollars would be earned before expenses. There was plenty at stake.

The government was however concerned about the impact of the huge floods of money that promised to flow ashore. It rightly recognized the great dangers that this could bring to traditional industry and the finely tuned Norwegian social structure. It very wisely decided to move cautiously, and hold the pace of exploration and development down. It further decided to strike the toughest bargain possible with the foreign companies. Norwegians soon became known as "blue-eyed Arabs". It was a time of perceived oil shortage and high prices: the companies had little option but to accept. The hard-nosed Texaco and Chevron did however pull out, and BP was a less than enthusiastic applicant.

The terms were truly outrageous, and could only have been countenanced in such unusual circumstances:

1. Statoil was granted a 50% interest in all licences, but its costs were borne by the other partners;

2. Statoil voted its full unpaid interest in the management of the venture, which with the support of the other Norwegian partners conveyed total control;

3. Statoil had the right to increase its interest to 70-80% in success;

4. Heavy drilling commitments were imposed (in some cases with as many as five, six and even more wells being required);

5. Checkerboard licensing was organized whereby offset blocks were retained for subsequent privileged allocation.

6. The government formed the groups, often deliberately putting together companies with differing technical views in moves calculated to maximize the management difficulties. (Normally, companies work together building up mutual understandings and relationships, and develop their knowledge for their mutual advantage, but this was denied them under a policy of divide-and-conquer).

Whereas the companies had invested huge amounts over years to assemble the data, knowledge and judgment that pointed to the prime blocks, Statoil was able to enter for free, and cream off the icing on the cake. It was born with a sugar-coated silver spoon in its mouth. And like many spoilt children became extraordinarily aggressive and uncompromising in its privilege.

But all was not quite how it seemed in those days of national socialism: the costs borne by the foreign companies in carrying Statoil were in fact deductible as expenses against their own taxable income. This in effect meant that the government was investing huge amounts of national money in a new state enterprise without having to seek parliamentary approval for the annual budget. Whereas the Minister of Health had to crawl on his hands and knees to extract budgetary support for a new hospital, Statoil had *carte blanche* to spend as much as it wished, free of control or constraint, in a hidden manner that virtually no one in the country appreciated. And spend it did. It was a form of daylight robbery: not against the foreign companies, but against the long suffering Norwegian taxpayer.

In command of this robbery were the mainly young administrators in the Directorate, the Ministry and Statoil itself. They were relishing the huge power they now wielded to insist that this reservoir be tested or that new draconian safety code observed.

Ironically, the enormous benefit that the foreign companies enjoyed in being able to deduct all the expenses of this Mad Hatters Tea Party was not always recognized, because exploration budgets were commonly administered apart from fiscal budgets to ensure that projects around the world were compared on equal technical merit. Norway soon earned for itself the reputation of being one of the most difficult countries in the world to work in, although in reality it was one of the best. They were strange times; and about twenty international companies hung in there, not wanting to be left out of what was perceived to be one of the world's last great oil provinces, as indeed it was.

Of course, Norway was Norway. The robbery was not perceived as such, and the robbers had the high-minded intent of protecting the national interest. The well known themes of foreign exploitation and abuse by multi-national companies were flowing through the veins of the young socialists. As I will explain, although the policy found a lot of oil, it did not protect the national interest in the long term.

In the short term, however, the Round was a resounding success as one giant discovery followed another. In this brief span of time most of Norway's oil and gas was found as illustrated in Figure 12-8.

The ball had started rolling, and it proved difficult to stop. At first, Norway and Statoil needed access to the industry's knowledge and expertise, and great efforts were made in the licence terms to provide for "technology transfer". It was a remarkably successful policy, and within a few years the dedicated and hard-working Norwegians had learnt it all. Actually, much of the business is not particularly high-tech: all sorts of humble people manage to drill oil wells, but money helps, especially if it isn't yours.

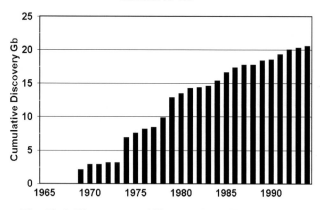

NORWAY
Ultimate 27 Gb

Fig. 12-8. Discovery of Norwegian oil over time.

Along with this, a number of Norwegian shipping companies went into oil drilling and even manufacturing rigs. They include Fred. Olsen, Wilhelmsen, Odfjell and Smedvig[7]. The national oil companies gave them enormous support by awarding contracts, often on better than commercial terms. GECO, a seismic contractor, was established by several enterprising Norwegians, largely from the expertise of the American GSI[8]. It too prospered with contracts for the huge surveys that were shot over the extensive Norwegian shelf.

The Fourth Round in 1979 was Norway's golden hour. With hindsight, it may be asked if Norway might then have been wiser to say "thank you and good-bye" to the foreign companies, who had brought their knowledge and expertise, and had trained up the three national companies to high levels of efficiency. The foreigners could have produced the fields to which they had contractual rights: honest Norway would never renege on them. But the exploration of the remainder of the shelf could have been left to the three national companies.

They could have gone about it slowly and seriously, and they could have aimed to deplete their resources gradually so that they might last as long as possible. Furthermore, they could have used the oil revenues cautiously, such that the economy would not become too dependent on oil. Remarkably enough, it would not have cost the nation any more, because the foreign contribution to exploration was largely deductible from the taxes that were levied on past production. Levied they were and at a high rate providing a marginal rate of 85%. Not only were the foreign companies able to deduct the expenses incurred in Norway, but they were able to claim the cost of the whole hierarchy of time-writers in regional offices, home offices and research establishments who reckoned they were in some way engaged in work linked to Norway, however peripherally. The taxman is an accountant. He is satisfied if the money was spent under the rules, but was not qualified to ask why it was spent.

Had this policy been adopted, the three national companies, and especially Statoil, would have been responsible and answerable for their exploration budgets.

But the government did not see it this way, and instead set forth to licence the rest of the shelf in successive rounds. Statoil continued until recently to have a mandatory 50% interest in everything, but the more excessive terms were gradually relaxed over time. Statoil's share of abortive exploration drilling probably cost the unconscious taxpayer almost ten billion dollars in lost revenue. Statoil was not to blame since it was, until recently, forced to take a 50% share in all licences, and the government imposed the drilling obligations. A further flaw with this arrangement was that the foreign companies had a contractual right to exploit whatever they found, even though their investments were in fact substantially paid for by the Norwegian taxpayer. It meant that Norway had lost control of the pace of development.

EYES TURN NORTH

The Fifth Round in 1980 found Norway's two other productive basins: Haltenbanken off mid-Norway and Hammerfest off North Norway. The outrageous Fourth Round terms were still in place, and surprisingly, the foreign companies accepted them, not wanting to be excluded. There is a sort of mindless momentum in the operation of major oil companies: if the Norway case had been successfully argued for the Fourth Round, the case for the Fifth was plain sailing: all the maps having been duly massaged to offer the same grand prospects. Besides oil prices were still high, generating large tax obligations, against which the exploration expense could be charged. In fact, the risks of the northern shelf were infinitely higher; and the world circumstances of perceived shortage after the Second Oil Shock in 1979 were changing. Some of the foreign companies may have come to understand the Norwegian system, and to realize just how much they were being subsidized, especially when taking into account the risk factor.

Haltenbanken turned out to be a fair success, delivering about ten billion barrels of oil, making it one of the world's last major *Petroleum Systems* to be found. The Hammerfest Basin never looked too promising, with the zone of oil generation being narrow and somewhat removed from the main structures. It proved to be gas-bearing, but gas in this remote location is unlikely to be commercial for a very long time to come.

STATOIL: A STATE WITHIN A STATE

Time passed, work continued, and Statoil grew ever larger. It built what in architectural terms must be one of the ugliest buildings in northern Europe as its corporate headquarters, and began recruiting staff – and how.

It now employs something like eleven thousand employees and enjoys an income of over ten billion dollars, almost all coming from the giant fields given to it in 1979. This puts it in the top dozen oil companies[9].

Arve Johnson, the aggressive Chairman, was thrown out when what were described as serious budgetary over-runs

Fig. 12-9. Statoil's character may be reflected in its architecture: large and square.

Photo courtesy of Statoil

on refurbishing the Mongstad Refinery were discovered. He was replaced by a young economist named Harald Norvik, who seems to want to run the Company as an efficient enterprise like any other, which of course it is not.

In 1987, came the penultimate throw of the dice, when the government decided to open up the huge Barents Sea, far within the Arctic Circle off northern Norway. It had a disputed border with the Soviet Union, which may have been one of the reasons for establishing a presence, especially as some discoveries had been made on the Russian side.

It was obvious from the first surveys that this was an entirely new province having virtually nothing in common with the North Sea. In particular, the well-known Upper Jurassic source-rock which made the North Sea work, appeared too shallow to generate oil[10]. It was necessary to hope for some older source to come to the rescue. The geological masseurs in their parlours soon went to work and came up with hypotheses that a Triassic organic clay in Svalbard might do the trick, or that very deep seated Devonian or Silurian rocks, which generate in Russia, might somehow not have been as burnt out as their great depth of burial suggested. In structural terms, there were some huge structures that deserved testing, even if the chance of success was remote.

It was already remarkable that the Almighty should have granted even God-fearing Norway one trend as prolific as the North Sea: it was too much to expect a second.

The environmentalists and fishermen, who always oppose the opening of new areas, were again against it. The government found a solution by classifying the exploration as a form of "research" which is a national passion. Several licences were let in this and subsequent rounds, and the data were shared so that everyone could know the outcome. That was not good. It appears that the whole region was uplifted by about 2000 m on the melting of an ice-cap in the Oligocene, about 25 million years ago. The resulting pressure differential caused the seals above the reservoirs to leak: in the same way as a fish brought from great depth almost explodes at the surface. There were numerous sniffs

of gas, but not much sign of oil, and there were no accumulations of sufficient size to warrant attention in this very remote area. In short, it was a bust. Again, the wisdom of bringing the foreign companies into it at all is to be questioned. The three national companies could equally have done the "research" at a slower pace. If they had held the negative results confidential, they could have perhaps farmed out their rights to uninformed foreign companies when they realized that it was no longer promising. If necessary, they could have offered new terms whereby the foreign companies spent money that was not deductible from Norwegian tax. Most such money could in fact be claimed as a cost against tax in the foreign companies' home countries.

Exploration continued in the North Sea, and a number of modest finds were made, without there being any particular surprises.

In 1995, they cast the last throw of the dice by opening the Voering Plateau, a deepwater area, north of Haltenbanken. The most obvious interpretation is that a thick sequence of Cretaceous shales rests directly on oceanic crust, in which case the essential Upper Jurassic source-rocks will be lacking. If so, it will be another bust. However, the imaginative geologists have managed to propose that some Upper Jurassic might have been preserved on very tenuous seismic evidence. If that should be confirmed, and if its organic carbon content has not been burnt out by deep burial, there is just a chance that some of the large structures could be filled with oil or gas[11]. It is certainly worth throwing this last dice to find out, however high the odds. The issue is: with whose money?

Figure 12-10 shows Norway's estimated endowment of oil; and Figure 12-11 shows the production profile based on these numbers. It shows that production will decline rapidly once past peak.

They have now culled the Norwegian shelf sixteen times with successive licence rounds, most of which were aimed to test the more promising prospects identified at the time. It stands to reason that there cannot be much left except smaller finds within known provinces. There is just a chance some new deepwater areas or sectors within the huge Barents Sea that for some reason escaped the regional effects that destroyed the prospects but the odds have lengthened considerably[12].

Norway	Gb
Produced	9.4
Reserves	14.6
Discovered-to-date	24.1
Yet-to-find	2.9
Yet-to-produce	17.6
Ultimate	27
Left after 2050	0.2

Fig. 12-10. Norway's oil endowment (1996).

NORWAY
Ultimate 27 Gb

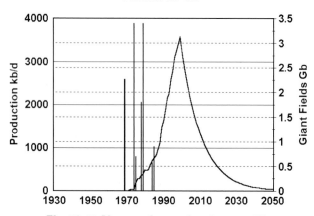

Fig. 12-11. Norway's production profile.

It is more difficult to assess its gas resources. As discussed in earlier chapters, gas source-rocks are more widely distributed in nature than are oil source-rocks, but gas needs a stronger seal to hold it in the reservoir. The production profile is also very different, being much more influenced by economics than is the case with oil. It is difficult to draw the economic thresholds. About 115 Tcf of gas has been discovered so far in Norway, which is equivalent to about 11 billion barrels of oil. In other words, about one-third of Norway's hydrocarbon resource is gas, but of this more than half lies in the super-giant Troll Field. I could imagine that another 10-20 Tcf might be found over the next twenty years, and some of the remote fields in the far north might become viable to produce by then. Gas prices are likely to rise in sympathy with oil prices, and Norway's gas is undoubtedly a major asset for the future. It will to some extent take the place of oil as its production declines. Managing the gas business is an ideal role for Statoil.

Why again is this chapter entitled: the finger points at Norway? It is because Norway developed its oil too quickly both for its own and the World's good. Production has risen rapidly in the past few years as the earlier discoveries reached peak. Norway with its small population could not absorb the oil, and therefore exported it. Today, Norway exports 2.8 Mb/d, 93% of its production, more than Kuwait. The North Sea contributed greatly to the so-called "glut" of 1986, when the share from the Middle East swing producers sank to 16% and prices collapsed, as I discussed in earlier chapters. The United Kingdom was equally to blame but had more of an excuse in that its 60 million inhabitants needed the oil. The OPEC producers tried to stiffen prices by cutting their production, however ineffectually. Norway and Britain pumped like there was no tomorrow. OPEC is not an altruistic organization, and its idea of increasing prices had no more than its own benefit in mind; but had it succeeded it would have curbed demand and thereby given the World a little precious breathing space to adjust to the coming shortfall. Norway and Britain, in ignorance of the world resource situation, may have thought they were contributing usefully by undermining the price of oil. But if so, it was at the expense of their future. It is more likely that they did not know what they were doing in terms of resource depletion.

State companies are being privatized everywhere outside the Middle East. It is claimed that this will improve efficiency. Perhaps it will. But these concerns are being returned to private hands when there is little more oil to find around the world. To privatize Statoil would be absolutely the worst thing to do: it would simply transfer to foreign investors the enormous and undervalued assets of the giant fields found in the early 1980s. They were a free gift by the taxpayer of some $100 billion[13] and should remain in national hands. It is at the same time absurd to allow Statoil to pretend to invest the Norwegian tax-payers' money overseas in new exploration when there is so little left to find. It will never make money doing that. It is not that Statoil does a bad job: just the wrong job.

Having licensed large interests in its fields to foreign companies, Norway is contractually bound to allow them to be exploited as soon as possible and as quickly as possible, barring some national emergency when it has powers to restrict production. Having gone to enormous pains to control the discovery of its oil, it effectively gave up control of the exploitation of this precious resource. It would have been wiser to have done it the other way round: to be lax in finding it and strict in producing it.

Statoil had privileged access to the prime positions, which was a benefit far exceeding the advantages of its favoured contract terms, but instead of delivering the proceeds to the widows and orphans in the fjords, it is bent on a programme of global expansion. The eleven thousand employees are well paid and want to continue that way. Nothing suits them better than global travel and all sorts of exciting experiences on the shores of the Caspian. It is impossible to go to a conference anywhere in the world and not find a Statoil man pronouncing on one subject or another. The management too can swell their breasts with the thought of their new global stature forgetting that their company's income owed little if anything to their endeavours, the assets having been a free gift from the taxpayer.

Norway has of course enjoyed the wealth that peak oil production brings. In the 1980s, Stavanger was a quiet place on a Saturday afternoon. It would be unkind to describe it as a new Babylon, even though beautiful, fresh-faced, teenage girls dismount from the pillion seats of motorcycles at midnight to head into the countless dives that surround the harbour. Some later become single-parents to be well husbanded by the State. In Oslo, a breed of clipped-speaking Yuppies enjoy the affluent life. Crime, only recently almost unknown, is prospering too: the country retreats are burgled and the yachts are stolen. It is becoming less of a fairy-land[14].

All of this is an understandable consequence of a heady flow of easy oil money, but it is a party that is nearly over, as oil production peaks, and heads sharply into decline. There is not much time to prepare for a new cold northern wind.

NEW POLICY OPTIONS

One can think of a few easy policy decisions that could be considered to save what is left.

1. Change the licensing system, such that the national companies are given exclusive access to any remaining prime tracts;

2. Disallow the deduction by foreign companies of exploration and related expenses from taxable income.

3. Open highly speculative areas to foreign companies to explore with their own money (itself mainly deductible in their home countries), accepting that they will be free to produce what they find. To attract them, it would be necessary to greatly ameliorate the terms. It might be enough to rely on royalty alone against which nothing can be deducted; scrapping tax on income altogether. It is a notional income because little is likely to be found. There is a big difference between looking for oil and finding it.

4. Stiffen, wherever possible and mainly in the area of deductible allowances, the taxes paid by foreign companies on existing production, offering at the same time to have Statoil buy these interests from willing sellers, no longer liking the new deal. The international companies are forced to have a short-term view of the future as their shares are traded on the World's markets, and might sell out for a quick buck. Several believe their own underestimates of future oil price rises.

5. Prohibit Statoil's overseas adventures, and trim it down to perform an efficient operator function on existing fields and those to be acquired in Norway.

6. Strictly control Statoil's budget and demand the highest possible dividend.

In fact, the State has already moved some way in this direction, as in latter years it has retained a direct economic interest in the oilfields it licensed, so as to recover some of its control. This economic interest is, however, administered by Statoil, no doubt at a direct and indirect fee and in a way advantageous to Statoil. A return has yet to come from the investment: it is not a good compromise. The State, as a shareholder, should aim Statoil to protect the national interest as part of a longer term policy, and not imagine that it will somehow grow to be another global "Shell", when the Shells of this world are heading for decline along with oil production.

Having recovered control of its oil and gas, Norway should do everything possible to slow the rate of production to make the resource last as long as possible. The Norwegian Petroleum Directorate already makes detailed resource estimates, which although I think over estimate the undiscovered potential, are valuable and no doubt can be revised as new knowledge comes in[15]. By recognizing that it is a finite resource, Norway should make a major effort to husband the proceeds for the days that are to come. An oil fund is already in place to build up a portfolio of overseas investments, in the same way as Kuwait has done. It is an excellent idea, but needs to be given every encouragement.

In addition, Norway is ideally placed both physically and mentally to dedicate its highly privileged oil income to research into all forms of renewable alternative energy. If the scenarios herein depicted are anywhere near correct, there is a growing future for such new forms of energy: from the windmill on the exposed headland; to the tidal and wave power plants using the energy released by the Atlantic rollers crashing against the rocky cliffs or on dedicated offshore structures; and to the lightweight hydrogen-powered or photo-voltaic vehicle. More than anything else, Norway could pioneer social engineering in tele-commuting; tele-cottaging; job-sharing; more simple non-consumeristic pleasures for less drudgery at the office; improved life styles. It is called voluntary simplicity: a subject now being studied at Oxford[16]. Norway is already more than half way there with its highly regulated society and the deeply held belief that life is for living rather than mindless consumerism. It could in fact set an example by preserving what it has.

Norway in recent years has excelled in many areas: its footballers win matches; its negotiators bring accord to the Israelis and Palestinians; its singers win contests; its skiers win slaloms, and its walkers win marathons. It has also done a marvellous job in conserving its fish stocks by adopting the simple but sensible policy of restricting the number of fishermen and keeping their boats small. In the 1980s, cod was almost wiped out by over fishing but now has recovered[17].

It is well set on the path to excellence but its greatest challenge is about to come. The World could learn a great deal both from how Norway harbors its oil and gas, and how wisely it uses the proceeds of its windfall. It was a windfall in the form of just a few hundred metres of Upper Jurassic Clay deposited 150 million years ago.

A POSITIVE POSTSCRIPT

There are some very promising recent developments. The oil fund has grown enormously with the higher oil prices, and is now the direct responsibility of a minister. It is having the attention it deserves, as the best hope of saving something for the future from the peak of oil production.

A new electric car, called the City Bee, has been designed and is beginning production[18]. As proposed above, it is exactly where Norway's research should be aimed.

New oil licensing terms have been introduced whereby the companies have the right to withdraw without drilling if the surveys are negative. It is at least a step in the right direction: reducing direct government interference in the conduct of exploration.

Statoil too is withdrawing from blocks whose prospects have been downgraded, most of which it would never have entered in the first place but for government policy.

There is much more to do, but these are promising and very important steps. Once the resource constraints are better understood, I am confident that the government will react appropriately. They may not understand yet but I hope that they soon will, and then the finger will no longer point at Norway.

DR. LEIK WOIE
The Preventive Doctor

Fig. 12-12. Dr Leik Woie, a Norwegian cardiologist interested in preventive medicine.

Q Leik: it may surprise you to be interviewed about oil. There are many oil experts in Norway, so why should I talk to you about your country's oil policy? I will answer that question myself: it is because it is not a technical or even an economic issue, but one of national interest on which your views are as valid as any other thinking citizen. Tell me how you came to be who you are?

A *That is impossible: no one either knows who they are or how they become who they are. But I have had certain experiences in the national life that make me very conscious of government policy and its effects. I have worked, as you know, in the health service.*

Q I can understand: as a doctor you face life and death decisions in your daily work; you have judgments and tests to make; but at the end of the day you have to rely on a sort of instinct to know what is best for your patient. Doctors are sometimes accused of being little more than the sales representatives of the drug companies, but with your interest in preventive medicine, you are clearly not guilty of that.

A *It is true, the phenomenal progress of modern drugs means that there is always something to prescribe, often at great expense. Long ago, I was shocked to treat young patients suffering from heart disease. I came to realise that in many cases their condition was almost self-induced by them having lived in the wrong way without knowing it. I began to see that preventive medicine would not only bring enormous benefits to many people, but would be very cost effective for the government. I tried to influence the politicians to move in this direction, but with only limited success. It is easier for ministries to pay drug bills on prescription than take the more difficult decisions that arise from preventive medicine. It is easier for them to say "you have to be sick before you are treated" whereas I say "you have to be treated to avoid becoming sick"*

Q I have often compared explorers with doctors. We use seismic surveys as you use X-rays; we do geochemistry in place of blood tests, and we use the drill instead of the catheter to find out what is buried under the seabed. We can make our scientific evaluations, but after years of experience, we develop an intuitive nose by which to get the diagnosis right. Does that make sense?

A I suppose it does, although I had not thought of it that way. To continue the analogy I can see that you begin to see relationships and patterns in the size of discoveries and the timing, which almost fit statistical theory. In medicine, we do exactly the same: we need to know the statistical chances but we need to know when to override the statistics and apply our judgment. Sometimes we save the patient in this way, but sometimes we fail, like you drill a dry hole. We may fail, but often that is no surprise. We knew that it is an almost hopeless case, and we are simply doing the best we can without much hope of success. We also know of near miraculous cures, which come about for reasons we don't understand. I suppose you also come upon the surprise good discovery.

Q Yes, there often are surprises, sometimes even good ones. But our patients are getting old: the oil basins on which we work are mature and over-mature: the hopes of giving them the elixir of life or the recipe for eternal youth have faded. We are dealing with an oil life-span, in much the same way as you deal with a human life-span. What about government policy?

A I certainly see the parallels. The demands on the health service must inexorably rise as the population ages. We need a realistic policy. Think of government tobacco policy. Tobacco should be banned on health grounds, yet those who smoke say that the tax they pay more than covers their health costs. Nothing is straightforward.

Q As an ordinary citizen what do you think about Norway's oil experience?

A First, I think it was a miracle that we should have been so blessed by having such a valuable resource. I think that the management of such a huge and important resource and benefit is a very serious thing. I think the earlier policy of a slow development to make it last as long as possible was the right one.

As a physician who cares about preventive medicine, I would like to be sure that my country does not suffer from a hangover and indigestion from the oil-feast we are enjoying. I am concerned that my grandchildren should think well of me in later years. I would prefer them to think of me as someone who stood out against the mindless consumption of their birthright rather than as someone who condoned the outrageous plunder of the resource.

It has come as a surprise to me that well informed oil people can still say that the resource is so large that it can be treated as infinite, when all the evidence indicates the contrary. The fact that Norway's larger fields were found first confirms the proposition even here. In medicine we use basic and fundamental research to answer questions. The same approach should be applied to assessing the world's oil resources: I am not impressed by the make-believe whether put out by politicians, economists, journalists, engineers or doctors

I am shocked when I hear that the people in charge fail to understand elementary arithmetic: production obviously eats into the reserves, and simple statistics show that the rate of discovery is falling. It is like drilling into a Swiss cheese, even without the advantage of seismic surveys, you will always hit the largest holes first.

Q What do you think about the role of Statoil? This huge enterprise was given free access to prime rights which belonged to the nation. Had the same rights been given to foreign companies, the Norwegian people would have received directly their rewards at a very high rate of tax. When the job was done, the foreign company would have left. But Statoil gets ever larger and is now taking the Norwegian taxpayers' investment into foreign areas where it no longer has the privilege it enjoys at home. Do you think the Norwegian taxpayer knows and agrees with what is being done?

A I think that most Norwegian taxpayers are satisfied with the work of Statoil, but I am afraid that he does not understand his indirect investment in Statoil. Drilling for oil in an unknown province is a risky business, and according to your research there is not much left to find. I also read in the newspapers that the findings so far in 1996 have been very disappointing. I am therefore sceptical that Statoil or other companies will succeed in finding big reserves in the Norwegian sector or overseas. The Barents Sea drilling has been an expensive experience.

As far as I understand, Statoil's technical achievements are excellent. Like all Norwegians I am a shareholder in Statoil and I would like to see my company dedicate itself to Norway's long term interest by concentrating on the Norwegian shelf. Can they extract more oil from the wells? Can they make marginal fields more profitable? If it is possible, they should also buy the interests now held by foreign companies in the existing fields, which are surely more profitable than anything likely to be found in the future. I do not think there will be another Statfjord.

I also believe very strongly that we in Norway must

realize that the oil income will not last forever. It may all be gone in one generation. We should be doing our best to use this asset to build a strong future in an age when energy is expensive throughout the world. There are all sorts of interesting things we could do, and we are exceptionally privileged both in our social structure and natural resources. We have a tradition in electrical power, which will come again into its own as the prime energy carrier. We should be using our present advantages to lead in developing things like the electric car. Above all, we need to prepare sensibly for the future, rather than live only for today when oil pays our bills.

I have looked into the Norwegian situation in some depth, partly because I know it well. Every country faces, and has faced, a dilemma in controlling its oil industry. It is different from most businesses because it is concerned with depleting a finite resource, which once gone is gone forever. It is also different because it has been difficult to tell in advance whether the resource exists or in what quantity.

Countries have to set terms to attract the companies to explore, and they are at least morally obliged to respect the terms if the companies should be rewarded with success. But whereas contract terms should be inviolable, all governments retain the right to change the tax regime, even retroactively.

Norway began with normal terms and tax that was an equitable basis for both parties at the beginning. Then when it became evident that a trend of major finds on the United Kingdom shelf was almost certain to extend into Norwegian waters, it imposed draconian terms that the industry accepted only because their imposition coincided with a time of world oil crisis. High taxes were imposed but expenses were allowed as a deduction. It led to distortions that are always associated with high taxes. Never were more Rolls Royces to be seen in the streets of London than when the socialists had an 80% marginal tax rate in the 1970s.

The creation of Statoil was a concealed tax-driven construction: Parliament would never have agreed to pour such colossal sums into the thing. As I have already emphasised several times, it was achieved indirectly by having the foreign companies carry its costs, but to deduct same against Norwegian tax. It sounds like a socialist conspiracy, but perhaps it just happened that way.

Britain had the same earlier experience. It opened the shelf to the companies, some working with state enterprises, like the Gas Council and the National Coal Board, who paid their way. The socialists then created BNOC and clawed back rights previously freely negotiated with the companies to give to it. Mrs Thatcher swept all that away. The oil interests of the Gas Council were floated off to become Enterprise, and BP bought BNOC. Everyone was thinking about how to share the oil profits: no one was thinking of depleting a finite resource or the next generation.

Petroleum Revenue Tax in Britain continued to effectively subsidize exploration by a free wheeling entrepreneurial industry made up both of major international companies and as many new small British companies, floated on the stock market. Some of them bought producing properties simply to acquire the benefit of paying PRT and enjoying the deductions that came with it. But eventually in 1994, the government closed that facility, when it realized that exploration was no longer yielding new revenue producing projects sufficient to justify the allowance. The tax deduction was in effect a form of national investment.

While some momentum continues, the removal of PRT probably heralds the end of the exploration business in Britain, and rightly so, since it has found pretty much all there is to find.

Many other countries develop their oil by Production Sharing Contracts. There are several variants but the essential formula provides that the State Company holds the rights and the foreign company takes the exploration risk, but can recover expenses from a certain percentage of the production. The split of the oil after recovery may be as high as 85:15. It comes to much the same thing at the end of the day, but is politically more attractive to the host country, as the foreign company is seen more as a contractor than as an owner of national resources. The facility to recover expenses through a share of production provides the same hidden subsidy for exploration and other indirect costs, as does tax, once production is established.

When the King of Spain let rights to his *conquistadores* to exploit the Spanish Empire, the tenancy lasted for a specific period of time. The Spanish currency was then divided in units of eight: "Pieces of Eight" as the pirates used to say. One of the eight was for the King: a $12\frac{1}{2}\%$ royalty. It was found to be a workable tax, which has been inherited by many oil contracts. In some ways, it is the best arrangement as it is not subject to distortion.

The whole issue of licensing terms will no doubt have to be reviewed throughout the world when supply shortages appear. The immediate response to shortage will be to stimulate further exploration, probably with all sorts of tax driven inducements. There may be another brief golden age of exploration. But if it does not deliver, what will they do then? Will they pump good money after bad? Or will they belatedly come to realize that it is a finite resource, and invest instead in finding ways to use less.

NOTES
(For references, see Bibliography)

1. Norway did well out of the settlement of the median line on geographic criteria. Had the boundary been drawn on the topography of the seabed, the deep Norwegian Trench would have excluded Norway from the prospective part of the North Sea. The Germans came out of the arrangement particularly badly.

2. A block is a designated area of the seabed bounded by 20′ of longitude and 15′ of latitude (about 525 km^2), and a licence conveys the right to explore for and produce petroleum under certain conditions.

3. See Kvendseth, S.S., 1988, for an excellent account of the story of Ekofisk.

4. The Jurassic Period occurred between 195 to 140 million year ago. The North Atlantic had not then opened, but a rift system, analogous to the present Red Sea, was beginning to form. Rivers flowed northwards into the Arctic Ocean, as a sort of precursor of the Rhine, and deposited huge deltas. The climate was warmer than today.

5. Companies tend to explore in packs, holding "undivided" rights to a concession. One of the companies is appointed "Operator" to manage the operation; the others being termed non-operators. Each votes his participating interest on important decisions. The arrangement gives rise to endless conflict. The Operator tends to think of his partners as worthless free-riders; whereas the non-operators tend to think of their Operator as a delinquent, overcharging on overhead, not giving their particular venture the priority it deserves, and being guilty of general ineptitude (see comments of Hardman p. 132).

6. A Round of Licensing is the term used to describe the periodic offering by the government of licences to explore and produce petroleum.

7. See Nerheim, G. and B. Utne, 1992, for a valuable description of the history of Norwegian shipping and its move into drilling.

8. It ran into difficulties in the mid 1980s, when oil prices collapsed, and was bought by the excellent French contractor, Schlumberger.

9. *Oil & Gas Journal,* 4.9.95.

10. Although it *appeared* too shallow, it may in fact have been too deep, prior to the 2000m of regional uplift that occurred in the Oligocene, which was not recognized until after drilling. It would explain the preponderance of gas. The remaining hopes are to find some part of the area that was not affected by this uplift for whatever local reason. Like a chicken, exploration often twitches after death.

11. There is a slim hope that an early structure might have been charged to be a form of "holding tank" for re-migration into later structures, as is the case West of the Shetlands. Such a prospect depends on the right timing of geological events which are rare and unpredictable. That it is very high risk cannot be denied.

12. The Norwegian Petroleum Directorate has recently (Feb. 1997) announced an implausible increase in reserves to 40.9 Gb for oil and condensate and 38 Gb for gas, and an undiscovered potential of 22 Gboe of oil and gas. The exact explanations need to be evaluated.

13. Assuming that about half of Norway's 17 billion barrels of oil belongs to Statoil, and is worth about \$15/b after necessary, direct and genuine production cost.

14. The Economist has reported that the world's highest per capita purchase of canned music is in Norway. Not all of it is Mozart.

15. See Brekke, H. and J-E. Kalheim, 1996. I believe the reason for the overestimation is the practice of risking every prospect independently. Every prospect will be perceived to have some chance, however high the risk, and the risked reserves of the many notional prospects on this huge shelf add up. In practice, the appraiser will rarely attribute a zero probability to a prospect although, in reality, prospects either succeed or fail; and huge areas may be entirely barren for the reasons discussed earlier.

16. By the Oxford Centre for the Environment, Ethics and Society at Mansfield College: a valuable initiative.

17. See *The Economist,* 19 October 1996 p.39.

18. See *The European Magazine* 5-11 December 1996.

Chapter 13

RENEWABLE ENERGY AND RESTRAINT

In the earlier chapters, I have explained how oil production is set to decline and become much more expensive. For many people, this sounds like bad news: a sort of doomsday message. There can, indeed, be no doubt that the transition to a new low-energy world will be difficult, perhaps very difficult. But it is not the end of the line. Almost by definition, Man will go on and find solutions. As often happens, there may even be a silver lining, such that many aspects of life may improve.

The economists like to say that if you want more oil, drill more wells. It is quite possible that the coming crisis will indeed spawn another exploration boom, as people in ignorance of the underlying resource constraints continue to live in the past and hope that what worked before will work again. It won't. Instead, some new thinking has to come into play.

There are two obvious solutions: use less energy and find alternative renewable sources. In fact, there is a more important third solution of better using what remains. I think the least promising approach is to try to find alternative energy sources to maintain the *status quo*. I do not believe that they can ever be a substitute for the fuels we use in such a profligate way today. That is not to say that they are irrelevant or unimportant: quite the contrary. They can contribute very usefully, and in some circumstances can make a great impact. I will summarise what they are. Clearly each approach has its own best application. It seems to me that alternative renewable energy and the idea of using less are complimentary processes that have to be considered together.

I am thinking mainly of *conventional oil,* so in this context alternative could include gas, *non-conventional* oil, nuclear power, coal as well as the more restricted forms of alternative energy such as solar power.

I am no expert in this subject but will try to imagine some scenarios and comment upon them from this vantage point.

FOOD

Without food, Man dies, so that has to be the priority. The first issue here is: how many mouths are there to feed. Human population has been growing since prehistoric times, reaching about one billion by 1850 when oil production began. It then expanded six-fold to present times in parallel with the growth in the production of oil which contributed greatly to ever more energy-intensive agriculture. It is difficult to avoid the conclusion that population will also peak soon after oil does so, although the mechanism by which that will be achieved does not bear thinking about. Some forecasts that ignore the oil factor project that population may continue to expand throughout the next Century[1]. That seems most unlikely.

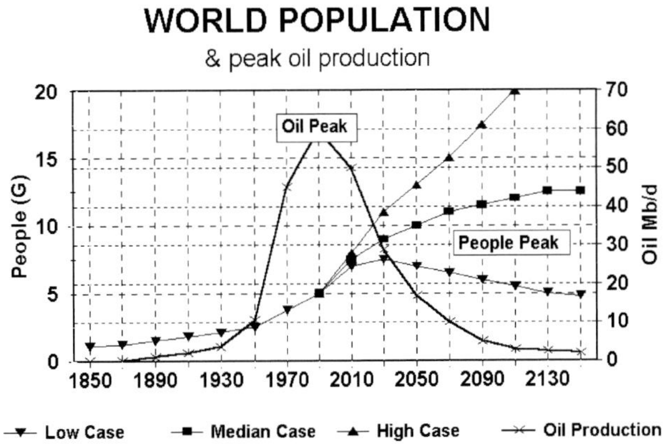

WORLD POPULATION
& peak oil production

Fig. 13-1. Population forecasts; peak oil precedes peak people. The low case seems to be not low enough.

Reproductive rates throughout the world have however been falling, some faster than others. Worldwide, the level is about three. In Western Europe, it has already declined from 2.8 in the early 1950s to 1.7, less than the 2.0 needed to maintain parity. But life-span has lengthened in parallel, so that the falling birth-rate has been compensated for by increased longevity. Longevity is, however, likely to stabilize, and the falling birth-rate will accordingly flow through to a declining population. Life expectancy in developed countries is now 71 for males and 79 for females compared with respectively 52 and 54 in undeveloped countries. A most remarkable decline has been in Russia where male life expectancy has fallen from 65 in 1987 to 57 in 1994, due apparently to alcoholism and environment factors.

Generally, reproductive rates in the Third World are much higher. However, in some countries, such as Colombia, they are already falling as people gradually come to prefer to have fewer children as social conditions change. The urbanization of society probably helps the move in this direction: children no longer being cheap labour to run the farm. In other countries, as in most of Africa, conditions are much worse, and declining birth-rate may be due more to pestilence, war and starvation. Longevity here too is likely to decline, so that a falling birth-rate will have a quicker impact.

Population disparities, however, raise the issue of migration. Already, large numbers of North Africans are crossing the Mediterranean in small boats to arrive as illegal immigrants on Spain's beaches. In 1996, a church in Paris was full of illegal African immigrants on hunger strike rather than accept repatriation. Other hopeful migrants are assembling in Poland at Europe's backdoor.

Most of the industrial countries, except Japan, built enormous long-term problems for themselves by admitting large numbers of *gast-arbeider:* Mexicans and Puerto Ricans in America; West Indians and Pakistanis in Britain; Algerians in France; and Turks in Germany. They proved themselves to be hard-working and useful members of society in an expanding economy fed by cheap oil, but they cease to be needed as they age and multiply in a stagnant and declining economy. Since they have no real roots where they live, they tend to degenerate into urban ghettos, living on crime and social security. They started out as fine people, but not all of them ended that way for no fault of their own.

But, by and large, migration is likely to be curtailed and controlled, especially as the global market declines as a consequence of the oil crisis, and as the world becomes a more protectionist place when nations break up into smaller regional units. The violent Basques and Ulstermen may already reflect a deeper, otherwise so far passive, trend in this direction. It is in fact enshrined in the Maastricht Treaty, which says that no decision shall be taken at any level higher than it has to be. The Mayor of Milhac, the village where we live, already has authority to grant us permission to live in his commune, and thereby France, for five years.

So, I think that populations in Western Europe, and the developed world generally, are likely to fall, probably by about twenty percent within a Century. If so, they would have more than enough agricultural capacity to feed themselves. In terms of energy, precious oil supplies would have to be dedicated to agriculture, probably by means of tax allowance on what will then be a high normal cost of this fuel. In addition, successful efforts could be made to reduce the energy content of food by a reduction of packaging and the production of frozen foods generally. The present system of high energy production with intensive use of fertilizers coupled with set-aside, by which subsidies are paid to take land out of production, would have to be reversed. Lower yields, consuming less energy

from more land, would be far preferable. The import of fresh foods by air would cease, and there would be more seasonal variety. More soup would be served.

TRANSPORT

Transport relies heavily on oil: indeed in the United States more than sixty percent of the oil used in the country is for transport[2]. It is here that the pending shock will be most severely felt.

There is however huge scope for savings in innumerable ways. To visit a large city, such as London, is to observe a scandalous waste of energy, with streets full of near stationary traffic burning up fuel at a prodigious rate in the most inefficient and polluting manner imaginable. That obviously has to end. If town centres were closed to private motoring, there would be room for efficient modern tramways and tracks for bicycles, which could become the normal mode of travel over the short distances involved. In Oxford, they have already introduced the rickshaw.

Private motoring is today heavily subsidized by the free marginal use of roads and parking, as well as by the widespread provision of company cars, whose cost is taken as an expense against corporate taxable income.

I remember when I was employed in London in 1979, I was given a large company car of which I had very little need. If I filled it up at a designated garage, I could have an unlimited supply of free gasoline, and I had a credit card for fuel on journeys farther away, which was however treated as a taxable benefit. It was an absurd arrangement, involving much unnecessary administration, all treated as a deductible operating expense. I learn that employees with company cars drive from London to meetings in Scotland in preference to going by train so as to increase their company mileage that increases their tax free personal allowance. It is madness.

Cars should be cheap to own, but the fuel should be expensive to encourage minimal usage and the development of more efficient vehicles. I drive a small Renault Clio, which is more than adequate for all normal needs, yet the roads are choked with enormous cars which must be more of a status symbol than an essential means of transport. A change of attitude could make enormous fuel savings at no hardship. High fuel costs would impose difficulties in relatively underpopulated rural areas, where people depend on cars, but that could be solved by appropriate allowances for those in need. The idea of the electric car is probably a good one, although electricity generation and transmission is itself not a very efficient process. Despite this drawback, it can in fact be more efficient than the internal combustion engine, which is in practice often not running at an optimal speed. The solar electric car is already a technical possibility, but it needs mass production to become economically competitive with existing vehicles. I suspect that there is much scope for improving engine design to reduce consumption. I read only the other day about a novel design by a Mr Negre in France[3]. It has three separate cylinders for respectively compressing the air-fuel mixture, exploding it and evacuating it, which are done in a single cylinder in the traditional four-stroke engine. It results in an efficiency improvement of 50%. Furthermore, it can run partly or exclusively on compressed air: fifteen litres being enough to drive it for one hour. The compressed air is partly generated by braking. Going back to the concept of energy concentration discussed in Chapter 1, one could imagine many ways of achieving low-level air compression: for example, by large numbers of small windmills running day and night. Perhaps useful energy could be concentrated in this way for use in novel new engines that may be developed. There is plenty of scope for inventiveness.

The railway network could be rediscovered and revitalized in developed countries. It should be used wherever possible to replace long-range trucking. The economics of building roads at public expense which subsidizes motoring needs to be re-evaluated in relation to railways. The economics should include a full evaluation of all the costs, including irreplaceable energy costs.

The airline business would go into near extinction. Very few people actually need to travel by air. Modern communications make most business travel unnecessary. It is also presently subsidized as a deductible operating expense. Tourism too would suffer.

Not much can be done to save marine bunkers for shipping, except by the reduction of world trade. The sailing ship to modern design may return.

The Third World will of course be the hardest hit, as they depend heavily on road transport, much in a dilapidated condition and very inefficient. It is difficult to envisage any solution for this problem.

ELECTRICITY

Electricity is a most convenient source of energy for lighting and operating power tools but it does suffer from huge losses in generation and transmission.

Much can be done to save electricity and to restrict its use to the purposes for which it is particularly suitable. There is great scope for savings. The widespread installation of modern lamp bulbs which use a fraction of the electricity consumed by traditional incandescent bulbs could make an enormous impact. Inverted tariffs could encourage more careful usage. I have worked in offices where the lighting was left on 24 hours a day because it prolonged the life of the bulbs, which cost more to replace than the fuel they used. People should walk up stairs rather than take the lift: it might also help their health.

Various alternative energy initiatives could contribute greatly: improved insulation; solar water and space heating; and even wider use of geothermal heat are obvious examples. Air conditioning and space heating are commonly excessive. I once worked for a company that offered its staff a pullover in preference to turning up the heat, although in this case the move was motivated by parsimony rather than energy saving.

Obviously electricity generation from oil needs to be phased out as soon as possible, to be replaced by nuclear, coal and natural gas fuels. Wind, tidal, wave, hydro and solar generation should be encouraged wherever possible.

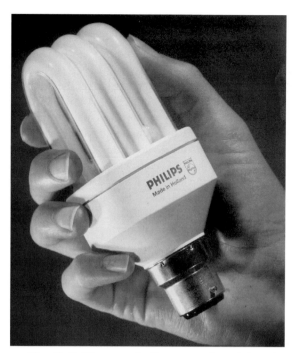

Fig. 13-2. The new bulb saves electricity.

Photo courtesy of Phillips Lighting

Again, the greatest difficulties are in the Third World. Half the world's population still has no access to modern energy, and their needs are increasing, quite apart from the growth in population itself. Much of the electricity as is available is generated from oil, and is widely subsidized already. The coming oil crisis will be a crippling blow in this area, especially in the sprawling urban centres of population which get larger every day. It seems a hopeless situation.

There is more hope in rural circumstances, where small scale alternative energy projects can have wide application. Even such simple steps as improving the efficiency of cooking stoves can have a great impact[4].

Writing about energy shortages is depressing: it seems so hopeless. But it may not be as bad as it seems. I am not after all speaking about the end of oil production for a very long time, nor is it going to get much more expensive to actually produce for a long time, even if it costs much more to buy. So, there is some breathing space, and the anticipated radical increases in price will concentrate the mind wonderfully.

Savings in energy usage have to be made, but since we are now so profligate they can probably be achieved relatively painlessly. It is a case of aiming in a new direction, rather than having to take any extreme measures. There is still some time to make the adjustment although not too much.

The most effective weapon is the tax system, which at present is inefficient and commonly gives unintended benefits. I have already discussed the hidden subsidies on exploration that are provided by allowing the expense to be deducted against high marginal tax rates. It is the same in many other parts of the economy. Business expenses, however incurred, are a form of subsidy that distorts the reality of the costs. One way to solve that would be to remove corporate tax altogether so that there would be nothing to deduct expenses from. Barker[5] of the prestigious Cambridge Econometrics has already shown that increasing fuel taxes while at the same time reducing national insurance contributions (a tax on employment) is a viable option for the United Kingdom. At the very least, the general principle that the real costs of doing things are properly metered and allocated should be observed more effectively. The environmentalists have already been pressing for changes in this direction, such that the cost of pollution should be charged to those who cause it[6],

including the consumer. It should be the same with energy use. Those who use it should pay for it in full. As Timson points[7] out, market forces aimed at reducing the cost of energy do not encourage energy efficiency.

I also think that the tax system should be used to impose shareholder loyalty by penalizing short-term trading. It goes to the heart of the matter. Ownership implies responsibility and a view of the future. World trade driven by a shiftless investment community, with thoughts of nothing but a quick buck, is a disastrous formula, encouraging the management excesses and abuses, which have become universal, and general feckless behaviour. So far, cheap oil has paid for it all, but not for much longer.

RON SWENSON
A Californian Expert on Renewable Energy

Fig. 13-3. Ron Swenson.

Q Ron: we have been in touch for some time. Obviously, the scope for alternative, or renewable, energy as it should better be called, depends heavily on the price and availability of conventional fuels. But first, how did you become interested in the subject?

A I had an experience in High School which showed me that the big defence contractors were operating a great boondoggle, so when I went to university, I began searching for solutions that contributed to "living-ry" not "weapon-ry". Then, from 1965 to 1968, I taught Cybernetic Systems at San Jose State University. For two months, Bucky Fuller was a visiting professor in one of the courses I taught, and he brought me to see the view that humanity can learn to do more with less, and thus discover that there is enough for everyone.

Q From what you say, renewable energy is already a viable option which in several areas successfully competes with under-priced conventional fuels.

A Hydro-electric, a renewable energy source, was at the forefront of electricity generation a hundred years ago, and still provides a large percentage. Wind energy is already an economic success, typically costing less than oil, coal and nuclear energy – that is if environmental costs are included in the accounting. Solar *thermal (by the Luz Company) has had a good economic performance. These are important signposts.*

Q It seems to me that the United States faces a very difficult energy future. Already it is importing more than 50% of its oil on a trend that can only rise. Prices are going up, and there is a lot of evidence that the world will soon face another oil price shock. How will America cope?

A *The meek will inherit the Earth. An oil crisis will not affect people who have never had it, as for example, the descendants of the Incas in the highlands of Peru who still have an indigenous lifestyle. But the United States citizen, who is very spoiled and very naive about the forces which affect our future, is ill-prepared. The likely reaction will be to blame a Middle East dictator, who is perceived to be diabolical, such as Saddam Hussein or the Ayatollah Khomeini. Then people will turn on each other. In '73 and '79 people were shot at the lines in gas stations. Then the environment will be ravaged, and only then will people wake up. The cities will be the worst hit.*

Q Solutions are there, but what is lacking is the understanding and awareness. Above all, we face the problem of lead-time. An oil shock arrives overnight, but it takes ten years to build a nuclear power station, and no one wants one in his backyard. It seems to me that the emphasis today must be in terms of education and spreading information.

A *I agree. I'm putting out the word as best I know how. Surprisingly, it's just as hard to wake up people in the solar industry – they've been underdog for so long, they don't know how to assert themselves. I do exhibits, talks; and we have an extensive website on renewables [www.ecotopia.com]. We are developing alternatives in high profile projects. Because young people like cars, solar car racing is an especially good way to educate and motivate. I've been involved in such a project in Mexico for the past four years. We competed in SunRayce '95 in the USA and participated in the World Solar Challenge 1996 in Australia.*

Q Tell us more about what the practical applications are.

A *Practical renewable energy is, above all else, a function of mass production. Wind turbines are already practical in areas of consistent high winds (Patagonia, the great plains of the USA, New Zealand etc.). Solar thermal electric has proved economical at large sites in the southern California desert. Photovoltaics are now often the best choice in remote sites, and will be economic in large fields when there is serious production. The new*

technology of photovoltaics in thin film or concentration gives substantially better energy performance and economics than the older style of flat-plate PV panels.

Q It seems to me that electricity, however convenient it is, is a very wasteful use of energy due to the huge losses in generation and transmission.

A *On the contrary, electricity is the future, especially for urban transport. It's clean, and delivers more overall efficiency, because a large power plant has a much higher efficiency than an automobile engine, which in traffic is often running at much less than its optimal performance. Long distance transmission is not required for photovoltaics, so it makes a lot of sense to generate power at the load.*

Q What about the environmental impact of renewable energy?

A *For photovoltaics and solar thermal concentrators in rural or desert areas, there is the impact of extensive land use, although depending on the structural design of the frames, land may remain relatively free underneath the solar panels. Obviously, if installed in forest areas, trees would have to be felled.*

Wind generators have had some impact on raptors (eagles, hawks), which are sometimes hit by the turning turbine blades. They are also sometimes considered unsightly and noisy if placed near residential areas.

In the case of solar, there is less need for long transmission lines, which are both inefficient and unsightly, because the receptors can be installed right at the load.

Q In America the automobile is regarded as almost a national birthright. The country is built around it. Gasoline prices are already absurdly low, about one-third as much as in Europe. Some people think it unfair that America should be burning up a disproportionate share of the world's resources. It is precisely in the transport sector that the crunch will be felt most severely, yet it seems the most difficult problem for which to find a solution in terms of alternative energy. Is that so?

A *In urban areas with short distances, solar panels on electric cars, or charging them from rooftop solar panels will be quite workable. For longer distances, solar powered railroads with a third rail or catenary lines are feasible. Buses and trucks with motive power based on renewable principles will be able to handle long distances with biomass fuels or hydrogen as an intermediary.*

Q The continental climate of the US means that many places face extremes of heat and cold. That seems a particularly good area for the application of renewable energy?

A *The climate for solar energy is good around the world. Seattle has hydropower, a form of solar energy; the Sahara Desert has direct sunlight for photovoltaics; Patagonia has wind; the Philippines has the potential for sustained and managed biomass production. The United States has it all too. There are bioregional differences, and so there will be some transmission from one region to another. I can imagine a shift in the industrial base from the coal-rich eastern states to the sun-drenched Southwest. Perhaps it will be a case of Phoenix vs. Pittsburgh vying for the industrialists' investment!*

Q I suppose what is needed is a dual approach: first to use less conventional fuel so it lasts longer and second to bring in as much renewable as possible.

America is a great believer in the capitalist system and market forces, but perhaps the depletion of a resource calls for government intervention as the problem cannot be handled by market forces alone. The market lives in the short term, whereas what is needed are longer term solutions. Have you had positive political responses?

A *The political response to date has been very primitive. Environmental issues have so far attracted much more attention than the depletion of the resource. The ozone threat has so far had the best response. The US has not paid enough attention to carbon dioxide emissions, and its failure to lead has meant that places such as China or India which rely on coal for energy generation have not been adequately pressed to control emissions that jeopardize everyone's future.*

NOTES

(For references, see Bibliography)

1.   See World Resources Institute, 1996, reporting UN estimates (p. 174).

2.   See US Department of Energy, 1989.

3.   *The Economist,* 16th October 1996.

4.   Bohnet, M., 1996, explains German experience on applying alternative energy in developing countries.

5.   Barker, T., 1995, from the prestigious Cambridge Econometrics unit gives a compelling argument.

6.   Hawken, P., 1993, surprisingly an executive of an American mail order company, who you would think was wedded to consumerism, makes some thoughtful and persuasive proposals.

7.   Timson, R., 1996, explains that the increasing application of market forces to British energy supply has done little to encourage more efficient usage.

Chapter 14

THROUGH THE WRINGER: THE INDUSTRY TODAY

The changes in the resource base and in the ownership of it, which have been discussed in earlier pages, have had a profound effect upon the industry. It is in a state of flux and transition: and the future is likely to call for even more radical adjustments. The evolution affects the size of the corporate vehicles, and how they conduct their business. It mirrors the wider political and market changes that are influencing the World more rapidly than we perceive. I have already reviewed the general historical trends, and will now concentrate more on the evolving corporate structure.

Much of the industry's early experience was in the free-wheeling environment of the United States, which was characterized by a large number of small exploration companies, often family controlled. Drilling was being pursued throughout the oil lands, and a new find, whether made by luck or skill, transformed the company that made it. Having made the discovery, it had to somehow find the capital to develop it, often at times of weak oil price. The situation was conducive to the merger, as the larger established companies swallowed up their smaller brethren. The larger companies did themselves engage in exploration, but it seems that they mainly grew by acquisition.

The great Rockefeller empire was the first and perhaps the greatest manifestation of this trend. The acquisitions were commonly achieved by often hostile share purchase, in which the critical step was to obtain a controlling interest. The original owners often survived as unhappy minority interests, with positions on the Board. There was the Trust structure, which soon acquired a pejorative meaning, by which controlling interests spread unseen tentacles and what were widely seen as unfair influences. Standard Oil was broken up in 1911 for precisely these reasons by anti-trust legislation.

While it broke up Standard into thirty-seven entities, the process of consolidation continued unabated, being in turn practised by the Standard daughters. I have already described how Standard of Indiana (Amoco) acquired a controlling interest in Pan American, then a growing concern with huge finds in Venezuela and Mexico (p. 67).

Things were very different in the Eastern Hemisphere. The reason was that, whereas in the United States the mineral rights were vested with the landowner, with whom the companies had to deal under normal commercial terms, in the rest of the world the rights were held by the State. The companies did not own the mineral rights outright but secured concessions for a given duration: they were tenants not owners, which gave the affair a certain short-term character with many political implications. Some of the early concessions were for huge tracts of territory, even entire countries. As a consequence, the concessionaire companies could not grow by acquisition, but had to rely on their own efforts. This was exactly the position of BP in Iran (p. 142).

However, this general framework was modified in various ways. Two companies might join forces, such as Chevron and Texaco to form Caltex, or BP and Gulf to form the Kuwait Oil Company. Such subsidiaries tended to have independent lives, either with a staff of their own or using seconded personnel from the parents. The Iraq Petroleum Company was a larger grouping of Shell, BP, CFP, Mobil, Exxon and Mr 5% (Gulbenkian), which resulted indirectly from how the Allies divided the spoils following the fall of the Turkish Empire in the First World War. In some cases, the joint ventures were integrated entities both producing and marketing oil as in the case of Caltex. In others, they were simply production companies supplying oil to their parents for marketing, and the point of profit was set wherever that was most advantageous to the parents. We must also not forget the important merger of Royal Dutch with Indonesian production and Shell Transport and Trading, a British marketing company to form the giant Shell Group, which was integrated from top to bottom.

The process of consolidation had been effectively completed by the Second World War with the emergence of seven huge integrated companies: Exxon, Shell, BP, Chevron, Mobil, Gulf and Texaco. The inclusion of CFP of France, with its important stake in Iraq, could make this eight.

But the domination by this group was short-lived: the expropriation of BP in Iran in 1951 was a mortal blow to the hegemony of the Seven Sisters. They did not succumb immediately, but their world would never be the same again.

The opening of Libya brought into prominence a new

breed of companies: the smaller daughters of Standard such as Conoco, Amoco, Arco and Marathon, together with independents such as Hunt, Amerada, Getty, Occidental, Sun, Murphy or Diamond Shamrock. They were expanding overseas in exploration, and had to sell their crude without a ready-made international marketing organization. They began to rock the cosy boat.

The post-war epoch also saw the great expansion of contracting. Hitherto, the major companies had done much of the work themselves but now found it expedient to contract it out. They no longer had exclusive expertise, and the contractors proved more efficient at applying it. This was another nail in the majors' coffins, although not perceived as such at the time. With the notable exception of the French company, Schlumberger, which concentrated on the specialist job of well-logging to identify where the oil zones were, most of the contractors were American. Haliburton and Western Atlas dominated well servicing. Geophysical Services Inc and Western were the main seismic contractors, along with CGG of France, GECO of Norway and Prakla of Germany[1]. The move offshore prompted the growth of several large drilling contractors owning the rigs. They included Zapata, ODECO, SEDCO, Santa Fe and Global Marine of the United States, who were later joined by several Norwegian companies, such as Dolphin, Smedvig, Willhelmsen and Odfjell. Two French contractors COMEX for diving and COFLEXIP for specialist drilling equipment deserve mention. The offshore further brought in two giant American construction companies: Brown and Root, and McDermot.

Fig. 14-1. Norwegian seismic vessel.

Photo courtesy of GELO

It is rather surprising that no notable British contractors arose from the high level of activity on the British shelf. They evidently did not compete with the Americans, and in their open economy lacked the hidden state subsidies of the Europeans.

The widespread expropriations in the Middle East and Venezuela in the 1970s accelerated the changes. Today, the once mighty Sisters own less than ten percent of the world's reserves[2].

New national and State-owned companies grew up almost everywhere except in North America. They soon controlled the world's supply of oil. Many became huge, inefficient, feather-bedded bureaucracies[3], and their wealth made them states within states, wielding immense political power. Conflicts often arose between the Minister responsible and his own State oil company as to who was actually in charge. Many governments rued the day they let these cuckoos into the nest, but of course could not admit publicly that it was easier to deal with the foreign companies. Whereas the foreign companies were able to take the taxes paid to producing governments as a charge against the home country tax, thus conveying a hidden subsidy from the consumer to the producer, the new State companies found themselves having to sell their oil in a cold market place. They further had to spend real money on exploration, whereas the foreign companies were beneficiaries of hidden tax subsidies. Expropriation was not a good idea. The shift in ownership to the State opened further doors to the contractor.

The new ownership led indirectly to the oil shocks of the 1970s and the "glut" of the mid 1980s. The global control previously exerted by the cosy majors was replaced by dog-eat-dog competition, exacerbated by the underlying "feast or famine" nature of the industry. OPEC tried and substantially failed to control the situation: it lacked the hidden subsidy that the foreign companies previously conveyed to the producing governments. As already described, the "glut" was partly due to Saudi Arabia opening its taps wide under the influence of the CIA.

Then the pendulum began to swing the other way, led by Mrs Thatcher when she privatized the British State company, BNOC, in the 1980s. Her example has been followed in several other countries, such as Argentina, but it has proved difficult to disband the State monolithic institutions once created. Venezuela and Brasil are likewise ending the exclusive positions of their state companies.

The 1980s were a trying time for the major companies: they too were monolithic institutions and slow to perceive that their roles were changing and that the good old days could never be restored. They reacted by policies of down-sizing[4]. Approximately, half the membership of the Petroleum Exploration Society of Great Britain was thrown out of work in the three years ending in 1990; and the purges were even more severe in the United States. Some of the luckier technical staff were however rehired as consultants or contractors, exchanging their careers for short-term livings and uncertain futures. The contractors themselves found the going hard.

In parallel with this downsizing, the major companies faced an embarrassment of riches, left over from their earlier hegemony. They began to buy their own stock: effectively saying that their past was worth more than their future. It is said that the greatest takeover of all time was when Exxon bought its own stock.

The mammoth self-serving bureaucracies of the past had to go. And about time too. I remember when I was in Norway during the early 1980s, I had to manipulate the budget fourteen times in the year, with virtually none of the adjustments being discretionary. It was a colossal waste of time. The centralized computer systems of the day exacerbated the situation. Global links to the screens in every office often carried messages no more important than "Congratulations to Astri on her new baby". It was a misuse of the information highway, and it took a huge staff to feed it.

In recent years there have been positive moves whereby the major company affiliates gained more independence and autonomy. Their managers began to have an authority to match their responsibility. The idea of the *Asset Team* began to take hold. Groups of technicians and middle management would work together on a specific area, behaving, at least in theory, as if they were independent companies or contractors. It was however no easy transition: there is a natural reluctance to delegate responsibility, and despite cries to the contrary, not everyone actually wants to receive it. Also, as the teams become more independent, they narrowed their horizons, and it became more difficult for the senior management to know how to allocate their efforts. They tried to overcome that with various mathematical risking techniques in the hope that there would be a level playing field between projects competing for investment.

These developments came in parallel with the unwelcome and often denied realization that there was much less to do anyway in terms of viable exploration. Emphasis shifted to engineering, where there was still scope for innovative approaches to improve the economics of production in small and difficult fields. It led to the so-called *Alliance*. Hitherto, the companies and the contractors had sat on opposite sides of the table, with the contractors bidding, and the companies selecting the lowest bid. Competitive bidding was supposed to lead to higher efficiency under the principles of the market economy, although it often did not always work like that. When I was letting seismic contracts in Norway, I often preferred the highest bid and the special relationship with the contractor on the principle that if you need an operation you don't normally search for the cheapest surgeon.

All this changed under the *Alliance,* when the company and selected contractors sat down together to jointly find the best solutions. It has called for new attitudes to make the system work and resolve the often conflicting interests of the parties in an optimal way. This evolution appears to reflect a further contraction in the role of the major company.

The national companies own most of the oil and their countries have the greatest potential for new discovery. They certainly do not need the major company for technical expertise, which is already readily available from the contractors, although they could use the major company's money, and perhaps his downstream facilities.

But looking ahead, I foresee, assuming that the arguments advanced in this book are broadly correct, that demand for oil will shortly be curbed by much higher prices, imposed by the producing governments. In such circumstances, the producing governments will be wealthier from higher revenues, and absolutely in the driving seat. It will spell the last nail in the majors' coffins. Some may merge, as BP and Mobil may be thinking of doing, but otherwise they will have little option but to continue to distribute their past assets to their shareholders. Even if the major companies don't actually merge they may work together, as for example BP and Statoil of Norway already are doing in another form of *Alliance* for international exploration.

There is however a large amount of oil yet to produce, albeit at a falling rate. I think that the contractor, working for the national company, will produce most of it. Exploration will accordingly lose its tax subsidy and wither away. Few people will do it with real money. Independent refiners will supply supermarket retailers.

For the investor, it will be a particularly interesting situation to try to benefit from the major company divestiture. The management will no doubt try to hang on to their benefits to the bitter end, and the shareholder will have to work hard to protect his interest as an owner. It will be a very different world.

CLIVE NEEDHAM
The challenge of contracting

Fig. 14-2. Clive Needham.

Q Clive, your personal experience has mirrored what the industry is going through now, and will increasingly do in the future: careers as such are over as everyone becomes a contractor or a consultant. But, first could you say something about how your career evolved before you made the change.

A I do agree that it appears that lifetime careers with one company are over, but this is true of many industries in the 90's. It is not exclusively a facet of oil field life. We have all become contractors now – even within oil companies: some have gone to a purchase and supply basis with internal customers.

The change is for the worse, in my view, because the romance has gone out of the oil business, particularly on the exploration side. I admit that a younger geologist may disagree and say that I am displaying alarming signs of an exponential slide into old bufferism. I'll leave that for the moment and recap briefly my career, which may explain the view that my early days in the business had their romance.

I graduated, none too auspiciously, in 1974 from Portsmouth Polytechnic. I only mention this because in those days the major oil companies did not recruit direct from such institutions, and so it was necessary as a first step to start out as a contractor. I worked for Geoservices, a French mud logging company (still in existence, unlike many of the oil companies that were around at the time), known those who either worked for it or employed it as "Frog Log". Mud logging in itself is not a glamorous activity. In essence, it is the means by which a well is monitored; its progress charted; and what it is finding – oil, gas or often nothing – determined.

For a new graduate, the job offered variety, travel and exposure to a wide range of interesting individuals. It provided a very good background and insight to the business: it is a good idea to do it for a while and then move on.

So, where is the romance? Seeing the Rockies from the top of the Regent 25 rig in Alberta, nearly 50 miles away across the top of the never-ending pine woods. Being asked in the mud logging unit by the company representative: "Do you guys have any beer in here?". "Of course not", we reply, horrified innocent lads that we were. "Well, you'd damn well better have some the next time I come to the unit". A complete travesty of all known Company regulations, but we were fifty miles from the nearest road.

Mud logging is not a career, it is experience. I joined Amoco in fairly typical manner for those times (1978). They had two rigs drilling and had lost most of their wellsite geologists through transfer and natural wastage. They were short staffed. I was working on their rig at the time and knew their wellsite procedure. So they took me on and I worked at twice the rate of pay, with half the time offshore that I did as a Frog Logger.

Working as a Wellsite Geologist for Amoco was great fun. I loved it because I had a lot of autonomy on the rig. In those days, you worked in co-operation with the Company rig superintendent, but you only reported to London by Telex, at 6.00 am every day. It was genuinely good fun. What spoiled the job was being transferred out of the field into the central London office and finding oneself reporting to pasty faced young lads who had never seen an oil rig but cut a dash in a suit at executive committee meetings.

So it was that I came to answer an advertisement by Getty Oil in the Sunday newspapers. I was interviewed at the then Getty offices in the grounds of Sutton Place, the Elizabethan home of the late John Paul Getty. This made a big impression, driving in through the wrought iron gates and up the winding drive, crossing the humpback bridge over the Wey river (complete with traffic lights) to the house itself. It was here that J.P. lived with his 'companions' and even a lion – and ran a multi-billion dollar business.

Getty was a great company to work for. Small in the UK, but with enormous income, courtesy of the Piper, Claymore and Heather fields and $30/b oil. Our job was simple – go out and find more of the stuff anywhere in Europe and Africa.

I worked for a splendid Chief Geologist from the US, Barry Faulkner, who was as rounded a character as he was in profile. Early one morning, he suspended in the office of a geophysicist colleague a tyre on a rope with a bunch of bananas attached to it – to indicate what he thought of him. I got the blame for this incident, but survived.

Well, we did look everywhere. After an early spell on the North Sea, I was put onto new ventures in the Mediterranean and had to travel to Spain, Malta, Sicily, Tunisia and other such spots in the pursuit of new acreage to explore. It certainly was great fun and a privilege; and it provided a huge range of experiences.

The business was different then. Note that I am not saying it was better in the sense of how effective it was, just that it was much less of an industrial process than it has become. Many people will say that is exactly why it had to change – the high and escalating price of oil hid the many inefficiencies and the poor finding record of many major companies, once the giant fields had been discovered in the North Sea. The price crash of the mid-eighties plus the hunt for oil on Wall Street rapidly led to a shake-out that has produced what we have today – a shrunken number of companies who have had to adapt to survival with oil prices below their real value of twenty years ago.

My time with Getty ended as Chief Geologist for a new venture in China, a grand title considering I was also the only geologist in the local office for a while. Getty had taken a view that they wanted to be in the first round of western companies to enter China and hence have the best opportunity to pick up the most prospective acreage. Unfortunately, geology doesn't work like that, and what had been statistically correct in the North Sea did not prove to be the case in the South China Sea (as for example BP learnt to its cost: four licences and seventeen straight dry holes as I recall).

The competition for acreage was such that Getty was forced into a joint venture with Japex for the licence they got (15/33) and a joint-venture company was formed with staff from both parents. The Pearl River Oil Operating Company (see what I mean about romance) operated from the nineteenth floor of the White Swan Hotel in Guangzhou and I had a Japanese boss. I rapidly learnt the cultural differences between East and West.

The critical factor was the interest, the excitement perhaps, of being involved in such a new venture in a place like China. That it came to an early end for me was indicative of the changes that were overtaking the industry at this time. Oil was cheaper and easier to find on Wall Street. You just hired a bunch of suits in a merchant bank and off exploring you went – just down the road. Texaco decided to capitalise on the well documented difficulties of Getty Oil at board and family level and appeared as a late entrant to a proposed merger with Pennzoil. There is no need to dwell on this further, except to note that Texaco's total market capitalisation a couple of years later was less than the sum it paid for Getty Oil. I felt that the time had come to go.

Q We then found ourselves together building up a new exploration office in Norway. I think we succeeded beyond all expectations. We turned around a company that had failed to secure new licences for many years, and put it at the head of the pack, even securing an operatorship, which is no mean feat in Norway. You were responsible for the technical work which was a key element in the growing success. What do you think were the most impressive technical developments of those days in Norway?

A We did indeed find ourselves together, working from above a garage in Hillevagveien, such was the commitment that Petrofina had put into re-investing its spectacular returns from Ekofisk. Petrofina S.A. was nick-named "the Anonymous Society" by the Norwegian authorities having had such a low profile in the country. We set about the task of improving the technical reputation of the Company and establishing credentials to justify the award of the sort of acreage that we aspired to. We did this by starting to compete on the basis of being 'the best non-operator in Norway' (should I acknowledge the authorship of this golden strategy – you often used to observe that I hadn't quite mastered the art of fawning!). Some thought this a negative strategy to adopt, an acceptance of always being second-best. However, it proved itself correct because the competition among the non-Norwegian companies for operatorship was so intense and it was better to deliberately aim for the No. 2 spot in the most attractive blocks.

The major tool we employed was regional geochemistry. There were a number of reasons for this. Firstly, much of the new ventures activity was aimed at large tracts of open acreage in the Barents Sea and off Mid Norway. At that stage of exploration in the mid-eighties, it was essential to be able to screen large areas in order to reject those of low potential before focusing the more expensive evaluation effort on the best targets.

It is a truism that oil can be found in areas with wildly varying structural styles and reservoirs, but it is never found if there is no source-rock. A source-rock is any rock formation that contains sufficient organic carbon to be able to generate and expel hydrocarbons under the combined effects of temperature and time. If oil is not generated, there can be no discoveries.

As source-rocks need to be volumetrically significant to generate a sufficient hydrocarbon 'flux' for migration and entrapment to occur, it makes sense to search first in unknown areas for evidence for the presence and effectiveness of source-rocks. In this way, large areas can be coarsely screened before looking for the correct combination of coincident source-rocks, reservoirs, structures and seals.

At that time, major technical advances were being made in two areas of research that combined to provide the oil industry with a very powerful tool for acreage evaluation. The first was the growing understanding and everyday use of geochemical modelling to look at the maturation path undergone by a source-rock in the course of its geological history. The second was the new understanding about the way in which the crust of the earth behaves when subject to the primary tectonic stresses that have shaped the oceans, continents and, critically, the sedimentary basins.

To take the second first, the understanding of extensional tectonics, literally the stretching of the continental crust in response to the separation of continents and the opening of oceans (e.g. the North Atlantic) had led to a general recognition of the relationship between the degree and rate of extension and the formation of the major sedimentary basins in which the bulk of the worlds oil is found. What was also recognised was the relationship between basin subsidence and the heat flow through the earth's crust from the thermonuclear heat of the interior to the surface.

A seminal paper was published in 1980 by Douglas Waples, acknowledging the earlier work of a Russian scientist, Lopatin, in relating the maturity of coals to both the temperature to which the coal bearing strata had been subject and (crucially) the duration of exposure. Waples extrapolated this thinking to hydrocarbon source-rocks (which coals are too) and stated that there is a good empirical fit between the behaviour of source-rocks under thermal stress over geological time and the second order reaction equation of the chemist Arrhenius. This simple exponential relationship has allowed a whole generation of geologists to undertake burial profile analysis for various points in a sedimentary basin, and pronounce with reasonable accuracy on the likelihood of hydrocarbon generation (and even assess the likelihood of gas versus oil) whilst knowing little more than the outline stratigraphy of the basin and a generalised figure for the temperature gradient (often taken to be 33-35 °C per kilometre of burial).

The regional exploration geologist is looking for a "well cooked" (note, not over-cooked) source-rock with the additional appetisers of reservoir and structure to contain the oil.

We (and all our competitors) therefore spent much time and effort on the understanding of the burial and thermal history of the relatively unexplored mid-Norway Atlantic margin and the very poorly understood Barents Sea.

It was also necessary to play the political game of 'supporting' Norwegian R & D and even industrial efforts. They were really little more than disguised subsidies by the State, as the offshore industry by this stage comprised only serious taxpayers who could recycle exploration related R&D to their profit and loss accounts. Thus many a wild idea was funded at the expense of the Norwegian exchequer. I seem to recall we had a few of our own, such as a heart foundation for corpulent oilmen. All was done in the name of providing an individual character to our applications for the prime acreage. Still, it was all good fun sifting the ideas that came forward and trying to judge their impact on the NPD and Ministry.

We succeeded at this game too. We got into our first choice of acreage in the Barents Sea, the ill-fated PL 136, Blocks 7219/9 and 7220/7. I remember this because it was the biggest disappointment of my technical career. It was a beautiful, giant (one billion barrel plus) prospect made up of a clear rotated fault-block with Jurassic reservoirs and the all important Upper Jurassic source-rock adjacent to it and modelled to be in the oil generating "window". I remember being in your office when we got the LWD resistivity curve faxed in from the rig over the top of the reservoir. The resistivity of the reservoir was lower than the overlying shales: no oil. Even worse, it turned out that there

had been oil, as residual oil was present, and that the structure had been filled and then it had leaked at some point in Late Tertiary time owing the late stage uplift of the western Barents margin, about which little was then known. The structure had previously been up to 1,500m deeper than its present depth of burial. When it had been uplifted, the relative reservoir pressure ruptured the seal and the oil migrated out again. It was rather like finding your well baited trap had caught your prey, he'd eaten the bait and left a rude note as he got out again.

Q We came to regard the endeavour as if it was ours and didn't curtsy enough to our masters. They resented our independence and success, and eventually pitched us out. At first, I was relieved to be free of the stress but then became unreasonably resentful at the injustice of it all. It left a scar, but I turned to consulting which in fact is a much better life. How did you react.

A In many ways, Norway was the high point of our careers. The problem is, to really succeed you have to be committed and you have to be have individual ideas. This can often result in a clash of ideology with a conservative and nervous management. Our management in distant Brussels awoke to the costs of the Barents Sea exploration campaign (about $25 million per well dry hole cost) and the fact we had failed to find anything other than residual oil. Then the finger started to be pointed and resentments recalled at what was seen as a cavalier approach to exploration and, particularly, to them. There is no doubt there is truth in that, but a cavalier approach, a bit of flair, was required to make us competitive with the likes of Conoco, BP, Mobil and Shell. Dull conformity was not going to work. Our failure was that of the Barents Sea itself - we implicitly said it was going to work and we failed to deliver the 'expected' success.

The first casualty of this was you. You were initially moved sideways to the role of 'Consultant' and then, wisely, decided to resign with pride intact. I was actually promoted initially, to be Exploration Manager. I was then moved to a new venture office in Ho Chi Minh City, Vietnam, where it was my turn to fall. The failing this time was telling the truth. Giving an accurate assessment of the exploration potential of the acreage that had been acquired in the Gulf of Thailand. Telling management the average reserve size of the identified prospects was one fifth of what they had been 'promised' proved a great disappointment. Though if the Exploration Manager cannot give an honestly optimistic view, I think they should have just put someone in the job who would say yes.

Q You are also back in the business, this time as a contractor. You face the same issues but from a different perspective. How would you sum up the main differences.

A I initially did not want to have anything further to do with this silly game – but it is addictive and I soon realised it was the only game in town that I had any capacity for. I started consulting for Western Atlas, a major oilfield service company, and then joined them as a employee. I believe the very fact that Western recruited me, a non-specialist in their areas of expertise (dominantly wireline and seismic data acquisition and processing) reflects the structural change in the business. Ten years ago, they would never have wanted me and I would not have wanted to work for them. They provided data for oil companies to work on and nothing more. This was a stable, symbiotic relationship that everyone understood and was happy with.

Two factors altered the arrangement: the restructuring of the oil companies and the greater complexity of the data now produced by service companies like Western. I can give two examples: 3D seismic data; and the workstations that integrate the seismic data with the well logs that depict the acoustic or resistivity properties of the formations in the wellbore.

Oil companies had decreased their technical staffs at a time when the volume of data they were receiving was increasing exponentially. There was only one outcome: they had to come to their suppliers for not only data, but also an interpretation of the data. Western Atlas increasingly find themselves asked for a technical assessment of the geology of the structures over which they have acquired data. Service companies are nothing if not aggressive and keen to expand their business, with the result that Western started to recruit people with oil company backgrounds who could speak the new (to the service company) language of interpretation.

It works quite well. An interesting facet is to see how the service companies cope with the new world of the oil companies for whom they now work. I will name no names, but some (who will thrive) have restructured themselves because they believe in it, are committed to and genuinely want to have a successful partnership with their service suppliers. They are willing to recognise that they do not have a monopoly on expertise and will accept the technical support of their supplier. Others, from the gadarene end of the oil company spectrum have cut staff, mouthed the words fed to them by their management consultants and have then carried on as before. They are failing to gain the benefits of downsizing and partnership with their service companies; and their staff appear increasingly demotivated as they have less and less people to do the work. Their success rates will fall and they will decline.

The main threat to the service companies is this failure of the oil companies to recognise the true meaning of their own restructuring. They want innovation, support and R&D to be supplied by the service sector. They make a great song and dance of concern for quality, safety, the environment and service over cost when seeking tenders for work. Yet they invariably seek to trim costs below the level of profit that will allow innovation to be funded.

As a species, oil companies do not yet fully comprehend the real requirements of partnering-alliancing for the future. They want the benefits, whilst preserving the master-servant culture of the past. There remains some way to go before mutual benefits will routinely flow from the new arrangements. Otherwise, the oil companies should staff up and do the job themselves – if the stock market will let them take on those costs.

Q All the evidence I can find suggests that the North Sea production will peak in a year or two and then decline rapidly. Every effort will be made to extend plateau production, but it simply means that the end when it comes will come quicker. There may soon be more work to do in removing exhausted platforms than in installing new ones. Does that make sense from your vantage point?

A I did some work in Norway during the course of the past year (1996) and it was a surprise to observe how far production decline had to go down the exponential decay curve before prompting a reaction. The old story of the dinosaur with a brain so small that it took seven minutes to react to a kick up the backside came to mind.

If stable (falling in real terms) oil prices are to remain, then most oil companies can face nothing but a prolonged decline, with an early exit from exploration for the least successful. In some ways, this started some years ago – remember Burmah. It could be argued that this is occurring in microcosm in Aberdeen even now. I should explain: when exploration commences in a new area and there is appreciable success, leading to production, the oil companies establish themselves and draw in a wide variety of service suppliers from the large – Western Atlas, Halliburton etc – down to a range of small specialists. As the area reaches maturity, the oil companies lose the efficiencies of scale in production: rising costs and declining productivity take their toll. They start to look elsewhere for their future because they find that declining production takes the same or more management and technical time as peak production. They can invest money elsewhere for a higher rate of return than they can hope to get from the declining years of a large field. Their interest wanes and other specialist companies – scavengers if you like, start to feed off the decaying carcasses of the older fields. This has happened in Aberdeen with the advent of Oryx and Talisman, to name but two. Who could have imagined Chevron moving out of Ninian House - but they have done so as they smoothly moved out of the Ninian Field and handed the responsibility for the field decline and abandonment to others.

The large players have already mentally gone from the mature area of the North Sea – how many have real executive decision making powers in Aberdeen? I mean by that, raising capital and employing it in a risk sense. I think it is a round zero. It is incidentally the one area where I can understand the fears of the Scottish National Party: Aberdeen is just a service industry town now. If the SNP *should come to power, they will not be dealing with the powers in the industry when they talk to people in Aberdeen. It is well down the slope towards withdrawal by the large oil companies. Sure, they will be there for many years to come, but the real decisions are being taken on other areas, elsewhere.*

This gives tremendous potential for the service companies to take on the management role of the declining resources. And why not, when they often actually do the work and fully understand the means by which it is to be carried out. After all, most of us watch television and control the channels but we know (or care) little for how the visual image is created. We are happy to pay others to take care of that, so long as the programme is satisfactory.

The future in the North Sea may well lie with contractors, but the real decisions will still be made by the sources of capital. Where the financial implications of the

decisions are in the realms of loose change, they can easily be left to others, with a spending authority cap.

The oil companies' problem is that they control such a small part of the world's remaining reserves that they will be increasingly marginalised as their reserve base declines. The evidence of this is around us now – Shell's cash mountain (have they really nothing to spend it on?), the flat market performance of some companies through a bull market (Petrofina). The only long term hope is in relationships with the Middle East, if that is possible, and a rise in the price of oil, justifying exploration and development in difficult areas – deep water, hostile environments etc.

The question of the future level of oil prices is, I know, a subject dear to your heart. I will stop here and let others with better crystal balls consider that.

NOTES

(For references, see Bibliography)

1. Prakla and GECO were later taken over by the French giant Schlumberger

2. Jennings, J.S., 1996, Chairman of Shell, actually reports 12%, but other estimates are as low as 7%. It much depends on how reserves are defined.

3. Petrobras in Brasil has for example a payroll of 41 000.

4. *The Economist,* 1996, comments on this trend.

Chapter 15

SYNTHESIS – WHAT IT ALL AMOUNTS TO

If you have come this far, you deserve a medal. If you have simply browsed, that will have served its purpose too if it has prompted you to think more about the implications of depleting a finite resource and the coming production peak. We are not used to depleting things, being confident that we can always run into the supermarket and replenish our stocks. And, we prefer not to think about the end of the one finite resource we do know about only too well: our own life-span. But in the 21st Century, we will come to experience the virtual depletion of oil, an energy source that has become central to our way of life. We will have to change the way we live. It is not too soon to start thinking about what that may entail.

In this last chapter, I will try to sum up the message of this book.

THE FORMATION AND ENTRAPMENT OF OIL AND GAS

Oil is derived for algae that proliferated from time to time in the Earth's long geological history. Gas comes from plant remains and is more widely distributed. On death, the organic material sank to the bed of the sea or lake from which it was derived, or was washed in from the surrounding land. In most cases, it was dissolved or destroyed, and only rarely in stagnant troughs was it preserved and concentrated. The resulting organic-rich layers were buried by other sediments, and with further subsidence became heated by the Earth's heat-flow. After a critical exposure to heat, the organic material was converted to oil and gas by chemical reactions. It is obvious why the circumstances for prolific oil generation occurred so rarely. The conditions for generation normally lie at depths of 2000 to 5000 m. There is not much oil to be found deeper.

Once formed, petroleum, whether consisting of oil or gas, which is under great underground pressure, begins to migrate upwards through the rocks in hair-line fractures until a porous and permeable layer, such as a sandstone, is encountered. It then follows the conduit, floating on the water that fills the pores. If the conduit leads straight to the surface, the petroleum will escape to the atmosphere. But in most geological basins, the rocks have been folded and faulted by earth movements. In such cases, the petroleum collects in the highest part of the folds or up against faults. It is contained in porous rocks, but will leak out over time unless well sealed by overlying impermeable strata, such as clay or salt. Much of the oil, that was once formed, has escaped. Older rocks therefore become less prospective because they have been exposed to leakage for longer.

Much of the petroleum is trapped in the cracks and crevices of the migration paths along which it moves, and much is lost at the surface. So only about one percent of the amount generated finds its way into traps that are large enough to be exploited. Oil accumulations at shallow depth on the margins of basins are oxidized and attacked by bacteria becoming tar and heavy viscous oil. Oil that is overheated on being buried too deeply is cracked to gas.

Gas contains dissolved liquid hydrocarbons that condense at the surface, and can be extracted by processing, being known as *Condensate* and *Natural Gas Liquids*. Oil also contains dissolved gas, which may separate out in the reservoir to form a gas-cap over an oil accumulation.

EXPLORATION

In earlier years, geologists searched the world for seepages of oil at the surface, and looked for promising structures in the vicinity to trap it. They found most of the prolific basins and many of the giant fields in this way.

Later, seismic surveys were developed to explore the subsurface. The technique involves the release of energy from an explosive charge, or in other ways, at the surface, and recording the echoes reflected back from rock interfaces far underground. By computing the time taken for the echo to return, it is possible to calculate the depth and configuration of the buried structures.

The offshore was opened after the Second World War, and marine seismic surveys were perfected, such that it became possible to map the continental shelves rapidly and inexpensively. Technological progress has greatly improved the resolution of seismic surveys, and the computer work-station has brought enormous computing power to the interpretation. Geologists and geophysicists can now investigate the oil zones in great detail, searching for thin and subtle reservoirs and finding ever smaller traps.

Exploration boreholes, known as wildcats, are drilled to test the geological interpretations and gather information. The technology of drilling has made enormous progress, such that it has become routine to drill 5000 m wells in the stormy waters of the North Sea. The process involves drilling a large diameter hole from the surface, commonly 30 inches in diameter, and then cementing steel casing into it to seal off the formations. The size of hole is progressively reduced and cased off. The drill string with a bit on the end of it rotates to make the hole, and a special mud, weighted up with the heavy mineral, barytes, is pumped through the drill string to lubricate the bit and remove the cuttings. The trick in drilling is to match the mud weight against the formation pressure: if it is too high, the mud escapes into the formation; if it is too low the formation encroaches on the well and causes the bit to stick.

Geologists examine the cuttings brought to the surface in the mudstream to identify the rocks the borehole is penetrating. Cores are taken where necessary. Sondes are also lowered down the borehole to record the electrical and radioactive properties of the rocks, making it possible to identify different rock types, measure porosity and determine which zones are oil or gas bearing.

The breakthrough offshore came with the development of the semi-submersible rig, in which a platform holding the drilling derrick is mounted on two submerged pontoons that lie beneath the wave base, providing a stable structure relatively unaffected by the weather.

Other important developments have been to find ways to drill highly deviated wells to reach far out from the platform. In extreme cases, the wellbore may be 90 degrees or more from the vertical. It can track a thin productive zone, which can be drained rapidly. Also, a single well may have several branches at depth. Various techniques to improve the permeability of the reservoir can be applied, such as injecting acid or fracturing it by injecting fluid under very high pressure.

To produce an oil zone, it is necessary to first seal it off from the overlying and underlying strata, and then let off explosive charges in the well to pierce holes in the casing. Normally, there is sufficient pressure in the reservoir to cause the oil to flow to the surface, although sometimes it has to be pumped.

In short, advances in technology have made exploration and production highly efficient. The geological processes responsible for oil accumulation are now very well understood. Of particular importance was the geochemical breakthrough of the 1980s that made it possible to identify and map the zones generating oil and gas. It not only showed which trends had potential but it allowed large tracts to be written off as non-prospective, once the critical information had been gathered.

RESERVE ESTIMATION

Before a wildcat is drilled, the geologists and engineers have to estimate the likely reserves of the prospect to determine if it has the potential to be commercially viable.

They map the volume of the trap, using seismic data and applying their best estimate of the likely reservoir conditions. The first well will reveal whether or not it is oil or gas bearing as well as much more information about the reservoir, but it is usually necessary to drill several appraisal wells to confirm the estimates.

There remains a range of uncertainty, the *Median Probability* case gives the best estimate of what is actually producible. However, for the purposes of planning the investment and in raising the finance, it is normal to take a more conservative *Low Case* estimate with a probability ranking of about 90 percent, and term the reserves *Proved Reserves*. This number is used for financial reporting purposes, and is the one normally recorded for official statistics. This so-called *Proved* conservative estimate is naturally subject to upward revision over the life of the field: the increase being termed *reserve growth*. It is, however, widely misunderstood, being taken as a dynamic akin to exploration, driven by improved technology, when in fact it is little more than the natural evolution from a *Low* to a *Median Case* estimate.

While, the reserves of a field will be known absolutely only on the day when it is finally abandoned, at which point they equate with the *Cumulative Production,* the estimation of reserves is a straightforward procedure in technical terms. The reporting of reserves is, by contrast, a political act reflecting more the needs of the reporting authority than the actual situation. In the absence of clear universal definitions, reporting procedures or audit, there is plenty of scope for latitude in reserve reporting. In part, companies treat reserves as a form of inventory which they book as best meets their financial and commercial needs.

Government statistics are often unreliable: the greatest distortion arising from the exaggeration of reserves by several OPEC countries in the late 1980s in order to secure higher production quotas.

CATEGORIES OF OIL

We may broadly classify oil into *conventional* and *non-conventional* categories. Most of the oil produced to date, as well as most to be produced over the next few decades, can be called *conventional*. It has a characteristic depletion profile with production starting at zero, and rising rapidly to one or more peaks before declining exponentially.

In addition, there are large amounts of what can be collectively termed *non-conventional* oil. It is made up of heavy oil and tar; oil dependent on enhanced recovery techniques that change its fluid properties in the reservoir by such methods as steam injection; oil in hostile environments; oil from late-stage infill drilling to tap small pockets missed by the primary wells; and oil in accumulations too small to be viable exploration targets. It has a different depletion profile, rising only slowly to a long low plateau before eventually declining in the far future. Although the boundary between the two categories may be blurred, it is very important to make the fundamental distinction.

Natural Gas Liquids are another source of confusion in

the statistics. They should be separated from oil because they relate to the gas domain, but in practice are often lumped together, being in some cases pumped through the same pipeline and metered together.

What concerns us most is when the production of *Conventional* oil will peak, because it is then that shortages of the cheap oil-based energy, on which our economy is based, will begin to appear. *Non-conventional* oil may well become important in the distant future. Some is already in production at today's prices, but the amounts are unlikely to be significant for a long time, if ever. Imagine having to drill patterns of five closely spaced wells, and inject steam into the peripheral wells to drive a little heavy oil out of the central well until it drains that area, and then move on to the next grid unit. It can be done and it may even make a profit, but the sheer scale of the operation is daunting. Imagine too the processing to convert this oil, which often has a high sulphur content, into a usable product. A great deal of energy is used making the steam, and the whole thing is damaging to the environment. It is critically important to distinguish this kind of activity from conventional oil production. Many of the high reserve estimates that are published fail to distinguish the two categories and give very misleading impressions.

HOW MUCH HAS BEEN FOUND AND WHEN

The search for oil has been going on for almost 150 years. During this time, almost everything that there is to know about the geological conditions responsible for it has been learnt. The World accessible to the international industry has been very thoroughly explored. Large parts of the World, made up of ancient shields, or oceanic rocks, are absolutely non-prospective. Almost all potential basins have now been identified, and investigated to some degree by seismic means and drilling. It is almost inconceivable that any new significant province remains to be discovered.

There are certain new tracts in very hostile environments that are under-evaluated, such as in Antarctica, the Falkland shelf, the Greenland icecap, off Iceland, and in several Arctic provinces. There is no particular reason to think that they are oil-bearing, still less that they can yield any significant amount. We can treat much of such notional oil as may be attributable to them as *non-conventional* insofar as it is in any event out of range for a long time to come.

The prospects in the Former Soviet Union and China are not well known in the West. This ignorance has tempted some to attribute a large undiscovered potential to these areas. I think that the Soviet explorers were certainly as intelligent as their western counterparts, and the systematic exploration of the Soviet system was probably efficient. It is said that oil is found in the head of the geologist, and I think that Russian heads were no thicker than ours. Their technology may not have been as advanced, but most of the oil in the West was found long ago when technology was even less advanced. I therefore think that all the large productive basins have been found as well as most of the

giant fields. There is however certainly scope to extend known trends offshore into the Caspian which was not investigated by the Soviets. The giant Tengiz Field with 8 Gb of reserves in a Devonian reef beneath an effective salt seal lies on the shores of the northern Caspian and provides encouragement for exploration in the adjoining waters, although there is a danger that with deeper burial the oil has been cracked to gas. There may be some scope left in very remote places like the Tarim Basin in the interior of China but again such oil is nudging the *non-conventional*.

Most deep and very deep water areas are non-prospective for geological reasons, but there may be a few new areas like the deep Gulf of Mexico or offshore Nigeria and Angola yet to bring in. It is however unlikely to be any great bonanza, even if now technologically feasible.

Figure 15-1 provides the essential data on discovery, showing that 784 billion barrels have been produced and that estimated *median probability* reserves from known fields stand at 836 billion barrels. Together, they add to a total discovery of 1.6 trillion barrels. This estimate of reserves excludes *Natural Gas Liquids,* and is respectively about 180 and 270 billion barrels lower than the numbers reported in industry journals[1]. The reason for the difference is that many countries, including Mexico and several large OPEC producers have released unreliable data and/or included *non-conventional* oil.

Produced to end 1996	784 Gb
Reserves	836
Discovered	1620
Yet-to-find	180
Yet-to-produce	1016
Ultimate	1800
Depletion Midpoint	2001
Depletion Rate	2.6%
Discovery Rate	<6 Gb/a

Fig. 15-1. The World's conventional oil endowment.

GIANT FIELDS
Initial reserves by discovery year

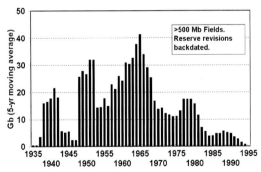

Fig. 15-2. Giant discovery.

About sixty percent of what has been discovered lies in just over three hundred giant fields, many in the Middle

East. Peak giant discovery was in the 1960s, and the discovery rate has fallen dramatically in recent years, see Figure 15-2.

Discovery as a whole also peaked in the 1960s being heavily influenced by the contribution of giant fields. More and more fields are being found but the average size is falling.

HOW MUCH REMAINS

The sum of the *Reserves* and the *Yet-to-Find* gives how much remains to produce. Put in other terms, it is *Cumulative Production* when it ends (namely *Ultimate* recovery) less *Cumulative Production* to-date.

Estimating how much is *yet-to-find* is not easy, but nor is it quite as difficult as it was. The World has been so extensively explored that we can be sure that almost all, if not all, of its prolific basins have now been identified. The bulk of what remains to be found lies in ever smaller fields within the established provinces.

We can estimate this amount by old-fashioned geological judgment relating the maturity of exploration with the underlying geology, and there is now sufficient data to use statistical approaches.

When the first well is drilled in a basin, nothing is known about the ultimate distribution of field size, but when the last well is drilled everything will be known. As we get close to the finishing line, we can begin to see it clearly.

Jean Laherrère has discovered a law of distribution stating that objects in a natural domain plot as a parabola when their size is compared with their rank on a log-log format. For example, the populations of the larger towns can be plotted to yield the population of a country down to the smallest settlement. It means that when the larger oilfields in a basin have been found, their size distribution can be used to predict what the *Ultimate* recovery will be. The difference between this and what has been discovered gives the *Yet-to-Find*.

Another approach is to plot cumulative discovery against the wildcats drilled. The plot is generally hyperbolic with the larger fields found first, and the asymptote equates with *Ultimate* recovery. Cumulative discovery may also be plotted over time.

Lastly, production peaks can be correlated with their related discovery peaks and extrapolated to zero, giving an indication of how much is yet to produce.

It is a case of using all of these techniques, as well as judgment, to come up with the best estimate, remembering always to distinguish *conventional* oil from *non-conventional* oil.

My best estimate computes that there are 180 billion barrels *yet-to-find*. One could round it to 200, but it is better to keep the exact number as calculated so that things add up properly. With reserves of 836 billion barrels, it means that there is a rounded one trillion barrels of *yet-to-produce*.

The distribution of the *yet-to-produce* is most uneven: about half of it lies in just five Middle East countries. The ten largest countries hold three-quarters of it.

PRODUCING WHAT REMAINS

The production of any finite commodity starts at zero, rises to one of more peaks and ends at zero. Think of your life-time spending pattern: you spend little in the cradle or the coffin, but have several peaks around middle age. It is the same with oil production in a country: peak comes around the midpoint of depletion. It could come a little before midpoint if there are a lot of giant fields found early; or it could come after midpoint if peak production were artificially restricted by prorating or quota. But the general coincidence of peak and depletion midpoint is valid for both theoretical and empirical reasons.

PRODUCTION SCENARIOS
1800 Gb Ultimate

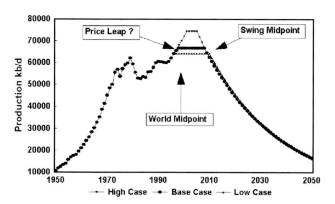

Fig. 15-3. Production profiles.

Most countries are now either past midpoint or close to it, except the United States which is far past it and the five Middle East countries which are far from it.

It means that the producing countries of the world can be divided into those past midpoint where production is set to continue to decline, and those which have not yet reached it where production can still increase. We may further consider the five Middle East countries as an extreme category of the pre-midpoint group, because their *yet-to-produce* is so large and the depletion rate so low. They can behave as swing producers, making up the difference between world demand and what the others can produce. This swing role can, however, apply only for a number of critical years before they too reach their midpoint. The swing producers are likely to control both the level of production and the price of oil.

SCENARIOS

World demand for oil rose rapidly until the 1970s, when it briefly declined before again beginning to increase at a slow rate. Most forecasters now predict rising demand, driven by the expanding economies of the East and the growing population. It is currently rising at just above two percent a year, the International Energy Agency predicting an increase of 2.5% in 1997.

One can envisage a large number of alternative scenarios

of supply and demand, but here we will be content with three, hoping that they will encompass in their spectrum the actual course of events. They all envisage a radical increase in the price of oil when the swing producers exercise their control. This price constraint is expected to lead to a plateau of production for a certain period of time before the resource constraints drive production down.

The base case scenario envisages that production rises at a conservative 2% a year until the swing producers control thirty percent of the world market, which will be in 1998 under this model. If production rises faster, as now seems quite possible, the thirty percent threshold will be reached sooner. Prices then increase by a factor of two or three, which curbs demand, giving a plateau of production at 67 Mb/d until 2008. *Plateau* may not be the right word because there will be many ridges and valleys in a very volatile market. It is expected to end when the swing share has risen to about fifty percent. It is assumed that there will then be physical shortage probably accompanied by a further radical increase in price. Production will then start to decline at the then world depletion rate.

SHOCK

In an ideal world, governments would properly study the resource base and understand the principles of depletion. They do not, and in democratic societies cannot, because they are elected for short terms and are therefore motivated to deliver short-term benefits to their electors. As a consequence, it is most unlikely that the governments of either the United States or the European Union will adopt an energy policy with the aim of preparing for the inevitable peak in oil production and subsequent scarcity.

It will therefore be left to the Middle East producers to alert the World to its predicament. They wont do so for an altruistic purpose, but simply to raise their revenues. Motive apart, their action will carry an important message. I don't think that their message will be delivered in small doses, nor can it be, given the efficiency of the new oil commodity markets. It will be the marginal barrel that sets the price. Quite a small shortfall could trigger a strong reaction. There will be nothing to counter it: oil will suddenly be in strong demand and the traders will hourly mark up its price as more buyers than sellers appear in the bullpen. Probably, as prices rise the buyers would at first hold back, but since their physical stocks are now so low, they could not do so for long. The market would move into *contango* whereby the futures would be above the present. That itself would deliver a message, which the sellers would pick up, holding back on physical delivery. It would spiral upwards as a crisis feeding on itself. Where would it end?

POST-SHOCK

The epoch immediately following the shock will likely see great volatility. There will be no shortage of comment, informed or otherwise, and it will be a field day for radio and television panel discussions. Mr Clinton may launch a few more missiles at someone. He might send in the marines to occupy Saudi Arabia. But it would all be posturing and gesture. If Exxon, backed by the marines, found itself controlling the world's oil supply, what would it do? Put up the price.

There would be a flurry of new exploration in the hope that old solutions would again come to the rescue: they won't.

But gradually, the realities will filter through. The Third World will be hit first: their oil-based energy consumption will begin to falter. We already have an example of what happened to Cuba when cheap Russian imports ended with the collapse of Communism.

"Cuba has become an undeveloped country. Bicycles are replacing automobiles. Horse-drawn carts are replacing delivery trucks. Oxen are replacing tractors. Factories are shut down and urban industrial workers resettled in rural areas to engage in labour intensive agriculture. Food consumption is shifting from meat and processed products to potatoes, bananas and other staples".[2]

It won't be so rosy in the developed world either.

A permanent doubling or more in the price of oil, followed by growing physical shortages, must lead to a major economic and political discontinuity in the way the world lives. It heralds the end of rampant and mindless consumerism in the more developed countries, and will bring great suffering to the Third World.

Every effort will be made to find alternative and renewable sources of energy. Nuclear power will be increased, although not fast enough to deal with the crisis. It will itself later become resource constrained by the finite quantities of uranium. Coal mining will be stepped up with adverse environmental consequences, especially in places like China where the power stations lack adequate smoke filters. The use of renewable energy will expand rapidly and successfully.

The greatest progress will however have to be made in terms of using less. The World will become a very different place with a smaller population. The transition will be difficult, and for some catastrophic, but at the end of the day the world may be a better and more sustainable place.

That seems to be a logical interpretation, but is it the correct one? I don't know, but I hope that the discussion which you have so patiently read will prompt you to think about it. I further hope that, having thought about it, you will make some provisions to protect yourselves as well as you can. I have discussed it with my broker, but I have to admit that we do not know what to do.

NOTES

(For references, see Bibliography)

1.   *Oil and Gas Journal* and *World Oil.*

2.   Falcoff, 1995, quoting Preeg & Levine, 1993.

BIBLIOGRAPHY

AAPG, 1991, The Arabian Plate – producing fields and undeveloped hydrocarbon discoveries; map published by *Amer. Assoc. Petro. Geol.*

AAPG Explorer, 1993, Future energy needs not being addressed; *AAPG Explorer,* June 1993.

Abelson, A., 1996, Crude awakening; *Barrons* 1.5.96.

Abi-Aad, N., 1995, Challenges facing the financing of oil production capacity in the Gulf; *Petroleum Review,* Inst. Pet. London. Feb 1995 83-86.

Abi-Aad, N., 1994, Middle East NGL vs pipeline transportation costs; preprint *APS Conf.*

Aburish, S.K., 1994, *The rise, corruption and coming fall of the House of Saud;* ISBN 0-7475-1468-2.

Abraham, K.S., 1993, UK adopts US view of industry as cash cow; *World Oil,* May 1993.

Abraham, K.S., 1994, Low oil prices killed the USSR; *Pet. Eng. Int.,* Sept. 1994.

Abraham, K.S., 1996, Sifting the stew of Iraqi sales, US politics and future supplies; *World Oil,* March 1996.

Abraham, K.S., 1997, Venezuela bets on heavy crude for long term; *World Oil,* Jan. 1997.

Adam, P., 1997, Modern asset pricing – a map of the future? *Petroleum Review* Feb. 1997.

Adams, G.A., 1989, *Policy and management of petroleum resources in Nigeria;* UN Seminar Policy and Management of Petroleum Resources, Oslo.

Adams, T., 1991, Middle East Reserves; *Oilfield Review* p.7-9.

Adelman, M.A., 1989, Problems in modeling world oil supply; *Energy Modeling Forum,* MIT Energy Lab Working paper MITEL -89-010WP.

Adelman, M.A., 199?, Modelling world oil supply; *The Energy Journal,* 14/1.

Adelman, M.A., 1995, Letter to Petroleum Economist; *Petroleum Economist,* July 1995.

Adelman, M.A., 1995, *The genie is out of the bottle: world oil since 1970;* MIT ISBN 0-262-01151-4.

Al-Fathi, S.A., 1994, *Opec oil supply outlook to the year 2010;* Wld. Petrol. Congr., Stavanger.

Alahakkone, R.R., 1990, Convenional supply of oil will slow down; *Oil & Gas Journ.,* Feb. 5 1990.

Aldinger, C., 1995, *World dependence on Gulf oil will grow;* Pentagon press release.

Alazard, N., J.H. Laherrère and A. Perrodon, 1992, Réserves et resources de pétrole et de gaz des pays Méditerranées; *Revue de l'Energie* 441, Sept. 1992.

Allen, R.V., 1966, The man who changed the game plan; *The National Interest,* 44 Summer 1996.

Amenson, H., 1988, Oil discovery principles, exploration strategy; *Oil & Gas Journ.* Oct 17, 1988.

Amenson, H., 1992, World developments since mid 1990. *AAPG Explorer,* Nov. 1992.

Amenson, H., 1993, US drilling: up is the only direction left; *World Oil,* Feb. 1993.

Amenson, H., 1993, Look for a moderate rise in US exploration; *World Oil,* Feb. 1993

Amenson, H., 1993, US oil and gas reserves resume downward trends; *World Oil,* Feb 1993.

Amenson, H., 1993, Drilling remains strong in many world regions; *World Oil,* Feb 1993.

ARAMCO, 1980, *Aramco and its world: Arabia and the Middle East;* (Ed. Nawwab I.I., P.C. Speers, & P.F. Hoye) Aramco. ISBN 0-9601164-2-7.

Arbatov, A., 1994, quoted in Abraham K.,1994, Economists project mixed view to 2000; *Petrol. Eng. Int.,* June 1994.

Arbatov, A., 1994, The foreign concerns of the Russian oil and gas complex; preprint APS Conf. Cyprus.

Arnold, R., G.A. Macready and T.W. Barrington, 1960, The first big oil hunt: Venezuela 1911-1916; Vantage Press, New York 353p *Lib.Congr.* 59-11872.

Bahree, B., 1994, World oil demand in 4th quarter to top earlier forecast, agency says; *Wall St. Journ.* July 8 1994.

Barber, A.H., C.J. George, L.H. Stiles and B.B. Thompson, 1983, Infill drilling to increase reserves – actual experience in nine fields in Texas, Oklahoma and Illinois; *Journ. Petrol. Technol.*

Barker, T., 1995, Taxing pollution instead of employment: greenhouse gas abatement through fiscal policy in the UK; *Energy & Environment* 6/1.

Barkeshli, F., 1996, Oil prospects in the Middle East and the future of the oil market; *Oxford Energy Forum,* 26 August 1996, 10-11.

Barrett, W.J., 1992, Calling all bottom fishers; *Forbes,* May 11th, 1992 .

Barron's, 1994, What next for oil prices; *Barrons,* Feb 28 1994.

Barry, R.A., 1993, *The management of international oil*

operations. PennWell Books, Tulsa.

Beardall, T.J., 1996, The world of reserve definitions – can there be one set for everyone; in Doré, A.G. and R. Sinding Larsen (Eds), *Quantification and prediction of hydrocarbon resources;* NPF Sp. pub. 6., Elsevier ISBN 0-444-82496-0.

Beardsley, T., 1994, Turning Green; Shell International projects a renewable energy future; *Scientific American,* 271/3, Sept. 1994.

Beck, R.J., 1992, Oil industry outlook 1992-1996; *Oil & Gas Journ.*

Beck, R.J., 1991, Second half economic gains to lift oil prices; *Oil & Gas Journ,* July 29.

Beck, R.J., 1993, World oil flow steady in 1992; stable market ahead in 1993; *Oil & Gas Journ.* Mar 8, 1993.

Beck, R.J., 1995, OGJ300 population shrinks but assets total grows; *Oil & Gas Journ.,* 4. Sept. 1995.

Bee, A.C., 1991, Long-run industry performance in exploration outside N. America; Abstract, *Amer. Assoc. Petrol. Geol.* 75/8, 1405 preprint.

Beeby-Thompson, A., 1961, *Oil pioneer;* Sidgwick & Jackson, London.

Beliveau, D., 1995, Heterogeneity, geostatistics, horizontal wells and blackjack poker; *Journ. Pet. Technol.,* Dec. 1995.

Bell Helicopter Textron, 1989, *World Energy Update.*

Bell Helicopter Textron, 1995, *World Energy Update.*

Bell, S., 1994, Distribution of resources knowledge helps locate undiscovered reserves; *Petrol. Eng. Int.*

BGR, 1994, *World Energy – a changing scene;* BGR Conference abstracts.

BGR, 1995, *Reserven, Ressourcen und Verfügbarkeit von Energierohstoffen 1995;* BGR Hannover 498p. ISBN 3-510-65170-7.

Bilkadi, Z., 1996, *Babylon to Baku;* Stanhope-Seta ISBN 0952-881608 230p.

Billo, S.M., 1990, Complex geology discussed by noted Arabian scientist: *Oil & Gas Journ,* Jan 29 1990.

BIP, 1995, *Quelle politique pour l'union europeanne;* BIP 7836.

Bird, K.J., F. Cole, D.G. Howell and B. Leslie, 1995, The future of oil and gas in northern Alaska; *Amer. Assoc. Petrol. Geol.,* 79/4 p.579.

Bjørlykke, K., 1995; From black shale to black gold; *Science Spectra 2,* 1995 44-49.

Blakey, E.S., 1985, *Oil on their shoes;* Amer. Assoc. Petrol. Geol. 192p ISBN 0-89181-804-9.

Blakey, E.S., 1991, *To the waters and the wild;* Amer. Assoc. Petrol. Geol. ISBN 0-89181-803-0.

Bohnet, M., 1996, Promotion of conventional and renewable sources of energy in developing countries; in Kürsten M. (Ed) *World Energy – Charging Scene:*

ISBN 3-510-65170-7

Bookout, J.F., 1989, Two centuries of fossil fuel energy; *Episodes* 12/4.

Bourdaire, J.M. *et al.* 1985, Reserves assessment under uncertainty – a new approach; *Oil & Gas Journ.* June 10. 135-140.

Bourdaire, J.M., 1993, Le pétrole dans l'economie mondiale; *Energie Universale,* Sept. 1993.

Boy de la Tour, X., 1994, Technologies petrolieres: les nouvelles frontieres; *Revue de l'energie* 456.

Brammer, R.,1995, *Oil Change:* Barron's, March 6th, 1995.

Brekke, H. and J.E. Kalheim, 1996, The Norwegian Petroleum Directorate's assessment of the undiscovered resources of the Norwegian continental shelf – background and methods; in Doré, A.G and R. Sinding Larsen (Eds), *Quantification and prediction of hydrocarbon resources;* NPF Sp. pub. 6., Elsevier ISBN 0-444-82496-0.

Breton, T.R. and J.C. Blaney, 1991, Production rise, consumption fall may turn Soviet oil exports higher; *Oil & Gas Journ.* Nov 18, 1991, 110-114.

British Petroleum Co., BP Statistical Review of World Energy; Published annually by BP, London.

British Petroleum Co., 1959, *Fifty years in pictures;* publ. BP, London 159 p.

Brown, D., 1977, *Some reminiscences of an industrialist;* Eastern Hive Publishing ISBN 0-87960-109-4.

Brown, T., 1991 Mexico oil assets called inflated; *The Arizona Republic,* Dec. 10, 1991.

Browne, E.J.P., 1991, *Upstream oil in the 1990s: the prospects for a new world order;* Oxford Energy Seminar, Sept. 1991. Publ. The British Petroleum Company, London.

Browne, E.J.P., 1991, The way ahead – hydrocarbons for the 1990s: *Amer. Assoc. Petrol. Geol.* London Conference. preprint.

Buderi, R., 1992, Oil's downhill skid may be ending; *Business Week,* Feb. 1992.

Bundesministerium für Wirtschaft, 1993, *Securing Germany's economic future;* Report 338.

Bundesministerium für Wirtschaft, 1995, *Sources of energy in 1995: reserves, resources and availability;* Report 383.

Business Week, 1993, The scramble for oil's last frontier; *Business Week,* Jan 11, 1993.

Byman, D., 1996, Let Iraq collapse!; *The National Interest* 45 48-60.

Cambridge Energy Research Associates, 1991, *The capacity race: the long term future of world oil supply;* James Capel Report 1991.

Campbell, C.J. and H. Bürgl, 1965, Section through the Eastern Cordillera of Colombia, South America; *Geol. Soc. Amer.* 76 567-590.

Campbell, C.J., and E. Ormaasen, 1987, The discovery of oil and gas in Norway: an historical synopsis; in Spencer A.M. (Ed) *Geology of the Norwegian oil and gas fields*, Norweg. Petrol. Soc. 493p. ISBN 086010-908-9.

Campbell, C.J., 1989, Oil price leap in the early nineties; *Noroil* 17/12, 35-38.

Campbell, C.J., 1991, *The golden century of oil 1950-2050: the depletion of a resource;* Kluwer Academic Publishers, Dordrecht, Netherlands; 345p. ISBN 0-7923-1442-5.

Campbell, C.J., 1992, The depletion of oil; *Marine & Petrol. Geol.,* v.9 Dec. 1992, 666-671.

Campbell, C.J., 1993, Assumptions, *AAPG Explorer,* Readers Forum, Sept. 1993.

Campbell, C.J., 1993, The Depletion of the world's oil; *Petrole et technique.* 383. 5-12 Paris

Campbell, C.J., 1994, *An oil depletion model: a resource constrained yardstick for production forecasting;* Rept., Petroconsultants S.A., Geneva.

Campbell, C.J., 1994, *The end of an era. What now?;* Abs. Sp. Lecture. Petrol. Explor. Soc. Gt. Britain Newsletter. Mar. 1994.

Campbell, C.J., 1994, Scraping the barrel; *The Economist* Aug. 6th 1994.

Campbell, C.J., 1994, Scrambling for oil; *Time Magazine* July 11.

Campbell, C.J., 1994, The imminent end of cheap oil-based energy; *SunWorld* 18/4, 17-19.

Campbell, C.J., 1994, Nytt prissjokk pa olje – og ny seilskuteetid; *Stavanger Aftenblad* 9 Sept, 1994.

Campbell, C.J., 1994, *Analysis of Norwegian exploration policy 1965-93 and proposals for change:* NORRET paper, Oslo.

Campbell, C.J., 1995, Taking stock; *SunWorld* 19/1 16-19.

Campbell, C.J., 1995, The Coming Crisis; *SunWorld* 19/2 16-19.

Campbell, C.J., 1995, The next oil price shock: the world's remaining oil and its depletion; *Energy Expl. & Exploit.,* 13/1 19-46.

Campbell, C.J., 1995, Proving the unprovable; *Petroleum Economist,* May 1995.

Campbell, C.J., 1995, Cassandra or prophet; *Petroleum Economist,* Oct. 1995.

Campbell, C.J., 1996, The resource constraints to oil production: the spectre of a pending chronic supply shortfall; in Kürsten M. (Ed) World energy – a changing scene; *Proc. 7th Int. Symposium BGR,* Hannover, E. Schweizerbart'sche Verlagsbuchhandlung, Stuttgart 227p. ISBN 3-510-65170-7.

Campbell, C.J., 1996, The status of world oil depletion at the end of 1995; *Energy Exploration and Exploitation;* 14/1. 1996.

Campbell, C.J., 1996, Oil Shock; *Energy World,* June 1996.

Campbell, C.J. and J.H. Laherrère, 1995, Gauging the North Sea; *Platt's Petroleum Insight,* April 10, 1995.

Campbell, C.J. and J.H. Laherrère, 1995, *The world's supply of oil 1930-2050;* Report Petroconsultants S.A., Geneva.

Campbell, C.J., 1996 *The global Hubbert Peak;* Internet: http://ecotopia.com/hubbert/campbell.htm

Campbell, C.J., 1996, World oil – Reserves, production, politics and prices; in Doré, A.G and R. Sinding Larsen (Eds), *Quantification and prediction of hydrocarbon resources*; NPF Sp. pub. 6., Elsevier ISBN 0-444-82496-0.

Campbell, C.J., 1997, Non-opec future production; *Oxford Energy Forum* 28 Feb. 1997.

Campbell, S.J.D. and N. Gravdal, 1995, Prediction and evaluation of high porosity chalks in the East Hod Field; *Pet. Geoscience* 1/1.

Carmalt, S.W. and B. St. John, 1986, Giant oil and gas fields; in Future petroleum provinces of the world; *Amer. Assoc. Petrol. Geol. Mem.* 40.

Cazalot, C.P., 1996, Texaco exploration and production; *Petrol. Eng. Int.,* May 1996, 17-22 .

Cazier, E.C., *et al.,* 1995, Petroleum geology of the Cusiana Field, Llanos Basin foothills, Colombia; *Amer. Assoc. Petrol. Geol.* 79/10 1444-1463.

Centre for Global Energy Studies, 1992, *The long-run price of oil:* 3.

Centre for Global Energy Studies, 1992, *The costs of future North Sea oil production;* 3.

Centre for Global Energy Studies, 1994 *Monthly oil report;* 3/5.

Centre for Global Energy Studies, 1994, *Oil production capacity in the Gulf,* II t.

Centre for Global Energy Studies, 1993, *Oil production capacity in the Gulf* II Saudi Arabia.

Chevron Corp., 1990, *Overview, energy outlook to the year 2000.*

Chevron Corp., 1993, Saudi memories; *Chevron World* winter/spring 1993, 14-18.

Clark, P., P. Coene and D. Logan, 1981, *A comparison of ten US oil and gas supply models;* Federal Reserves Institute, Washington.

Cleveland, C.J., 1992, Yield per effort for additions to crude oil reserves in the Lower 48 United States 1946-1989; *Amer. Assoc. Petrol. Geol.* 76/7, 948-958.

Cleveland, C.J. and R.K. Kaufmann, Forecasting ultimate oil recovery and its rate of production: incorporating economic forces into the models of M. King Hubbert; *The Energy Journal* 12/2.

Colvin, M., 1994, Britain's Gulf War ally helped Saddam build nuclear bomb; *Sunday Times* July 24, 1994.

Cook, E., 1996, Marking a milestone in ozone protection: learning from the CFC phase-out; *Issues and Ideas,*

World Resources Inst.

Cook, W.J., 1993, Why Opec doesn't matter anymore; *US News & World Report,* 13/12/93.

Cornelius, C.D., 1987, Classification of natural bitumen: a physical and chemical approach; in Meyer, R.F. (Ed) Exploration for heavy crude oil and bitumen; *AAPG Studies in geology 25.*

Cornot-Gandolphe, S., 1994, *The main issues in gas processing, liquifaction LNG transport and storage;* preprint APS Conf., Cyprus.

Corzine, R., 1993, Warning over Saudi output; *Financial Times,* 4/11/93.

CPN, 1994, A forecast of China's petroleum resources; *China Pet. Newsletter* 1/9, April 27 1994.

Crandell, J.D. *et al.,* 1993, '93 E&P spending: shifting back to the US, *World Oil,* Feb. 1993.

Crandell, J.D., 1994, Depends on oil prices; *World Oil,* Feb 1994.

Creswell, J., 1996, Global demand fuels need for more drilling; *Press & Journ.,* Aberdeen.

Crow, P., 1991, Oil and bullets; *Oil & Gas Journ.,* Oct 21. 1991.

Dallas Morning News, 1993, Arco says Alaskan fields could increase reserves 60%; *Dallas Morning News,* April 14 1993.

Davies, P., 1994, Oil supply and demand in the 1990s; *World Petrol. Congr.* Topic 16.

Davidson, D.D. and W. Rees-Mogg, 1988, *Blood in the streets – investment profits in a world gone mad;* Sidgwick & Jackson, London, 385p, ISBN 0-283-99601-3.

Davidson, J.K., 1992, *Tectonic control of world's oil reserves – Australia's position;* APEA Conference, Perth 1992 preprint.

Davidson, J.K., 1995, Globally synchronous compressional pulses in extensional basins: implications for hydrocarbon exploration : *APEA Journ.* 1995 169-188.

Deitzman, W.D. *et al.* 1983, *The petroleum resources of the Middle East;* US Dept. of Energy / Energy Information Administration Report. DOE/EIA-0395 May 19 83 169p.

Dell, P.R., 1994, Global energy market : future supply potentials; *Energy Exploration and Exploitation,* 12/1 59-72.

Demaison, G. and G.T. Moore, 1988, Anoxic enviorments and oil source-bed genesis; in Beaumont, E.A. and N.H. Foster (Eds) *Geochemistry;* AAPG Treatise of Petroleum Geology Reprint Series No. 8 ISBN 0-89181-407-8.

Demaison, G. and B.J. Huizinga, 1991, Genetic classification of petroleum systems; *Amer. Assoc. Petrol. Geol.* 75 1626-43.

Demaison, G. and A. Perrodon, 1994, *Petroleum systems and exploration strategy;* AAPG Course, Mexico. Oct. 1994.

Department of Energy, 1989, *Federal oil research: a strategy for maximizing the producibility of known US oil;* US Dept of Energy DOE/FE 0139.

Department of Energy, 1995, *US crude oil, natural gas and natural gas liquids reserves;* 1994 Annual Rept.

Department of Trade and Industry; 1996, "Brown book" extract on reserves.

DeSorcy, G.J., *et al.,* 1993, Definitions and guidelines for classification of oil and gas reserves; *Journ. Canadian Petrol. Technol.* 32/5, 0-21.

Dicea, O., 1995, The structure and hydrocarbon geology of the Romanian East Carpathian border from seismic data; *Petrol. Geoscience,* 1/2, 135-144.

Doré, A.G. and E.R. Lundin, 1996, Cenozoic compressional structures on the NE Atlantic margin: nature, origin and potential significance for petroleum exploration; *Petroleum Geoscience,* 2/4, 299-311.

Dromgoole, P. *et al.,* 1994, *Oil supply stretch;* BP background paper.

Duncan, R.C., 1995, *The energy depletion arch: new constraints enhance the Hubbert model:* Public Interest Environmental Law Conf., Univ. of Oregon.

Duncan, R.C., 1996, *Mexico's petroleum exports: safe collateral for a $50 billion loan?* Sp. Report, Inst. on Energy & Man, Seattle.

Duncan, R.C., 1996, D-Model; pers.comm.

Duncan, R.C., 1996, *The Olduvai Theory: sliding towards the post-industrial stone age;* Inst. Energy & Man, Seattle.

Duncan, R.C., 1996, *Fossil fuel prospects for the twenty-first century;* Inst. Energy & Man, Seattle.

Dunn, P.D. and H. DuMoulin, 1996, The future of fossil fuel resources; *Energy World,* 240.

Economist, The, 1993, Pollution; *The Economist,* June 13th, 1993.

Economist, The, 1993, A shocking speculation about the price of oil; *The Economist,* Sept.15, 1993. 87-88.

Economist, The, 1994, Power to the people – a survey of energy; June 18th. 1994.

Economist, The, 1995, Illusion of change; *The Economist* 12/8/95.

Economist, The, 1995, The future of energy; Oct.7 1995.

Economist, The, 1996, From major to minor; May 18th.

Economist, The, 1996, Pipe dreams in central Asia; May 4th.

Economist, The, 1966, Kurdistan, which one do you mean? Aug. 10th.

Einhorn, C.S., 1994, *Well oiled;* Barron's. June 27.

Egg, M., 1995, Secrets of the Ice Man; *Science Spectra,* 2.

1995.

Ellis, P.A., 1991, New technology for gas finding: how important has it been? *Oil & Gas Journ.* Sept 30, 1991.

Energy Economist, 1993, The 1996 oil shock?; 139/17. May 1993.

Energy Economist, 1993, Saudi Arabian sands; 145/14. Nov. 1993.

Energy Information Administration, 1985, International energy outlook 1985; DOE/EIA 0484(85).

Energy Information Administration, 1988, Annual energy review; DOE/EIA 0384 (88).

Energy Information Administration, 1990, US oil and gas reserves by year of field discovery; DOE/EIA 0534.

Energy Information Administration, 1993, International oil and gas exploration and development 1991.

Esau, I., 1992, Azeri initiative; *Offshore Eng.* Aug. 1992.

Eureka, 1991, *Total-Colombian discovery;* July 1991.

European Commission, 1995, *For a European Union Energy Policy;* Green Paper.

Falcoff, M.,1995, The Cuba in our mind; *National Interest.* Summer 1995.

Farago, L., 1973, *The Game of Foxes;* Bantam Books, New York 878p.

Fernandez, R., 1994, Energy finance and geopolitics; preprint APS Conf., Cyprus

Ferrier, R.W., 1982, *The history of the British Petroleum Company: Volume 1, the developing years 1901-1932;* Cambridge University Press 801p ISBN 0-521-24647-4.

Ferriter, J.P., 1994, International energy cooperation; *Petrole et Technique* 389.

Fisher, W.L., 1991, Future supply potential of US oil and natural gas; *Geophysics: The leading edge of exploration,* Dec 1991.

Fisher, W.L., 1993, Oil and gas prices are headed higher; *Amer. Oil & Gas Reporter,* Jan 1993.

Fleay, B., 1995, *The decline of the age of oil, petrol politics: Australia's road ahead;* Pluto Press 152p ISBN 1-86403-021-6.

Fleay, B., and J.H. Laherrère, 1997, Sustainable energy policy for Australia; submission to the Department of Primary Industry and Energy Green Paper 1996; Paper 1/97 Institute for Science and Technology Policy, Murdoch University, W. Australia.

Flittie, C.G. and B.M. Robertson, 199?, William Elliott Humphrey; *Amer. Assoc. Petrol. Geol.*

Flores, G., 1987, *Arc of the sun;* Lion & Thorne, Tulsa. 229p. ISBN 0-914381-05-9.

Foley, G. and A. van Buren, 1978, *Nuclear or not: choices for our energy future;* ISBN 0-435-54770-4.

Folinsbee, R., 1977, World view from Alph to Zipf; *Geol. Soc. Amer.* 88 897-907.

Forbes, 1992, *The next oil crisis?* Forbes Feb. 17, 1992.

Foreman, N.E., 1997, The bear awakens: resurgence of FSU oil and gas; *World Oil,* Feb. 1997.

Foreman, N.E., 1996, Opec influence grows with world output in next decade; *World Oil,* Feb 1996.

Friedman, A., 1990, Opec needs $60 bn extra for productivity increase; *Financial Times,* Feb 8th 1990.

Friedman, T.L., 1994, Opec's lonely at the tap, but China's getting thirsty; *New York Times,* Apr 3 1994.

Fromkin, D., 1989, *A peace to end all peace;* Avon Books 635p ISBN 0-380-71300-4.

Fuller, J.G.C., 1971, The geological attitude; *Amer. Assoc. Petrol. Geol.* 55/11.

Fuller, J.G.C., 1993, The oil industry today; Brit. Assoc. Advancement of Sci. Published as The British Association Lectures 1993 by The Geological Society, London.

Garb, F.A., 1985, Oil and gas reserve classification, estimation and evaluation; *Journ. Petrol. Technol.* March 373-390.

Gauthier, D.L. *et al.* 1995, *1995 national assessment of United States oil and gas resources – results, methodology and supporting data:* US Geological Survey CD-ROM.

George, A., 1996, *Reality and risk in the Middle East;* preprint Oil & Gas Project Finance in the Middle East Conference, Dubai May 12-13.

George, D., 1993 Caspian Sea reserves could rival those of Saudi Arabia; *Offshore/Oilman* July 1993.

Geoscientist, 1996, World oil supplies, 6/1.

Ghadhban, T.A. *et al.,* 1995, *Iraq oil industry: present conditions and future prospects;* Report Iraq Oil Ministry.

Ghorban, N., 1994, *The evaluation of recent gas export pipeline proposals in the Middle East;* preprint, APS Conf. Cyprus.

Gibson, W.R. and C.F. Garvey, 1992, 3D seismic has renewed the search for stratigraphic traps; *World Oil,* Sept. 1992.

Giddens, P.H., 1955, *Standard Oil Company (Indiana) – oil pioneer in the Middle West;* Appleton-Century-Crofts, New York

Giraud, A., 1993, *1973-1993 – L'eclairage du passe, l'approche du futur;* Profils IFP 94/1.

Gold, T., 1988, Origin of petroleum; two opposing theories and a test in Sweden; *Geojournal Library* 9 85-92.

Goldsmith, J, 1994, *The trap;* Macmillan 216p.

Goodman, M. and N.C. Chriss, 1979, Mexico oil estimates inflated, experts say; *Los Angeles Times,* May 18 1979.

Goodwin, S., 1980, Hubbert's Curve; *Country Journal,* Nov 1980.

Grace, J.D., R.H. Caldwell and D.I. Heather, 1993,

Comparative reserve definitions: USA, Europe and the former Soviet Union; *Journ. Petrol. Technol.* Sept. 1993 866-872.

Grassl, H., 1994, *Scientific assessment of climate forcing implication for energy policy;* preprint BGR Conference, Hannover.

Gray, D., 1989, North Sea outlook and its sensitivity to price; *Petrol. Revue.* Jan 1989.

Greenhouse, S., 1992, Chevron taps ex-Soviet reserves; *Ft. Worth Star Telegram,* May 19 1992.

Greenwald, J., 1994, Black gold, *Time,* June 20.

Greer, C., 1994, My challenge to the World: an interview with Mikhail Gorbachev; *Times-Picayune Parade,* N. Orleans. Jan 4 1994.

Groeneveld, R., 1966, Auto LPG: global review and criteria for success; *Pet. Review,* May .

Grove, V., 1994, Brussels is a madness. I will fight from within; *Times* June 10.

Gulbenkian, N., 1965, *Portrait in oil;* Simon & Schuster, New York.

Haines, L., 1994, *Oil price recovery: for real or a dream;* NewsWell.

Haldorsen, H.H., 1996, Choosing between rocks, hard places and a lot more: the economic interface; in Doré, A.G and R. Sinding Larsen (Eds), *Quantification and prediction of hydrocarbon resources;* NPF Sp. pub. 6., Elsevier ISBN 0-444-82496-0.

Hart's Petroleum Engineer International, 1996, New paradigm: mining for oil; Sept. p.15.

Harvey, H., 1993, Innovative policies to promote renewable energy; *Advances in Solar Energy,* Amer. Solar Energy Soc.

Haun, J.D. (Ed), 1975, Methods of estimating the volume of undiscovered oil and gas resources; *Amer. Assoc. Petrol. Geol. Studies in Geology* No.1.

Hawken, P., 1993, *The Ecology of Commerce – a declaration of sustainability;* Publ. Harper Business, New York 250p. ISBN 0-88730-655-1.

Halbouty, M.T., 1970, Geology of giant petroleum fields; *Amer. Assoc. Petrol. Geol. Mem.* 14.

Harris, R., 1994, *Production Analysis shows growth in worldwide oil and gas reserves: World Watch,* Petroconsultants.

Hatfield, C.B., Will an oil shortage return soon? *Geotimes* Nov. 1995.

Hewins, R., 19?, *Mr Five Per Cent: the biography of Calouste Gulbenkian;* Hutchinson.

Hiller, K., 1996, *Depletion midpoint and the consequences for oil supplies;* pre-print WPC.

Hoagland, J., 1994, The US-Saudi line is off the hook; *Int. Herald Tribune,* 31.1.94 .

Hobbs, G.W., 1995, Oil, gas, coal, uranium, tar sand resource trends on rise; *Oil & Gas Journ.,* Sept 4 1995.

Hobson, G.D., 1980, Musing on migration; *Journ. Petrol. Geol.* 3/2.

Hobson, G.D., 1991, Field size distribution – an exercise in doodling? *Journ. Petrol. Geol.* 14/1.

Hollis, R., 1996, Stability in the Middle East – three scenarios; *Pet. Review* May p. 205.

Home, P.C., 1987, Ula; in Spencer, A.M. *et al.* (Eds); *Geology of the Norwegian oil and gas fields;* Graham & Trotman, 493p, ISBN 086010-908-9.

Horton, S. and N. Mamedov, 1996, Investment in Azerbaijan's upstream requires attention to legal details, *World Oil,* April 1996.

Hotelling, H., 1931, The Economics of exhaustible Resources; *Journ. Politic. Economy* 1931.

Howgego, H., 1996, Outwardly simple, inwardly rich; *Oxford Today,* 8/3 p.40.

Hubbert, M.K., 1949, Energy from fossil fuels; *Science* 109, 103-109.

Hubbert, M.K., 1956, Nuclear energy and the fossil fuels; Amer. Petrol. Inst. Drilling & Production Practice. Proc. Spring Meeting, San Antonio, Texas. 7-25.

Hubbert, M.K., 1962, *Energy resources, a report to the Committee on Natural Resources;* Nat. Acad. Sci. Publ. 1000D.

Hubbert, M.K., 1969, Energy resources; in Cloud P. (Ed) *Resources and Man;* W.H.Freeman.

Hubbert, M.K., 1971, Energy resources of the Earth; in Energy and Power, W.H. Freeman.

Hubbert, M.K., 1980, Oil and gas supply modeling; in Gass S.I., (Ed). proceedings of symposium, US Dept. of Commerce, June 18-20,1980.

Hubbert, M.K., 1981, The world's evolving energy system; *Amer. J. of Physics,* 49/11, 1007-1029.

Hubbert, M.K., 1982, *Technique of prediction as applied to the production of oil & gas;* in NBS Special Publication 631. US Dept.Commerce/National Bureau of Standards, 16-141.

Imbert, P., J.L. Pittion and A.K. Yeates, 1996, Heavier hydrocarbons, cooler environment found in deepwater; *Offshore,* April.

Independent Petroleum Association of America, 1993, *The promise of Oil and Gas in America;* final report of IPAA Potential Resources Task Force.

Ibrahim, Y.M., 1990, Widespread unrest threatens world oil supply, experts say; *Int. Herald Tribune,* 6/3/90.

Institute of Petroleum, 1993, Valuable Saudi upstream data published; *Pet. Review,* Dec 1993.

Institute of Petroleum, 1995, *The UK continental shelf in 2010: is this the shape of the future?* Report.

International Energy Agency, 1993, *Oil Market Report,* 7 Sept. 1993.

International Energy Agency, 1994, *World Energy Outlook 1994 Edition.*

International Energy Agency, 1995, *World Energy Outlook 1995 Edition.*

Investors Chronicle, 1994, Driven by supply; 23/12/94.

Ismail, I.A.H., 1994, Untapped reserves, world demand spur production expansion; *Oil & Gas Journ.* May 2, 1994. 95-102.

Ismail, I.A.H., 1994, *The world oil production perspective; the future role of Opec and Non-Opec;* Preprint, APS Conf., Cyprus.

Ismail, I.A.H., 1994, Future growth in OPEC oil production capacity and the impact of environmental measures; *Energy Exploration and Exploitation* 12/1 17-58.

Ivanhoe, L.F., 1976, Evaluating prospective basins; in three parts, *Oil & Gas Journ.* Dec 13. 1976.

Ivanhoe, L.F., 1980, World's prospective petroleum areas; *Oil & Gas Journ.* April 28 1980.

Ivanhoe, L.F., 1984, Oil discovery indices and projected discoveries; *Oil & Gas Journ.* 11/19/84.

Ivanhoe, L.F., 1985, Potential of world's significant oil provinces; *Oil & Gas Journ.* 18/11/85.

Ivanhoe, L.F., 1886, Oil discovery index rates and projected discoveries of the free world; in Oil & Gas Assessment. *Amer. Assoc. Petrol. Geol. Studies in Geology* #21, 159-178.

Ivanhoe, L.F., 1987, Permanent oil shock; *Amer. Assoc. Petrol. Geol.* 71/5.

Ivanhoe, L.F., 1988, Future crude oil supply and prices; *Oil & Gas Journ.* July 25 111-112.

Ivanhoe, L.F., 1990, Liquid fuels fill vital part of US Economy; *Oil & Gas Journ.* Apr. 23, 106-109.

Ivanhoe, L.F., 1990, Competition increases to obtain oil imports; *Oil & Gas Journ.* Oct 29. 1990.

Ivanhoe, L.F., 1991, Oil, gas dominant sources of energy in US; *Oil & Gas Journ.* Sept 30 1991.

Ivanhoe, L.F., 1995, Future world oil supplies:there is a finite limit; *World Oil,* Oct. 1995.

Ivanhoe, L.F., and Leckie, G.G., 1991, Data on field size useful to supply planners; *Oil & Gas Journ,* April 29 1991.

Ivanhoe, L.F., and Leckie, G.G., 1993, Global oil, gas fields, sizes tallied, analyzed; *Oil & Gas Journ.* Feb 15 1993, 87-91.

Ivanhoe, L.F., 1995, *Oil reserves and semantics;* M. King Hubbert Center for Petrol. Supply studies.

Ivanhoe, L.F., 1996, World oil supply; pers. comm.

Ivanhoe, L.F., 1996, Updated Hubbert curves analyze world oil supply; *World Oil.* Nov. 91-93.

Jacque, M. 1994, Reserves mondiales de petrole; *geochronique* 49.

James, M., 1953, *The Texaco Story – the first fifty years 1902-1952:* Publ. The Texas Company . 115p.

Jefferson, M., 1994, World energy prospects to 2010; *Petrole et Techn.* 389.

Jenkins, D.A.L., 1987, An undetected major province is unlikely; *Petrol. Revue* Dec 1987 p. 16.

Jennings, J.S., 1996, The millenium and beyond; *Energy World* 240 June.

Kaletsky, A., 1996, Time has come to review the demand side disaster; *Times* 25.1.96.

Kassler, P., 1994, *Two global energy scenarios for the next thirty years and beyond;* World Petrol. Congr. Stavanger.

Kaufmann, R., 1991, Oil production in the Lower 48 States: *Res. & Energy* 13.

Kaufmann, R. and C.J. Cleveland, 1991, Policies to increase US oil production likely to fail, damage the economy, and damage the environment; *Ann. Rev. Energy Environ.* 1991.

Kaufmann, R., W. Gruen and R. Montesi, 199?, *Drilling Rates and expected oil prices: the own price elasticity of US oil supply;* Centre for Energy & Environmental studies, paper.

Kenney, J.F., 1996, Impeding shortages of petroleum re-evaluated; *Energy World* 250 June 1992.

Kelsey, T., 1995, Can 'mother' stop the sons of Islam? *Sunday Times,* 31.12.95.

Khalimov, E.M., 1993, Classification of oil reserves and resources in the former Soviet Union; *Amer. Assoc. Petrol. Geol.* 77/9 1636 (abstract).

Klemme, H.D., 1983, Field size distribution related to basin characteristics; *Oil & Gas Journ.* 1983 169-176.

Klemme, H.D. and Ulmishek, G.F., 1991, Effective petroleum source rocks of the world: stratigraphic, distribution and controlling depositions factors; *Amer. Assoc. Petrol. Geol.* 75/12, 1908-185.

Knott, D., 1991, Calm surface, deep currents: is industry storing up problems?; *Offshore,* Dec 1991, 25-27.

Knott, D., 1994, Opec, once all-powerful, faces a cloudy tomorrow; *Oil & Gas Journ.* Aug 22, 1994.

Knott, D., 1996, Reserves debate; *Oil & Gas Journ.* Jan 29.

Krayushkin, V.A., *et al.,* 1994, *Recent application of the modern theory of abiogenic hydrocarbon origins: drilling and development of oil and gas fields in the Dneiper-Donets Basin;* 7th Int. Symposium on the observation of the continental crust through drilling, Sante Fe, New Mexico, proc. 1994.

Kulke, H., I.Taner and A. Mayerhoff, 1994, *China; in Regional Petroleum Geology of the World.* Borntrager, Stuttgart.

Kulke, H. 1994, *Sweden, in Regional Petroleum Geology of the World.* Borntrager, Stuttgart.

Kvint, V., 1990, Eastern Siberia could become another

Saudi Arabia; *Forbes,* Sept 17 1990.

Laherrère, J.H., 1990, Hydrocarbon classification rules proposed; *Oil & Gas Journ.* Aug. 13. p.62.

Laherrère, J.H., 1990, *Les Reserves d'hydrocarbures;* BIP 6629.

Laherrère, J.H., 1992, *Reserves mondiales restantes et à découvrir:* ATFP Conf. Paris 18.4.91.

Laherrère, J.H., 1993, Le petrole, une ressource sure, des reserves incertaines; *Petrol et Technique,* 383, Oct.

Laherrère, J.H., A. Perrodon and G. Demaison, 1994, *Undiscovered petroleum potential: a new approach based on distribution of ultimate resources;* Rept. Petroconsultants S.A., Geneva.

Laherrère, J.H., 1994, Published figures and political reserves; *World Oil,* Jan 1994. p. 33.

Laherrère, J.H., 1994, Study charts US reserves yet to be discovered; *American Oil & Gas Reporter* 37/9 99-104.

Laherrère, J.H., 1994, Nouvelle approche des reserves ultimate – application aux reserves de gaz des Etas-Unis; *Petrole et Technique,* Paris 392. 29-33.

Laherrère, J.H., 1994, Reverves mondiales de petrole: quel chiffre croire?; *Bull. Inform. Petrol.* 7727, 7728, 7729.

Laherrère, J.H., 1995, World oil reserves: which number to believe?; *OPEC bull.* Feb.

Laherrère, J.H., 1996, Distributions de type <fractal parabolique> dans la nature; *C.R. Acad. Sci.* Paris 322 IIa 535-541.

Lamar, L., 1992, World energy statistics; *Shale Shaker,* May/June 1992.

Laurier, D., *Le gaz: exploration, gisements, stockages;* ATFP Conference, Paris.

Leckie, G.G., 1993, Hydrocarbon reserves and discoveries 1952 to 1991; *Energy Exploration & Exploitation,* 11/1, 1993.

Lee, P.J and P.R. Price, 1991, Successes in 1980s bode well for W. Canada search; *Oil & Gas Journ.* April 22.

Lee, P., 1992, *Shifting sands;* Fort Worth Star Telegram, Jan 15.

Lelkes, A., 1996, Hungary's resources; pers, comm.

Lelkes, A, 1996 , Russian depletion study; pers.comm.

Lenzner, R. and J.M. Clash, 1994, Wrong again; *Forbes* June 6, 1994.

Lenzner, R., 1994, The case for hard assets; *Forbes,* June 20 1994.

Leonard, R.C., 1984, Generation and migration of hydrocarbonson southern Norwegian shelf; *Amer. Assoc. Petrol. Geol.* 68 796.

Leonard, R.C., 1993, *Distribution of subsurface pressure in the Norwegian Central Graben and applications for exploration;* Petrol. Geol. of NW Europe Proc. 4th Conf.

Lia, A., 1966, Elf fine tunes its strategy; *Pet. Eng. Int.,* May 1996.

Lippman, T.W., 1990. Saudis come up with major oil find; *Washington Post,* Oct. 15.

Lippman, T.W., 1992, Saudis and US teamed on oil issues; *Fort Worth Star Telegram,* July 24, 1992.

Long, D., 1996, Oil stocks; *Oxford Energy Forum,* 27, 3-5.

Longhurst, H.,1959, *Adventure in oil: the story of British Petroleum;* Sidgwick and Jackson, London, 286p.

Los Angeles Times, 1991, Mexico lied about Proven Oil Reserves, report says; Dec. 10, 1991.

Lübben, H. von and J. Leiner; 1988, Öl: perspektiven im Upstream bereich; Erdol, *Erdgas,* Kohle 104/5 Mai 1988.

Lynch, M.C., 1992, *The fog of commerce: the failure of long-term oil market forecasting;* MIT Center for Int. Studies Sept. 92.

Mabro, R., 1996, The world's oil supply 1930-2050 – a review article; *Journ. Energy Literature* II.1.96.

Macgregor, D.S., 1996, Factors controlling the destruction or preservation of giant light oilfields; *Petroleum Greoscience* 2. 197-21.

Mack, T., 1991, Are the big oil stocks a buy now?; *Forbes,* Feb. 18, 1991.

Mack, T., 1992, The last frontier; *Forbes,* May 8th 1992.

Mack, T., 1994, History is full of giants that failed to adapt; *Forbes* 28 Feb 1994.

MacKenzie, J.J., 1994; Transportation in the People's Republic of China: beginning the transition to sustainability; preprint World Resources Inst.

MacKenzie, J.J., 1995, *Oil as a finite resource: the impending decline in global oil production:* World Resources Inst.

MacKenzie, J.J., and K. Courrier, 1996, Cutting gas tax will make things worse; *Los Angeles Times,* May 8.

Macleay, J., 1996, Academic warns of black days ahead for oil; *The Australian* 10.10.96.

Macovei, G., 1938, *Les gisements de pétrole;* Masson. Paris.

Malone, A., 1996, Revenge of the Saudi exile; *Sunday Times,* 7 Jan 96.

Mansfield P., 1992, *A history of the Middle East;* Penguin Books 373p.ISBN 0-14-01.6989X.

Marino, J., 1992, Operations set to grow in Colombia; *BPXpress* Jan-Feb 1992.

Martell, H., 1989, *Exploration and Resources, Venezuela;* UN Seminar Policy and Management of Petroleum Resources, Oslo.

Martin, A.J., 1985, *Prediction of strategic reserves in prospect for the world oil industry;* Eds. T. Niblock and R. Lawless. Univ. of Durham 16-39.

Martinez, A.R. *et al.* 1987, Study group report: classification and nomenclature system for petroleum and petroleum reserves; *Wld Petrol. Congr.* 325-342.

Mast, R.F. *et al.,* 1989, *Estimates of undiscovered conventional oil and gas resources in the United States – a part of the nation's energy endowment;* US Dept of Interior.

Masters, C.D., 1987, Global oil assessments and the search for non-OPEC oil; *OPEC Review,* Summer 1987, 153-169.

Masters, C.D., 1991, World resources of crude oil and natural gas; Review and Forecast Paper, Topic 25, p.1-14. *Proc. Wld. Petrol. Congr.,* Buenos Aires 1991.

Masters, C.D., D.H. Root and E.D. Attanasi, 1991, Resource constraints in petroleum production potential; *Science* 253.

Masters, C.D., 1993, US Geological Survey petroleum resource assessment procedures; *Amer. Assoc. Petrol. Geol.* 77/3 452-453 (with other relevant references).

Masters, C.D., 1994, World Petroleum analysis and assessment; *Wld. Petrol. Congr.* Stavanger.

Masters, C.D., 1994, Bibliography of the world energy resources program; *USGS Open File* 94-556.

Mathews, J., 1996, World oil market faces instability and change; *ABQ Journ.* 28.2.96.

McCabe, P.J., *et al.,* 1993, *The future of energy gases;* USGS Circ.1115.

McRae, H., 1994, *The world in 2020 : power, culture and prosperity, a vision of the future;* Haper Collins 302p ISBN 0-00638382-3.

McRae, H., 1993., Oil looks slippery; *Independent* 7/5/93.

McRae, H., 1993, No end of cheap oil; *Independent* 11/11/93.

Megill, R.E., Discoveries lag oil consumption; *AAPG Explorer* Aug.1993.

Megill, R.E., 1994, Another look at finding cost; *AAPG Explorer* July 1994.

Mellbye, P., 1994, Norway's role, satisfying increasing demand for natural gas in Europe; *World Petrol Congr.* Stavanger.

Menezes, F. Ramos de, 1994, Worldwide criteria versus regional criteria; *Wld Petrol. Congr.* Stavanger.

Merrill Lynch, 1994, *Energy Monthly;* report.

Miller, R.G., 1992, The global oil system: the relationship between oil generation, loss half-life and the world crude oil resource; *Amer. Assoc. Petrol. Geol.* 76/4 489-500.

Mitchell, J., 1995, *The geopolitics of energy;* EU Report.

Mitchell, J., 1996, *The new geopolitics of energy;* Royal. Inst. Int. Affairs April 1996.

Mohammed, A.H., 1989, *Crude oil production, refining, petrochemicals and research activities in Iraq;* UN Seminar : Policy and Management of Petroleum Resources, Oslo.

Moody-Stuart, M., 1993, *Resources and resourcefulness;* AAPG Int. Conf., The Hague.

Moody-Stuart, M., 1996, *Winners and losers – meeting the upstream challenges of the 21st Century;* AAPG/SVG Congress, Caracas. 9 Sept. 1996. (Shell publication).

Morange, A., A. Perrodon and F. Héritier, 1992, *Les grandees heures de l'exploration pétrolière du Groupe Elf Aquitaine;* Bol. des Centres de Recherches exploration – Production Elf Aquitaine ISBN 2-901026-36-2.

Mortished, C., 1995, Shell thinks, then does the unthinkable; *Times* , March 3rd.

Mortished, C., 1996, Iraq oil could be back on sale soon; *Times,* March 11th.

Mortished, C.,1997, Tempus – BP; *Times,* February 6th.

Mosley, L., 1973, *Power play: oil in the Middle East;* Random House.

Narimanov, A. and A. Palaz, 1994, Baku region rich with oil, history; *AAPG Explorer* Oct. 1994 p.40.

Nasr, S., 1944, The economic opportunities and risks in the Middle East; preprint APS Conf.

Nation, L., 1992, Bass Straits oil still a mainstay; *AAPG Explorer* July 1992.

Nation, L., 1993, Delegates told oil prices must rise; *AAPG Explorer.*

Nation, L., 1995, Hodel sees looming energy crisis; *AAPG Explorer* May 1995 p.28.

Naturhistorischen Museum, 1987, *Hans G. Kugler 1893-1986;* Memorial, Basel.

NatWest, 1994, *Oil market outlook;* NatWest Securities report.

Nasmyth, J., 1996, A story of oil price reporting 1945-1985; *Pipeline,* May 1996.

N.E.R.A., 1993, Oil and natural gas price outlook; *Energy Outlook,* Nat. Econ. Research. Assoc. Feb. 15 1993.

Nehring, R., 1978, *Giant oil fields and world oil resources;* CIA report R-2284-CIA.

Nehring, R., 1979, *The outlook for conventional petroleum resources;* Paper P-6413 Rand Corp.21p.

Newswell, 1991, *Higher oil prices to spur 1991 economic growth in southwest.*

Norwegian Petroleum Directorate, 1993, *Improved oil recovery.*

O'Dell, S., 1994, *Prospects for non-opec oil supply;* 13th CERI Conf.

Odell, P.R., 1996, Middle East domination or regionalisation; Erdol, Erdgas, Kohle 4.

Odell, P.R., 1996, Britain's North Sea oil and gas production – a critical review; *Energy Exploration &*

Exploitation 14/1.

Odum, H. and E., Odum, 1981, *Energy basis for man and nature;* McGraw Hill, New York.

Offshore, 1989, Non-opec producers limited in ability to take advantage of rising oil demand; *Offshore,* June 1989

Offshore, 1989, *Poor US performances underlie reserve buys, international move;* June 1989.

Offshore, 1996, *Improved recovery grows Norwegian reserves;* April 1996.

Oil & Gas Journal, World Production Reports; December each year.

Oil & Gas Journal, 1989, IPAA, US oil production headed for biggest slide since the 1970s; Nov 6th 1989.

Oil & Gas Journal, 1989, Crude/condensate production slips in Soviet Union; Nov. 27 1989.

Oil & Gas Journal, 1994, Steady rise in oil, gas demand ahead; June 6 1994.

Oil & Gas Journal, 1994, Opec draws praise for restraint; Aug. 18 1994.

Oil & Gas Journal, 1994, Worldwide oil flow up, reserves steady in 1994; 26/12/94.

Oil & Gas Journal, 1995, OGJ Newsletter;2/1/95.

Oilman, The, 1996; UK reserves and potential; 18 March p 3-5.

Orphanos, A., 1995, Looking for oil prices to gush; *Fortune,* 15/5/95.

Pace, B.W., 1996, Petroleum economics; Course material Imperial College.

Parra, A.A., 1995, *Opec: the question of new oil:* 5th CGES Annual Conf. London.

Parent, L., 1989, Natural gas: life after the bubble; *World Oil,* Feb. 1989.

Patricelli, J.A. and C.L. McMichael, 1995, An integrated deterministic/ probabilistic approach to reserve estimations; *Journ. Petrol. Technol.* 47/1 49-53

Patricelli, J.A. and C.L. McMichael, 1996, Authors' reply to discussion of an integrated deterministic/probabilstic approach to reserve definition; *Journ. Pet. Technol.* Dec. 1996 .

Pauwels, J-P. and F. Possemiers, 1996, Oil supply and demand in the XXIst century; *Revue de l'energie,* 477.

Pees, S.T., 1989, Guidebook, history of the petroleum industry symposium; *Amer. Assoc. Petrol. Geol.*

Perdue, J.M., 1966, Canadian operators boost heavy oil production; *Pet. Eng. Int.,* May 1996.

Perrin, F., 1994, *L'impact de la baisse des prix sur les strategies d'exploration-production;* 7th Seminaire petrolier et gazier Int., Paris.

Perrodon, A., 1985, *Histoire des grandes découvertes pétrolières;* Elf Aquitaine – Masson ISBN 2-225-80659-4.

Perrodon, A., 1988, Hydrocarbons in Beaumont, E.A. and N.H. Foster (Eds) *Geochemistry. Treatise of petroleum geology,* Amer. Assoc. Petrol. Geol. Reprint Series No.83-26 (reproduced from Bull. Centre Rech. Elf Mem. 5 1983).

Perrodon, A. and J. Zabek., 1990, Paris Basin: in Leighton, M.W. *et al.* eds. Interior cratonic basins. *Amer. Assoc. Petrol. Geol. Mem.* 51 819p.

Perrodon, A., 1991, *Vers les reserves ultimes;* Centres Rech. Explor.-Prod. Elf-Aquitaine 15/2 253-369.

Perrodon, A., 1992, Petroleum systems, models and applications; *Journ. Petrol. Geol.* 15/3, 319-326.

Perrodon, A., 1993, *Historique des recherches petrolieres en Algerie;* 118 Congr. Nat. des soc. hist et scient, Pau, 323-340.

Perrodon, A., 1995, Petroleum systems and global tectonics: *Journ. Petrol. Geol.* 18/4 471-476.

Petrie Parkman, 1992, Llanos foothills Trend/ Cusiana Field's potential reserves and production economics assessed; Petroleum Research v IV EPPO1 Feb 22 1992.

Petroconsultants S.A.,1993, *World Production & Reserve Statistics; oil and gas 1992;* Petroconsultants, London.

Petroconsultants S.A., 1993, *Strategic petroleum insights;* Report.

Petroconsultants S.A., 1994, *Oil production forecast;* Report.

Petroleum Economist, 1993, Size and success win votes; June 1993.

Petroleum Engineer Int., 1993, The Norwegian North Sea contains about 75 billion bbl; June 1993.

Petroleum Engineer Int., 1994, Worldwide oil and gas activity forecast; Jan 1994.

Petroleum Review, 1996, Report on travel trends and fuel consumption in the United Kingdom; May 1996 p 202.

Petroleum Review, 1996, Ample energy reserves for the future; Aug. 1996.

Petroleum Review, 1996, First stages in production of Australian synfuels gets underway; Aug. 1996.

Phipps, S.C., 1993, Declining oil giants, significant contributors to US production; *Oil & Gas Journ.* Oct. 4. 1993.

Pickens, T.B., 1987, *Boone;* Houghton Miffin, Boston ISBN 0-450-42978-4.

Pitts, J.P., 1992, Tighter infill means more oil; *Midland Reporter* Jan. 19 1992.

PIW, 1994, PIW ranks the world's top 50 oil companies; *Petroleum Int. Weekly* 12/12/94.

Plassart, P., 1996, Norvège: derniers puits de Pétrole avant le desert; *Le Nouvelle Economiste* 1050.

Popescu, B.,(Ed) 1994, *Hydrocarbons of Eastern Central Europe;* Springer Verlag ISBN 3-540-55014-3.

Popescu, B., 1996, World oil suppliers at risk; *EEI*

Newsletter 9.

Porter, E., 1995, *Are we running out of oil?* API Discussion Paper 081.

Pratt, W., 1952, Toward a philosophy of oil finding; *Amer. Assoc. Petrol. Geol.,* 26/12 2231-36.

Preusse, A., 1966, Coalbed methane production – an additional utilization of hard coal deposits; in Kürsten M. (Ed) *World Energy – a changing scene;* E. Schweizerbart'sche Verlagshandlung, Stuttgart ISBN 3-510-65170-7.

Protti, G.J., 1994, Canada's upstream petroleum industry; *Journ. Canada. Pet. Tech.* May 1994.

Rahmani, B. Mossavar-, 1983, The Opec Multiplier, *Foreign Policy* 52.

Randol, W., 1995, No gushers; *Barron's* 6/2/95.

Rasmusen, H.J., 1996, Bright future for natural gas; *Oil Gas European* 2/1/96.

Rees-Mogg, W., 1992, *Picnics on Vesuvius: steps towards the millenium;* Sidgwisk & Jackson 396p ISBN 0-283-06147-2.

Reifenberg, A., 1996, April crude oil futures top $21-a-barrel; *Wall St. Journ* March 15.

Reuters, 1990, Oil stocks in west at low point; *Int. Herald Tribune* 7/3/90.

Riva, J.P., 1991, Dominant Middle East oil reserves critically important to world supply; *Oil & Gas Journ.,* Sept 23 1991.

Riva, J.P., 1992, *Petroleum in the Muslim Republics of the Commonwealth of Independent States: more oil for Opec;* US Congressional Research Report 92-684 SPR.

Riva, J.P., 1992, *The domestic oil status and a projection of future production;* US Congressional Research Report 92-826 SPR.

Riva, J.P., 1993, Large oil resource awaits exploitation in former Soviet Union's Muslim republics; *Oil & Gas Journ.,* Jan 4 1993.

Riva, J.P., 1996, World production after year 2000: business as usual or apocalypse; *Geopolitics of Energy* 18/9 September 2-6.

Roadifer, R.E., 1984, Size distribution of world's largest oil, tar accumulations; in Mayer R.E. (ed) Exploration for heavy crude oil and natural bitumen; *Amer. Assoc. Pet. Geol Studies in Geology* #25.

Roadifer, R.E., 1986, Size distribution of world's largest oil, tar accumulations; *Oil & Gas Journ.* Feb. 26. 1986 93-98.

Roberts, J., 1992, Saudi Ambitions; *Petrol. Revue.* July 1992.

Roberts, J., 1995, *Visions & Mirages – the Middle East in a new era;* Mainstream ISBN 1-85158-429-3.

Roberts, J., 1996, IEA studies Middle East; *Petrol. Review,* Feb. 1996.

Robertson, J., 1989, *Future Wealth;* Cassell 178p. ISBN 0-304-31930-9.

Robinson, D., 1992, US oil firms find new lenders; *Petrol. Econ.* Dec 1992.

Rodenburg, E., *The decline of oil; World Resources – a guide to the global environment.* ISBN 0-19-521160-X.

Roeber, J., 1985, Term or spot : the search for stability: *Middle East Economic Survey* 28/23.

Roeber, J., 1994, *The future of crude oil markets east and west of Suez:* preprint APS Conf. Cyprus.

Roeber, J., 1994, *The evolution of trading instruments;* Brookings.

Roeber, J., 1994, Oil industry structure and evolving markets; *Energy Journal* 15.

Roger, J.V., 1994, Use and implementation of SPE and WPC petroleum reserve definitions; *Wld. Petrol. Congr.,* Stavanger.

Root, D., E. Attenasi, and R.M. Turner, 1987, *Statistics of petroleum exploration in the non-communist world outside the United States and Canada;* USGS Circ. 981.

Root, D., E. Attenasi, and R.M. Turner, 1989, Data and assumptions for three possible production schedules for non-Opec countries; Memorandum US Dept. of Interior 26 Sept. 1989.

Rossant, J. and P. Burrows, 1994, Pain at the pump; *Business Weekly* July 4.

Rothenberg, M., 1994, Risk factors in Azerbaijan; *World Oil,* Feb. 1994.

Rothschild, E.S., 1992, The roots of Bush's oil policy; *The Texas Observer,* Feb. 14, 1992.

Sampson, A., 1988, *The seven sisters: the great oil companies and the world they created;* Coronet, London.

Sauer, J.W., 1991, Oil markets: what now?; *World Oil,* Feb. 1991.

Sauer, J.W., 1993, Crude oil prices: why the experts are baffled; *World Oil,* Feb. 1993.

Scanlon, A.F., 1994, *Long term energy investment trends – weaknesses and opportunities;* Preprint, APS Conf. Cyprus.

Schollnberger, W., 1996, A balanced scorecard for petroleum exploration; *Oil Gas European,* 2/1/96.

Schuler, G.H.M., 1991, A history lesson: oil and munitions are an explosive mix; *Oil & Gas Journ.* Nov.18, 1991.

Schweizer, P., 1994, *Victory: the Reagan administration's secret strategy that hastened the collapse of the Soviet Union;* Atlantic Monthly Press, New York 284p ISBN 0-87113-567-1.

Science et Vie, 1995, *Energie: un fantastique tresor cache au fond de mers;* April 1995.

Scott, R.W., 1995, Bloody fiasco; *World Oil,* Feb 1995.

See, M., 1996, Oil mining field test to start in East Texas;

World Oil, Nov. 1996.

Sell, G., 1938, Statistics of petroleum and allied substances; *The Science of Petroleum* v1 1938.

Shafranik, Y.K., 1993, *Fuel and energy complex of Russia: modern conditions and perspectives:* printed lecture to Univ. of Leiden May 1993.

Shammas, P., 1994, *LNG business – is it sustainable beyond the 2010s;* Preprint APS Conf.

Shell, 1995, *Oil market developments;* Confidential report, 10 July 1995.

Shell, 1995, *The evolution of the world's energy system 1860-2060.*

Shilling, A.G., 1993, The poor get poorer; *Forbes* 8/11/93.

Shirley, K., 1992, Colombia finds wows explorers; *AAPG Explorer,* Aug. 1992.

Sierra, J., 1994 European energy supply security; *Petrole et Tech.* 389.

Simeoni, C., 1994, *Prospects for gas export pipelines from North Africa and Middle East;* Preprint APS Conf., Cyprus.

Simon, B., 1990, Oil project approaches last hurdle; *Financial Times,* Feb 2, 1990.

Simmons, M.R., 1990, Our upcoming domestic embargo; Panel discussion, Alaska

Simmons, M.R., 1994, It's not like '86; *World Oil,* Feb. 1994.

Simmons, M.R., 1995, Strong market indicators; *World Oil,* Feb. 1995.

Simmons, M.R., 1995, Despite sloppy prices, fundamentals tighten; *Pet. Eng. Int.* Sept 1995.

Simmons, M.R., 1995, *1995 global wellhead review and drilling review; a new era for the oil service industry;* Simmons & Co report.

Simmons, M.R., 1996, Robust demand strengthens outlook; *World Oil,* Feb. 1996.

Simmons, M.R., 1997, Are our oil markets too tight? *World Oil,* Feb. 1997

Snow, N., 1995, Nazar's dismissal does not mean change in Saudi oil policy; *Pet. Eng. Int.* Sept. 1995.

Solomon, C., 1993. The hunt for oil; *Wall Street Journ.* Aug. 25 1993.

Soros, G., 1995, *Soros on Soros: staying ahead of the curve;* John Wiley & Sons, ISBN 0-471-12014-6. 326p.

Statoil, 1996, *Oil price scenarios towards 2020;* Statoil Report.

Steeg, H., 1994, *World energy outlook to the year 2010;* 7th Int. Oil & Gas Seminar, Paris

Stosur, G.J. and R.W. Luhning, 1994, Worldwide EOR activity in the low price environment; *Pet. Eng. Int.* Aug. 1994, p. 46.

Stosur, G.J., 1996, Enhanced recovery – the international

perspective; in Doré, A.G and R. Sinding Larsen (Eds), *Quantification and prediction of hydrocarbon resources;* NPF Sp. pub. 6., Elsevier ISBN 0-444-82496-0.

Sullivan, A., 1992, Iraq isn't expected to resume oil exports soon despite progress in talks with UN; *Wall St. Journ.,* Jan 13 1992.

Sunday Times, 1993, King Fahd; *Sunday Times* 10/10/93.

Sunday Times, 1994, Uncle Sam gets heavy with Fahd; *Sunday Times,* 9/10/94.

Takin, M., 1972, Iranian geology and continental drift in the Middle East; *Nature* 235.

Takin M., 1988, Energy cycles: can they be avoided? *Opec Bulletin,* Oct. 1988.

Takin, M., 1989, The high cost of misunderstanding Opec; 14th Congr. Wld Energy Conf.

Takin, M., 1990, Opec, Japan and the Middle East; *OPEC Bull.* April 1990.

Takin, M., 1993, OPEC, Japan and the Middle East; *OPEC Bull.* 4/2 (March-April 1993) 17-34.

Takin, M., 1994, How much gas is there in the Middle East; *Pet. Review,* July 1994.

Takin, M., 1996, Many new ventures in the Middle East focus on old oil, gas fields; *Oil & Gas Journ.* May 27.

Takin, M., 1996, Future oil and gas: can Iran deliver?; *World Oil,* Nov 96-106.

Tanner, J., 1990, Looming shock: Mideast peace could trigger a sharp drop in crude oil prices: *Wall St. Journ.* Dec. 19.

Tanner, J., 1992, Agency says world oil demand rose in 4th quarter and sees 1% rise in 1992; *Wall St. Journ.,* Jan 15 1992.

Tanner, J., 1992, Iran, in need of revenue, lifts oil output; *Wall St. Journ,* 19/10/92.

Taylor, E., 1996, Future Chill; *Tyler Morning Telegraph* 12 Feb.

Tchuruk S., 1994, Les relations entre les societes et les gouvernements dans l'industrie mondiale du petrole et du gas; *Petrole et Technique* 389, July 1994.

Tempest, P., 1994, *Oil geopolitics: some new dilemmas;* Preprint APS Conf.

Teitelbaum, R.S.,1995, Your last big play in oil; *Fortune,* 30 Oct. 1995.

Tickell, C., 1996, Climate & history; *Oxford Today* 8/2.

Times, The, 1994, State of terror; Editorial, April 29 1994.

Tissot, B and D.H. Welte, 1978, *Petroleum formation and occurrence;* Springer verlag, New York

Townes, H.L., 1993, The hydrocarbon era, world population growth and oil use – a continuing geological challenge; *Amer. Assoc. Petrol. Geol.* 77/5, 723-730.

Tomitate, T., 1994, World oil perspectives and outlook for

supply-demand in Asia-Pacific region; World Petrol. Congr. Topic 16.

Tull, S., 1997, Habitat of oil and gas in the Former Soviet Union; *Geoscientist* 7/1.

Tugendhat, C. and A. Hamilton, 1968, *Oil – the biggest business;* Eyre Methuen. ISBN 0-413-33290-X.

Tyler, N. and N.J. Banta, 1989, *Oil and gas resources remaining in the Permian Basin: targeted for additional hydrocarbon recovery;* Bureau of Economic Geology, Univ. of Texas. Circular 89-4. 20p.

Ulmishek, G.F., R.R. Charpentier, and C.C. Barton, 1993, The global oil system: the relationship between oil generation, loss, half-life and the world crude oil resource: discussion; *Amer. Assoc. Petrol. Geol.* 77/5 896-899.

Ulmishek, G.F. and C.D. Masters, 1993, Oil, gas resources estimated in the former Soviet Union; *Oil & Gas Journ,* Dec 13. 59-62.

USGS, 1995, 1995 national assessment of United States oil and gas resources; USGS Circular 1118.

Verleger, P., 1996, The role of futures markets; *Oxford Energy Forum* 27 5-7.

Vlierboom, F.W., B. Collini and J.E. Zumberge, 1986, The occurrence of petroleum in sedimentary rocks of the meteor impact crater of Lake Siljan, Sweden; 12th Europ. Assoc. Organic. Geochem. International meeting, Julich Sept 1985; *Org. Geochem* 10 153-161.

Volsett, J., A. Abrahamsen and K. Lindbo, 1994, An enhanced resource classification – a tool for decisive exploration; preprint, *Wld Petrol, Congr.* Stavanger.

Wardt, J.P. de, 1996, Operational realities in the '90s; *World Oil,* Feb. 1996.

Warman, H.R., 1972, The future of oil; *Geographical Journ.* 138/3 287-297.

Warman, H.R., 1973, The future availability of oil; proc. Conf. World Energy Supplies, by *Financial Times,* London, 11p.

Washington Inst., 1995, Saudi succession uncertain; News Release.

Wellmer, F-W., 1994, Rerserven und reservenlebensdauer von energierohstoffen; *Energie Dialog,* July 1994.

Weyant, J.P. and D.M. Kline, 1982, Opec and the oil glut: outlook for oil export revenues in the 1980s and 1990s; *Opec Review,* Winter 1982 334-365.

Will, G., 1996, Heavy oil – the jewel in western Canada's oil play; *Petroleum Review,* Nov. 505-507.

Willingham, B.J., 1994, Energy shortage looms again; *The Nat. Times Mag,* Oct/Nov.

World Oil, 1994, World not running out of oil; March 1994 p. 9.

World Oil, 1994, Worldwide drilling stable in most areas; Feb. 1994.

World Oil, 1994, US oil and gas reserves hit 15-year lows: Feb 1994.

World Oil, 1994, US drilling: industry needs to think positive; Feb 1994.

World Oil, 1996, Global oil output inches forward; Feb. 1996.

Yamani, A.Z., 1995, Oil's global role – the outlook to 2005; *MEES 38/33.*

Yasunaga, Y., 1994, *Japan's new energy policy and strategy;* Preprint APS Conf.

Yergin, D., 1991, *The Prize: the epic quest for oil, money and power;* Simon & Schuster, New York, 877p ISBN 0-671-50248-4.

Yergin, D., 1988, Energy security in the 1990s; *Petroleum Review,* Nov. 1988.

Zakaria, F., 1996, Thank goodness for a villain; *Newsweek* Sept 18.

Zellner, W., 1994, Steamed about natural gas; *Business Week,* 10/10/94 .

Ziegler, P.A., 1993, Plate-moving mechanisms: their relative importance; *Journ. Geol. Soc.,* 150/5.

Ziegler, W.H., 1975, Outline of the geological history of the North Sea., in Woodland, A.W. (Ed.) *Petroleum and continental shelf of North-West Europe,* Applied Science. ISBN 0-85334-648-8.

Ziegler, W.H., 1990, World oil and gas reserves; unpublished report.

Zischka, A., 1933, *La guerre secrète pour le pétrole;* Payot, Paris.

APPENDIX I

Units, Abbreviations, Conversion Factors & Glossary

UNITS

S.I. (Système International) units are the recommended system replacing the Imperial system, which is however still entrenched in the oil industry.

ABBREVIATIONS

b = barrel (US) = 159 litres = 42 US gals (Also B)
boe = barrel of oil equivalent (see below for conversion)
cf = cubic feet = 0.0283 m³
Ma = million years
toe = ton of oil equivalent
Giant oilfield = >500 Mb initial reserves

k (kilo)	= 10^3	= thousand	(Also M)
M (mega)	= 10^6	= million	(Also MM and m)
G (giga)	= 10^9	= billion (US)	(Also MMM, B, bn)
T (tera)	= 10^{12}	= trillion (US)	

CONVERSION FACTORS

sq. mile	sq. km	acre	hectare
1	2.59	640	259
0.386	1	247	100

barrel	cu. m	tonne
1	0.159	0.136
6.29	1	0.855
7.33	1.17	1

(Depends on oil gravity: above applies to Arabian Light 33.5° API)

Gas
1 m³ = 35.3 cf
1 cf = 0.0283 m³

Energy equivalents
1000 m³ gas = 1 toe = 7.168 boe (in calorific terms)
10 000 cf gas = 1 boe (in value @$1.8/kcf, $18/b)
1 Tcf = 0.1 Gboe
1 b/d = 50 t/a (32° API oil)

A SHORT GLOSSARY OF OIL AND TECHNICAL TERMS USED IN THIS BOOK

Acreage: the amount of land held under concession.

Alidade: an instrument for measuring distances, used with a planetable for surveying.

Anticline: an arch-like geological structure at the top of a fold in which petroleum tends to accumulate.

Appraisal (or delineation) wells: boreholes drilled to confirm the extent of a discovery.

Basement: non-prospective rocks beneath a sedimentary basin: often granite or metamorphic rocks.

Basin: a geological depression filled with sedimentary rocks, often containing petroleum.

Backwardation: a term in trading when the price of oil for future delivery is below present delivery.

Block: a commonly rectangular licence area or concession.

Cap-rock: an impermeable rock the prevents the escape of oil from a trap (also seal).

Choke: a restriction placed in a well to reduce flow. Also "choking-back" production.

Condensate: a liquid hydrocarbon that condenses from gas at surface conditions

Contango: the converse of backwardation.

Dip: the angle at which strata are inclined from the horizontal.

Downstream: the marketing and refining end of the oil business.

Evaporites: salt and similar rocks deposited by the evaporation of sea-water, commonly forming effective seals preventing the escape of oil from a reservour.

Farm-in: the converse of farm-out

Farm-out: a transfer of ownership in a petroleum concession to a new company, which will commonly drill one well at its expense to earn a 50% undivided interest. Oil rights change hands frequently, and many oilfields have been found by the incoming company.

Fault: a geological fracture such that the rocks on one side are offset relative to those on the other. Petroleum often accumulates against faults.

Flaring: the practice of setting light to unwanted gas, for which no current market exists.

Flysch: a coarse grained sediment resulting from the rapid erosion of a mountain chain, as found in front of the Alps.

Independent company: a general term for oil exploration and production companies other than the large integrated companies.

Licence: used largely in Europe, meaning concession.

Migration: the process by which petroleum moves through the earth from the rock (source-rock) in which it was generated.

Nappe: a large horizontally displaced mass of rocks, typical of Alpine geology.

NGL: Natural Gas Liquid, a liquid obtained from gas naturally or by processing.

Oil window: the depth at which oil is generated in a particular area.

Operator: a member of a group of companies in a joint venture charged with managing the operations. The other companies are known as Non-operators.

Palaeontology: the study of fossils, which are used to date and correlate geological formations.

Plane-table: a table mounted on a tripod, used in the traditional method of geological survey.

Play: an oil play is a group of prospects having similar geological characteristics.

Pore-throat: the communication between pores in a reservoir rock.

Prospect: an undrilled geological structure which is perceived to have the characteristics necessary for it to contain oil or gas.

Round: an offering at a particular time of concessions by governments.

Salt-plug: salt, being mobile and less dense than other rocks, may flow and penetrate the overlying sequence of rocks. Oil and gas sometimes accumulate around salt-plugs, also known as diapirs.

Sedimentary environment: the physical environment in which the rocks were laid down, such as shallow-water, deltaic, fluvial etc.

Seismic surveys: a survey procedure that involves the release of energy at the surface and recording the reflections from buried structures. It is now the most widely used exploration tool. The results are called seismic data. The surveys are conducted on a grid, each being known as a seismic-line. The surveys are said to be "shot".

Semi-submersible rig: a rig mounted on two submerged pontoons that float in calmer waters beneath the wave base.

Signature bonus: a cash payment on signature of a concession.

Stratigraphy: the study of the sequence and nature of sedimentary rocks.

Stratigraphic trap (strat-trap): a trap for oil or gas formed by stratigraphic factors alone, such as where a sandstone occurs as a lenticular body.

Stripper well: a well producing a few barrels a day from a nearly depleted oilfield, mainly in the United States.

Sub-sea completion: a technique of installing the wellhead and production equipment on the seabed as opposed to on a platform above the waves.

Tank-farm: an area containing large oil tanks.

Thrust: a low-angled fault.

Transcurrent fault: a fault on which movement has been mainly horizontal.

Upstream: the exploration and production end of the oil business.

Wildcat: an exploration borehole, which if successful finds a new oil or gas field.

APPENDIX II
World Oil Endowment

Country	kb/d 1996	Cum. Prod.	World Oil	O&GJ	Adjust +/−	Factor	Median Prob. (P50)	Discovered	Yet-to-Find	Yet-to-Produce	Ultimate	Prod. of Disc.	Disc. in Giants	Peak Disc/ Prod. yrs	MP Dep.	Dep. Rate	Peak Prod.
Saudi Arabia	7841	77.3	259.0	259.0	−90.50	1.20	202.2	279.5	20.5	222.7	300	28%	94%	57	2013	3.0%	2008
FSU	7044	128.1	189.68	57.00	−17.63	2.50	98.4	226.5	48.5	146.9	275	57%	47%	24	2000	1.8%	1988
USA	6478	173.2	22.16	22.35	0.00	1.25	27.9	201.1	8.9	36.8	210	86%	29%	12	1973	6.0%	1971
Iran	3675	46.5	57.70	93.00	−44.20	1.20	58.6	105.1	14.9	73.5	120	44%	87%	13	2007	2.6%	1974
Iraq	600	22.7	99.17	112.00	−52.90	1.20	70.9	93.6	21.4	92.3	115	24%	90%	55	2017	2.7%	2008
Kuwait	1818	27.0	92.72	94.00	−39.02	1.10	60.5	87.5	7.5	68.0	85	31%	92%	33	2013	2.7%	1971
Venezuela	2955	47.9	65.57	64.88	−25.95	0.90	35.0	83.0	7.0	42.1	90	58%	74%	29	1993	2.5%	1970
Abu Dhabi	1846	14.0	62.21	92.20	−40.25	1.20	62.3	76.3	3.7	66.0	80	18%	73%	44	2017	2.8%	2008
China	3127	21.4	30.96	24.00	0.55	1.20	29.5	50.9	4.1	33.6	55	42%	45%	42	2001	4.5%	2001
Mexico	2854	23.3	48.80	48.80	−27.50	1.15	24.5	47.8	2.2	26.7	50	49%	62%	29	1998	3.7%	1998
Libya	1403	19.9	34.74	29.50	−10.51	1.20	22.8	42.7	2.3	25.1	45	47%	70%	17	2000	2.7%	1978
Nigeria	2014	17.8	21.27	15.52	0.00	1.20	18.6	36.4	3.6	22.2	40	49%	25%	12	1999	3.8%	1979
UK	2633	14.2	4.54	4.52	0.00	3.00	13.6	27.7	2.3	15.8	30	51%	46%	23	1997	6.4%	1997
Canada	1820	19.8	5.55	4.89	0.00	1.20	5.9	25.7	2.3	8.2	28	77%	44%	14	1987	7.5%	1972
Norway	3086	9.4	24.18	11.23	0.00	1.30	14.6	24.1	2.9	17.6	27	39%	53%	18	1999	8.6%	1999
Indonesia	1516	17.4	5.95	4.98	0.00	1.30	6.5	23.9	1.1	7.6	25	73%	49%	22	1986	6.8%	1977
Algeria	816	10.4	9.98	9.20	−1.99	1.30	9.4	19.7	3.3	12.6	23	52%	60%	22	1999	2.7%	1978
Brasil	780	4.3	6.22	4.80	0.00	1.30	6.2	10.6	1.4	7.7	12	41%	72%	17	2001	3.1%	2001
N. Zone	484	5.4	4.55	5.00	−0.77	1.10	4.7	10.0	1.5	6.1	12	54%	95%	20	1999	1.8%	1979
Oman	883	5.0	3.30	5.14	−0.33	1.25	6.0	11.0	0.5	6.5	12	45%	37%	26	1998	4.8%	1998
Egypt	923	6.9	3.80	3.70	0.00	1.00	3.7	10.6	0.4	4.1	11	65%	42%	27	1992	5.6%	1996
India	643	4.1	5.28	4.98	0.00	1.20	6.0	10.1	0.4	6.4	11	41%	30%	26	2000	3.6%	2000
Qatar	480	5.3	4.95	3.70	−0.49	1.30	4.2	9.5	0.5	4.7	10	56%	72%	11	1994	3.6%	1973
Argentina	753	6.5	2.39	2.84	0.00	0.90	2.6	9.1	0.9	3.5	10	72%	34%	36	1989	6.5%	1996
Colombia	621	4.2	5.50	2.80	0.00	1.50	4.2	8.4	1.6	5.8	10	50%	49%	16	1999	3.8%	1999
Angola	816	3.0	3.13	5.41	−0.58	1.10	5.32	8.30	0.70	6.02	9.00	36%	16%	18	2000	7.2%	2000
Malaysia	650	3.6	5.20	4.30	−0.08	1.40	4.73	8.37	0.13	4.86	8.50	43%	13%	25	1998	4.7%	1998
Romania	136	5.5	1.01	1.61	−0.10	0.85	1.28	6.76	0.24	1.52	7.00	81%	12%	23	1971	2.9%	1976
Australia	535	4.5	3.46	1.56	−0.20	1.30	1.77	6.24	0.76	2.54	7.00	72%	33%	26	1991	7.1%	1990
Ecuador	386	2.4	3.39	2.12	−0.14	1.30	2.57	5.00	0.50	3.07	5.50	49%	45%	29	1998	4.4%	1998
Syria	602	2.6	2.56	2.50	−0.44	1.20	2.47	5.10	0.40	2.87	5.50	52%	39%	31	1997	7.1%	1997
Dubai	320	3.1	0.89	4.00	−2.60	1.10	1.54	4.66	0.34	1.88	5.00	67%	64%	25	1991	5.9%	1991
Yemen	338	0.8	2.98	4.00	−0.66	0.90	3.01	3.80	0.20	3.20	4.00	21%	13%	17	2004	3.7%	2004
Brunei	150	2.6	1.12	1.35	−0.35	1.30	1.30	3.93	0.07	1.37	4.00	67%	80%	15	1985	3.8%	1978
Trinidad	128	2.9	0.57	0.55	0.00	1.10	0.61	3.54	0.21	0.82	3.75	73%	17%	24	1978	5.4%	1978
Gabon	370	2.1	1.47	1.34	−0.27	1.10	1.18	3.32	0.43	1.61	3.75	64%	24%	34	1994	6.0%	1996
Peru	121	2.1	0.69	0.81	−0.04	1.10	0.84	2.95	0.55	1.39	3.50	71%	17%	23	1988	3.1%	1983
Congo	203	1.0	1.35	1.51	−0.32	1.20	1.42	2.37	0.13	1.55	2.50	40%	21%	24	2000	6.4%	2000
Germany	57	1.8	0.38	0.39	0.00	1.15	0.44	2.24	0.06	0.50	2.30	80%	0%	−7	1974	3.6%	1966
Denmark	206	0.7	0.96	0.96	0.00	1.25	1.20	1.85	0.15	1.34	2.00	35%	0%	29	2000	9.3%	2000
Tunisia	87	1.0	0.38	0.31	0.00	1.20	0.37	1.40	0.35	0.72	1.75	74%	57%	17	1991	4.2%	1981
Italy	101	0.7	0.70	0.65	0.00	1.20	0.78	1.45	0.30	1.08	1.75	46%	0%	20	2001	5.2%	2001
Viet Nam	171	0.3	0.66	0.60	0.00	1.20	0.72	1.03	0.47	1.19	1.50	30%	48%	16	2001	5.4%	2001
Bahrein	104	1.0	0.05	0.21	−0.11	1.50	0.14	1.15	0.05	0.19	1.20	88%	87%	61	1976	16.4%	1993
Sharjah	50	0.3	0.38	1.50	−1.10	1.10	0.44	0.79	0.41	0.85	1.20	44%	0%	26	2006	2.1%	2006
Cameroon	90	0.8	0.00	0.40	−0.28	1.10	0.13	0.95	0.25	0.39	1.20	86%	0%	10	1991	8.6%	1986
Papua	107	0.2	0.65	0.28	0.00	1.30	0.36	0.54	0.66	1.01	1.20	34%	0%	18	2004	3.7%	2004
Netherlands	60	0.7	0.13	0.09	0.00	1.40	0.12	0.85	0.25	0.37	1.10	86%	0%	6	1989	5.6%	1987
Turkey	69	0.7	0.24	0.26	0.00	1.20	0.31	1.01	0.09	0.40	1.10	69%	0%	27	1991	5.9%	1991
Hungary	31	0.7	0.12	0.12	0.00	1.10	0.13	0.78	0.22	0.35	1.00	83%	0%	30	1986	2.8%	1994
France	44	0.7	0.14	0.12	0.00	1.30	0.15	0.81	0.09	0.24	0.90	81%	0%	30	1986	4.5%	1988
Austria	21	0.7	0.07	0.08	0.00	1.25	0.10	0.83	0.07	0.17	0.90	88%	62%	6	1970	4.0%	1955
Albania	9	0.5	0.14	0.17	−0.03	1.10	0.15	0.68	0.12	0.27	0.80	78%	0%	55	1986	6.1%	1983
Thailand	59	0.2	0.28	0.30	0.00	1.20	0.35	0.56	0.19	0.54	0.75	37%	0%	22	2003	3.8%	2003
Pakistan	54	0.3	0.17	0.21	0.00	1.20	0.25	0.59	0.11	0.36	0.70	58%	0%	9	1996	5.2%	1992
Bolivia	30	0.4	0.11	0.13	0.00	1.30	0.17	0.53	0.17	0.34	0.70	67%	0%	8	1996	3.1%	1974
Chile	9	0.4	0.12	0.30	−0.05	0.75	0.18	0.59	0.01	0.20	0.60	69%	0%	22	1983	1.6%	1982
Philippines	1	0.0	0.31	0.21	0.00	1.50	0.32	0.37	0.23	0.55	0.60	13%	0%	9	2009	0.2%	2009
By Region																	
ME Gulf	16264	193.0	575.35	655.20	−267.64	1.18	459	646.3	75.2	534.4	722	29%	91%	57	2013	2.8%	2008
Eurasia	10347	156.2	221.90	82.89	−17.21	1.97	129	285.7	53.1	182.6	339	55%	46%	25	2000	2.3%	1988
N. America	8297	193.0	27.71	27.24	0.00	1.24	34	226.8	11.2	45.0	238	85%	45%	13	1975	2.9%	1972
L. America	8637	94.5	133.33	128.02	−53.69	1.03	77	171.4	14.6	91.5	186	55%	64%	38	1995	3.0%	1998
Africa	6720	62.8	76.12	66.88	−13.95	1.19	63	125.7	11.5	74.4	137	50%	47%	34	1999	3.5%	1999
W. Europe	6207	28.9	31.09	18.02	0.00	1.72	31	59.8	6.1	37.1	66	48%	44%	24	1998	6.8%	1999
East	3886	33.4	23.08	18.76	−0.54	1.22	22	55.7	4.1	26.4	60	60%	39%	26	1994	4.4%	1996
ME Other	2844	18.9	15.34	21.31	−5.73	1.16	18	37.0	2.5	20.6	40	51%	47%	9	1997	5.0%	1996
Other	283	3.0	0.00	0.00	0.00	0.00	2	5.4	7.8	10.2	13	55%	0%	0	2016	4.0%	2016
WORLD	63486	784	1104	1018	−359	1.27	836	1620	180	1016	1800	48%	62%	41	2001	2.6%	2001

Table header: WORLD — CONVENTIONAL OIL ENDOWMENT — Ref. date end 1996 — Revised 20/02/97 — Gb (billion barrels). Production: kb/d 1996, Cum. Prod. Reserves: Reported (World Oil, O&GJ), Adjust +/−, Factor, Median Prob. (P50). Discovered: Yet-to-Find, Yet-to-Produce | Ultimate. Prod. of Disc., Disc. in Giants, Peak Disc/Prod. yrs, MP Dep., Dep. Rate, Peak Prod.

The Coming Oil Crisis

APPENDIX III
Production Profiles

SAUDI ARABIA
Ultimate 300 Gb

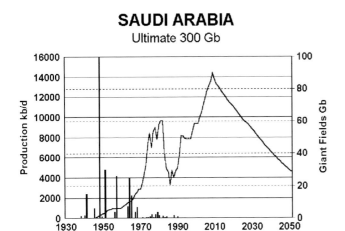

F.S.U.
Ultimate 275 Gb

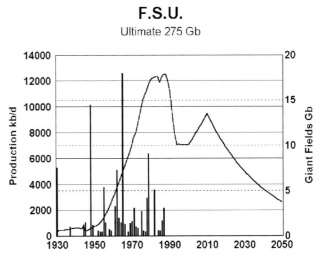

IRAN
Ultimate 120 Gb

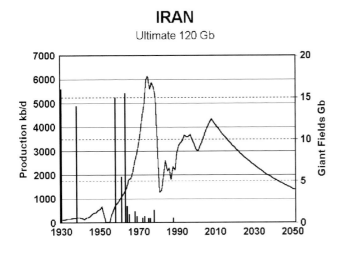

CHINA
Ultimate 55 Gb

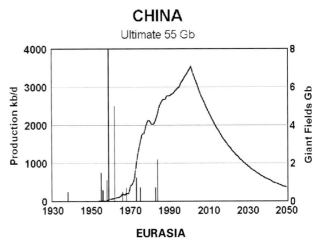

EURASIA

IRAQ
Ultimate 115 Gb

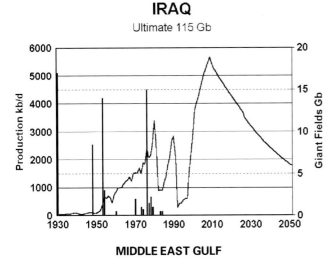

MIDDLE EAST GULF

N AMERICA

U.S.A.
Ultimate 210 Gb

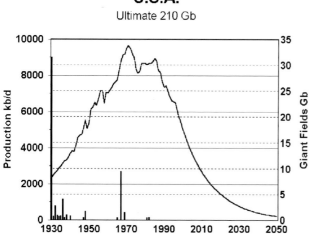

CANADA
Ultimate 28Gb

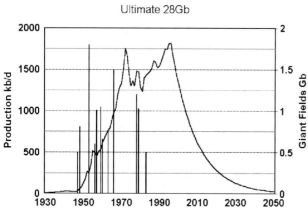

L. AMERICA

VENEZUELA
Ultimate 90 Gb

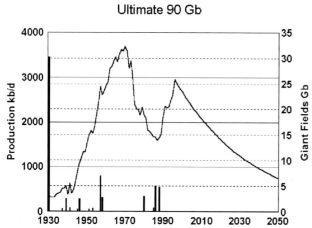

MEXICO
Ultimate 50 Gb

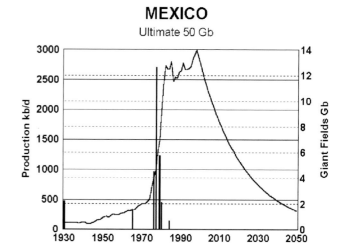

BRASIL
Ultimate 12 Gb

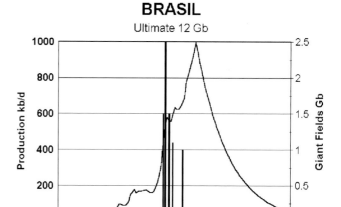

COLOMBIA
Ultimate 10 Gb

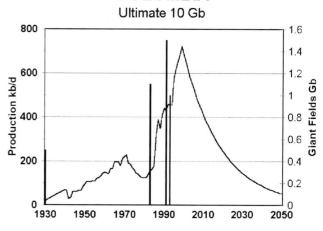

AFRICA

W. EUROPE

LIBYA
Ultimate 45 Gb

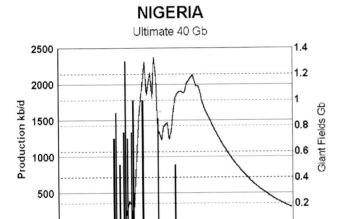

UNITED KINGDOM
Ultimate 30 Gb

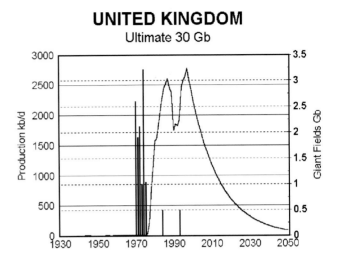

NIGERIA
Ultimate 40 Gb

NORWAY
Ultimate 27 Gb

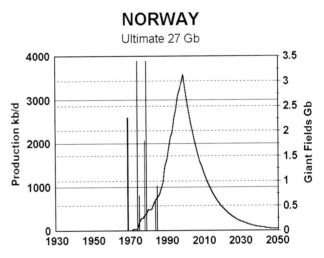

ALGERIA
Ultimate 23 Gb

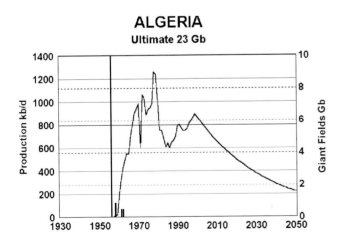

INDONESIA
Ultimate 25 Gb

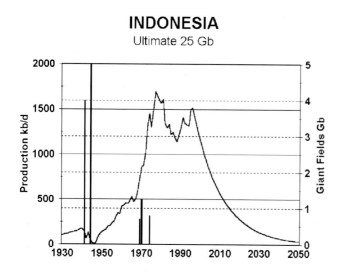

OMAN
Ultimate 11.5 Gb

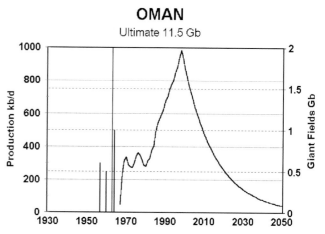

INDIA
Ultimate 10.5 Gb

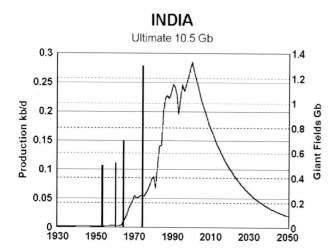

QATAR
Ultimate 10 Gb

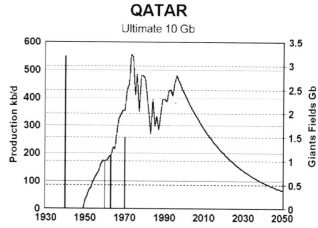

MALAYSIA
Ultimate 8.5 Gb

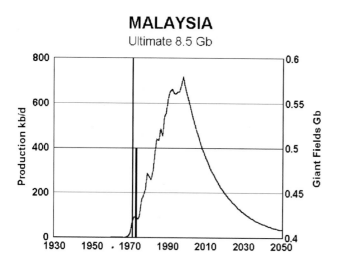

SYRIA
Ultimate 5,5 Gb

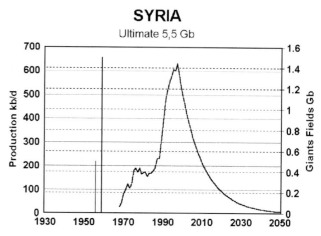

APPENDIX IV
Depletion Plots

Base Case Scenario			WORLD				
Cum.Prod.	783.84	MIDPOINT		PRODUCTION		DISCOVERY	
Reserves	835.96	Amount	900.00	Peak	2001	Midpoint	1960
Discovered	1619.80	Date	2001	3yr-trend	0.02	RESERVES	
Yet-to-Find	180.20	Years	5	Disc-Prod	41	O&GJ	1018.33
Yet-to-Produce	1016.16	GIANTS		DEP. RATE		Adjust	−358.75
Ultimate	1800.00	Amount	1011.29	Current	0.02	World Oil	1103.92
Cum. Prod. 2050	1619.16	%	62%	Midpoint	0.03	Factor	1.27
Reference date	12/96	Last	1993	Diff.	−0.42	Revised	03/08

	Production		Yet-to-Produce	Dep.	Giants	Cum.
Date	kb/d	Gb/y	Gb	Rate	Gb	
Pre-1930		17	1782.64		73	
1930	3724	1.36	1781.28	0.08%	104	104.34
1931	3955	1.44	1779.84	0.08%	2	106.08
1932	4115	1.50	1778.33	0.08%	3	108.84
1933	4305	1.57	1776.76	0.09%	1	109.80
1934	4614	1.68	1775.08	0.09%	1	110.56
1935	4843	1.77	1773.31	0.10%	1	111.42
1936	5059	1.85	1771.46	0.10%	4	115.45
1937	5326	1.94	1769.52	0.11%	2	117.55
1938	5458	1.99	1767.53	0.11%	77	194.56
1939	5730	2.09	1765.44	0.12%	3	197.16
1940	5762	2.10	1763.33	0.12%	6	203.05
1941	6087	2.22	1761.11	0.13%	20	222.80
1942	5775	2.11	1759.00	0.12%	0	222.80
1943	6231	2.27	1756.73	0.13%	0	222.80
1944	7110	2.60	1754.13	0.15%	6	229.05
1945	7156	2.61	1751.52	0.15%	8	237.08
1946	7605	2.78	1748.75	0.16%	3	239.58
1947	8380	3.06	1745.69	0.17%	1	240.61
1948	9480	3.46	1742.23	0.20%	126	366.65
1949	8420	3.44	1738.79	0.20%	2	368.87
1950	10562	3.86	1734.93	0.22%	0	368.87
1951	11824	4.32	1730.62	0.25%	31	399.42
1952	12495	4.56	1726.06	0.26%	1	400.02
1953	13314	4.86	1721.20	0.28%	18	417.52
1954	13837	5.05	1716.15	0.29%	12	429.03
1955	15587	5.69	1710.46	0.33%	16	445.23
1956	16737	6.11	1704.35	0.36%	17	462.49
1957	17588	6.42	1697.93	0.38%	35	497.10
1958	17942	6.55	1691.38	0.39%	37	533.70
1959	19439	7.10	1684.29	0.42%	28	561.90
1960	21055	7.69	1676.60	0.46%	5	566.88
1961	22464	8.20	1668.40	0.49%	22	588.64
1962	24353	8.89	1659.51	0.53%	27	615.44
1963	26058	9.51	1650.00	0.57%	31	646.45
1964	28328	10.34	1639.66	0.63%	52	698.90
1965	30200	11.02	1628.64	0.67%	38	736.65
1966	32747	11.95	1616.69	0.73%	13	750.00
1967	35210	12.85	1603.84	0.79%	20	770.10
1968	38572	14.08	189.76	0.88%	12	782.25
1969	41253	15.06	1574.70	0.95%	7	788.97
1970	45085	16.46	1558.24	1.05%	10	799.17
1971	48310	17.63	1540.61	1.13%	9	808.19
1972	49894	18.21	1522.40	1.18%	4	811.91
1973	55541	20.27	1502.13	1.33%	8	819.68
1974	56900	20.77	1481.36	1.38%	10	829.74
1975	53701	19.60	1461.76	1.32%	7	837.03
1976	57154	20.86	1440.90	1.43%	19	856.18
1977	58901	21.50	1419.40	1.49%	5	861.23
1978	59667	21.78	1397.62	1.53%	13	874.60
1979	62058	22.65	1374.94	1.62%	18	892.76
1980	59351	21.66	1353.31	1.58%	6	898.56
1981	55529	20.27	1333.04	1.50%	1	899.97
1982	52800	19.27	1313.77	1.45%	7	906.98
1983	52638	19.21	1294.55	1.46%	2	909.31
1984	53610	19.57	1274.98	1.51%	7	915.82

	Production		Yet-to-Produce	Dep.	Giants	Cum.
Date	kb/d	Gb/y	Gb	Rate	Gb	
1985	53224	19.48	1255.56	1.52%	3	918.53
1986	55700	20.33	1235.23	1.62%	7	925.23
1987	55890	20.40	1214.83	1.65%	3	928.34
1988	57686	21.06	1193.77	1.73%	7	935.74
1989	59893	21.86	1171.91	1.83%	1	936.44
1990	60518	22.09	1149.82	1.88%	2	937.94
1991	60272	22.00	1127.82	1.91%	0	937.94
1992	60112	21.94	1105.88	1.95%	0	937.94
1993	59850	21.85	1084.04	1.98%	1	938.44
1994	60470	22.07	1061.96	2.04%	0	938.44
1995	62009	22.63	1039.33	2.13%	0	938.44
1996	63486	23.17	1016.16	2.23%	0	938.44
1987	65431	23.88	992.28	2.35%		
1998	66735	24.36	967.92	2.45%		
1999	66731	24.36	943.56	2.52%		
2000	66728	24.36	919.21	2.58%		
2001	66725	24.35	894.85	2.65%	Plateau	
2002	66722	24.35	870.50	2.72%		
2003	66719	24.35	846.15	28.0%		
2004	66716	24.35	821.79	2.88%		
2005	66646	24.33	797.47	2.96%		
2006	66579	24.30	773.17	3.05%		
2007	66726	24.35	748.81	3.15%		
2008	66722	24.35	724.46	3.25%		
2009	64547	23.56	700.90	3.25%		
2010	62447	22.79	678.11	3.25%		
2011	60415	22.05	656.05	3.25%		
2012	58449	21.33	634.72	3.25%		
2013	56547	20.64	614.08	3.25%		
2014	54707	19.97	594.11	3.25%		
2015	52927	19.32	574.79	3.25%		
2016	51205	18.69	556.11	3.25%		
2017	49539	18.08	538.02	3.25%		
2018	47927	17.49	520.53	3.25%		
2019	46368	16.92	503.61	3.25%		
2020	44859	16.37	487.23	3.25%		
2021	43400	15.84	471.39	3.25%		
2022	41988	15.33	456.07	3.25%		
2023	40622	14.83	441.24	3.25%		
2024	39300	14.34	426.89	3.25%		
2025	38021	13.88	413.02	3.25%		
2026	36785	13.43	399.59	3.25%		
2027	35588	12.99	386.60	3.25%		
2028	34430	12.57	374.03	3.25%		
2029	33310	12.16	361.88	3.25%		
2030	32226	11.76	350.11	3.25%		
2031	31178	11.38	338.73	3.25%		
2032	30164	11.01	327.72	3.25%		
2033	29182	10.65	317.07	3.25%		
2034	38233	10.31	306.77	3.25%		
2035	27315	9.97	296.80	3.25%		
2036	36426	9.65	287.15	3.25%		
2037	25566	9.33	277.82	3.25%		
2038	24735	9.03	268.79	3.25%		
2039	23930	8.73	260.06	3.25%		
2040	23152	8.45	251.61	3.25%		
2041	22399	8.18	243.43	3.25%		
2042	21670	7.91	235.52	3.25%		
2043	20965	7.65	227.87	3.25%		
2044	20283	7.40	220.47	3.25%		
2045	19623	7.16	213.30	3.25%		
2046	18985	6.93	206.37	3.25%		
2047	18367	6.70	199.67	3.25%		
2048	17770	6.49	193.18	3.25%		
2049	17192	6.28	186.91	3.25%		
2050	16633	6.07	180.84	3.25%		

The Coming Oil Crisis

REGION	Scenario: base case		MIDDLE EAST GULF			
Cum.Prod.	193.04	MIDPOINT	PRODUCTION		DISCOVERY	
Reserves	459.15	Amount 360.75	Peak	2008	Midpoint	1951
Discovered	646.25	Date 2013	3yr-trend	1%	RESERVES	
Yet-to-Find	75.25	Years 17	Disc-Prod	57	O&GJ	655.20
Yet-to-Produce	534.40	GIANTS	DEP. RATE		Adjust	−267.64
Ultimate	721.50	Amount 552.06	Current	1.1%	World Oil	575.35
Cum. Prod. 2050	605.74	% 85%	Midpoint	2.8%	Factor	1.18
Reference date	12/96	Last 1990	Diff.	−1.7	Revised	03/08

Date	Production kb/d	Gb/y	Yet-to-Produce Gb	Dep. Rate	Giants Gb	Cum.
Pre-1930		0.98	720.52		33	
1930	151	0.06	720.46	0.01%	33	33
1931	168	0.06	720.40	0.01%	0	33
1932	185	0.07	720.34	0.01%	0	33
1933	202	0.07	720.26	0.01%	0	33
1934	219	0.08	720.18	0.01%	0	33
1935	236	0.09	720.10	0.01%	0	33
1936	253	0.09	720.00	0.01%	0	33
1937	280	0.10	719.90	0.01%	0	33
1938	306	0.11	719.79	0.02%	76	108.5
1939	309	0.11	719.68	0.02%	0	108.5
1940	261	0.10	719.58	0.01%	2	110.34
1941	185	0.07	719.51	0.015	15	125.34
1942	264	0.10	719.42	0.01%	0	125.34
1943	287	0.10	719.31	0.01%	0	125.34
1944	386	0.14	719.17	0.02%	0	125.34
1945	512	0.19	718.99	0.03%	6	131.34
1946	680	0.25	718.74	0.03%	0	131.34
1947	813	0.30	718.44	0.04%	0	131.34
1948	1109	0.40	718.04	0.06%	109	239.84
1949	1372	0.50	717.54	0.07%	1	240.84
1950	1703	0.62	716.91	0.09%	0	240.84
1951	1849	0.67	716.24	0.09%	30	270.84
1952	1997	0.73	715.51	0.10%	0	270.84
1953	2372	0.87	714.64	0.12%	14	284.84
1954	2581	0.94	713.70	0.13%	11	295.84
1955	3086	1.13	712.58	0.16%	9	304.64
1956	3191	1.16	711.41	0.16%	4	308.74
1957	3330	1.22	710.20	0.17%	26	334.74
1958	4016	1.47	708.73	0.21%	29	363.74
1959	4346	1.59	707.14	0.22%	9	372.74
1960	5022	1.83	705.31	0.26%	1	373.24
1961	5395	1.97	703.34	0.28%	6	378.77
1962	5956	2.17	701.17	0.31%	14	392.77
1963	6493	2.37	698.80	0.34%	24	416.27
1964	7417	2.71	696.09	0.39%	47	463.27
1965	7938	2.90	693.19	0.42%	15	478.24
1966	8852	3.23	689.96	0.47%	7	485.17
1967	9468	3.46	686.51	0.50%	5	490.07
1968	10562	3.86	682.65	0.56%	8	498.42
1969	11331	4.14	678.52	0.61%	1	499.22
1970	12579	4.59	673.92	0.68%	2	501.22
1971	15009	5.48	668.45	0.81%	1	501.895
1972	15950	5.82	662.62	0.87%	1	502.495
1973	19989	7.30	655.33	1.10%	3	505.655
1974	21193	7.74	647.59	1.18%	1	507.005
1975	18900	6.90	640.69	1.07%	2	508.555
1976	20375	7.44	633.26	1.16%	19	527.105
1977	20485	7.48	625.78	1.18%	3	530.205
1978	19320	7.05	618.73	1.13%	6	536.355
1979	19740	7.21	611.52	1.16%	5	540.955
1980	16830	6.14	605.38	1.00	3	543.755
1981	14340	5.23	600.15	0.86%	0	543.755
1982	11167	4.08	596.07	0.68%	1	545.055
1983	10450	3.81	592.26	0.64%	1	546.155
1984	10024	3.66	588.60	0.62%	2	547.755
1985	8857	3.23	585.36	0.55%	0	547.755

Date	Production kb/d	Gb/y	Yet-to-Produce Gb	Dep. Rate	Giants Gb	Cum.
1986	10790	3.94	581.43	0.67%	0	547.755
1987	10956	4.00	577.43	0.69%	0	547.755
1988	12179	4.45	572.98	0.77%	3	550.355
1989	13998	5.11	567.87	0.89%	1	551.055
1990	14648	5.35	562.53	0.94%	1	552.055
1991	14011	5.11	557.41	0.91%	0	552.055
1992	15080	5.50	551.91	0.99%	0	552.055
1993	15875	5.79	546.11	1.05%	0	552.055
1994	15970	5.83	540.28	1.07%	0	552.055
1995	16125	5.89	534.40	1.089%	0	552.055
1996	16264	5.94	528.46	1.11%	0	552.055
1997	18238	6.66	521.81	1.26%		
1998	19915	7.27	514.54	1.39%		
1999	20594	7.52	507.02	1.46%		
2000	21855	7.98	499.04	1.57%		
2001	23225	8.48	490.57	1.70%		
2002	24602	8.98	481.59	1.83%		
2003	26137	9.54	472.05	1.98%		
2004	27818	10.15	461.89	2..15%		
2005	29205	10.66	451.23	2.31%		
2006	30482	11.13	440.11	2.47%		
2007	31664	11.56	428.55	2.63%		
2008	32749	11.95	416.60	2.79%	Peak	
2009	31570	11.52	405.07	2.77%		
2010	30392	11.09	393.98	2.74%		
2011	29781	10.87	383.11	2.76%		
2012	29159	10.64	372.47	2.78%		
2013	28526	10.41	362.06	2.80%		
2014	27885	10.18	351.88	2.81%		
2015	27239	9.94	341.93	2.83%		
2016	26588	9.70	332.23	2.84%		
2017	26002	9.49	322.74	2.86%		
2018	25416	9.28	313.46	2.87%		
2019	24831	9.06	304.40	2.89%		
2020	24248	8.85	295.55	2.91%		
2021	23668	8.64	286.91	2.92%		
2022	23093	8.43	278.48	2.94%		
2023	22522	8.22	270.26	2.95%		
2024	21958	8.01	262.25	2.97%		
2025	21400	7.81	254.43	2.98%		
2026	20850	7.61	246.82	2.99%		
2027	20307	7.41	239.41	3.00%		
2028	19773	7.22	232.19	3.01%		
2029	19247	7.03	225.17	3.03%		
2030	18731	6.84	218.33	3.04%		
2031	18223	6.65	211.68	3.05%		
2032	17725	6.47	205.21	3.06%		
2033	17236	6.29	198.92	3.07%		
2034	19758	6.12	192.80	3.07%		
2035	16289	5.95	186.86	3.08%		
2036	15830	5.78	181.08	3.09%		
2037	15381	5.61	175.47	3.10%		
2038	14942	5.45	170.01	3.11%		
2039	14513	5.30	164.72	3.12%		
2040	14094	5.14	159.57	3.12%		
2041	13685	4.99	154.58	3.13%		
2042	13285	4.85	149.73	3.14%		
2043	12896	4.71	145.02	3.14%		
2044	12516	4.57	140.45	3.15%		
2045	12145	4.43	136.02	3.16%		
2046	11784	4.30	131.72	3.16%		
2047	11432	4.17	127.55	3.17%		
2048	11090	4.05	123.50	3.17%		
2049	10756	3.93	119.57	3.18%		
2050	10432	3.81	115.76	3.18%		

REGION				EURASIA			
Cum.Prod.	156.20	MIDPOINT		PRODUCTION		DISCOVERY	
Reserves	129.45	Amount	169.40	Peak	1988	Midpoint	1963
Discovered	285.65	Date	2000	3yr-trend	-2%	RESERVES	
Yet-to-Find	53.15	Years	4	Disc-Prod	25	O&GJ	82.89
Yet-to-Produce	182.60	GIANTS		DEP. RATE		Adjust	-17.21
Ultimate	338.80	Amount	130.75	Current	2.0%	World Oil	221.90
Cum. Prod. 2050	306.04	%		46% Midpoint	2.3	Factor	1.97
Reference date	12/96	Last	1890	Diff.	-0.3	Revised	03/08

Date	Production kb/d	Gb/y	Yet-to-Produce Gb	Dep. Rate	Giants Gb	Cum.
Pre-1930		1.48	337.32		8.35	
1930	466	0.17	337.15	0.05%	8.35	8.35
1931	590	0.22	336.94	0.06%	0.00	8.35
1932	578	0.21	336.73	0.06%	0.00	8.35
1933	579	0.21	336.52	0.06%	0.00	8.35
1934	656	0.24	336.28	0.07%	0.00	8.35
1935	675	0.25	336.03	0.07%	0.00	8.35
1936	694	0.25	335.78	0.08%	0.00	8.35
1937	685	0.25	335.53	0.07%	1.00	9.35
1938	708	0.26	335.27	0.08%	0.50	9.85
1939	736	0.27	335.00	0.08%	0.00	9.85
1940	738	0.27	334.73	0.08%	0.00	9.85
1941	787	0.29	334.44	0.09%	0.00	9.85
1942	769	0.28	334.16	0.08%	0.00	9.85
1943	692	0.25	333.91	0.08%	0.00	9.85
1944	858	0.31	336.60	0.09%	1.25	11.101
1945	539	0.20	333.40	0.06%	1.53	12.631
1946	553	0.20	333.20	0.06%	0.00	12.631
1947	629	0.23	332.97	0.07%	0.00	12.631
1948	725	0.26	332.71	0.08%	14.51	27.141
1949	781	0.29	332.42	0.09%	1.23	28.366
1950	855	0.31	332.11	0.09%	0.00	28.366
1951	900	0.33	331.78	0.10%	0.00	28.366
1952	1041	0.38	331.40	0.11%	0.60	28.966
1953	1251	0.46	330.94	0.14%	0.50	29.466
1954	1406	0.51	330.43	0.16%	0.51	29.976
1955	1659	0.61	329.82	0.18%	6.90	36.876
1956	1942	0.71	329.12	0.21%	2.06	38.936
1957	2232	0.81	328.30	0.25%	0.00	38.936
1958	2537	0.93	327.38	0.28%	1.86	40.795
1959	2892	1.06	326.32	0.32%	8.60	49.395
1960	3293	1.20	325.12	0.37%	0.00	49.395
1961	3678	1.34	323.78	0.41%	3.29	52.68
1962	4102	1.50	322.28	0.46%	12.30	64.98
1963	4518	1.65	320.63	0.51%	2.01	66.99
1964	4910	1.79	318.84	0.56%	1.50	68.49
1965	5345	1.95	316.89	0.61%	18.00	86.494
1966	5855	2.14	314.75	0.67%	1.80	88.294
1967	6225	2.27	312.48	0.72%	0.00	88.294
1968	6706	2.45	310.03	0.78%	1.20	89.494
1969	7077	2.58	307.45	0.83%	1.44	90.934
1970	7733	2.82	304.62	0.92%	1.60	92.534
1971	8417	3.07	301.55	1.01%	3.10	95.634
1972	8885	3.24	298.31	1.08%	1.00	96.634
1973	9850	3.60	294.71	1.21%	2.11	98.744
1974	10922	3.99	290.73	1.35%	0.00	98.744
1975	11834	4.32	286.41	1.49%	3.50	102.247
1976	12626	4.61	281.80	1.61%	0.60	102.847
1977	13111	4.79	277.01	1.70%	0.50	103.347
1978	13578	4.96	272.06	1.79%	4.22	107.567
1979	14159	5.17	266.86	1.90%	9.14	116.702
1980	14526	5.30	261.59	1.99%	0.00	116.702
1981	14549	5.31	256.28	2.03%	0.00	116.702
1982	14640	5.34	250.93	2.09%	5.12	121.822
1983	14707	5.37	245.57	2.14%	0.73	122.552
1984	14876	5.43	240.14	2.21%	2.83	125.383

Date	Production kb/d	Gb/y	Yet-to-Produce Gb	Dep. Rate	Giants Gb	Cum.	
1985	14701	5.37	234.77	2.23%	0.56	125.943	
1986	15245	5.56	229.21	2.37%	1.70	127.643	
1987	15481	5.65	223.56	2.47%	3.11	130.753	
1988	15519	5.66	217.89	2.53%	0.00	130.753	Peak
1989	15189	5.54	212.35	2.54%	0.00	130.753	
1990	14397	5.26	207.09	2.47%	0.00	130.753	
1991	13305	4.86	202.24	2.34%	0.00	130.753	
1992	11964	4.37	197.87	2.16%	0.00	130.753	
1993	10968	4.00	193.87	2.02%	0.00	130.753	
1994	10211	3.73	190.14	1.92%	0.00	130.753	
1995	10308	3.76	186.38	1.979%	0.00	130.753	
1996	10347	3.78	182.60	2.03%	0.00	130.753	
1997	10425	3.81	178.79	2.08%	0.00		
1998	10508	3.84	174.96	2.15%	0.00		
1999	10596	3.87	171.09	2.21%	0.00		
2000	10674	3.90	167.20	2.28%	0.00		
2001	10967	4.00	163.19	2.39%			
2002	11019	4.02	159.17	2.46%			
2003	11086	4.05	155.12	2.54%			
2004	11166	4.08	151.05	2.63%			
2005	11260	4.11	146.94	2.72%			
2006	11367	4.15	142.79	2.82%			
2007	11488	4.19	138.60	2.94%			
2008	11622	4.24	134.36	3.06%			
2009	11770	4.30	130.06	3.20%			
2010	11931	4.35	125.70	3.35%			
2011	11525	4.21	121.50	3.35%			
2012	11131	4.06	117.44	3.34%			
2013	10751	3.92	113.51	3.34%			
2014	10385	3.79	109.72	3.34%			
2015	10031	3.66	106.06	3.34%			
2016	9690	3.54	102.52	3.33%			
2017	9361	3.42	99.11	3.33%			
2018	9043	3.30	95.80	3.33%			
2019	8736	3.19	92.62	3.33%			
2020	8440	3.08	89.54	3.33%			
2021	8154	2.98	86.56	3.32%			
2022	7878	2.88	83.68	3.32%			
2023	7611	2.78	80.91	3.32%			
2024	7354	2.68	78.22	3.32%			
2025	7105	2.59	75.63	3.32%			
2026	6866	2.51	73.12	3.31%			
2027	6634	2.42	70.70	3.31%			
2028	6411	2.34	68.36	3.31%			
2029	6195	2.26	66.10	3.31%			
2030	5986	2.18	63.92	3.31%			
2031	5785	2.11	61.80	3.30%			
2032	5591	2.04	59.76	3.30%			
2033	5403	1.97	57.79	3.30%			
2034	5222	1.91	55.89	3.30%			
2035	2047	1.84	54.04	3.30%			
2036	4878	1.78	52.26	3.29%			
2037	4714	1.72	50.54	3.29%			
2038	4557	1.66	48.88	3.29%			
2039	4404	1.61	47.27	3.29%			
2040	4257	1.55	45.72	3.29%			
2041	4115	1.50	44.22	3.29%			
2042	3978	1.45	42.76	3.28%			
2043	3845	1.40	41.36	3.28%			
2044	3717	1.36	40.00	3.28%			
2045	3593	1.31	38.69	3.28%			
2046	3474	1.27	37.42	3.28%			
2047	3358	1.23	36.20	3.28%			
2048	3247	1.19	35.01	3.27%			
2049	3139	1.15	33.87	3.27%			
2050	3035	1.11	32.76	3.27%			

REGION — WESTERN EUROPE

Cum.Prod.	28.88	MIDPOINT		PRODUCTION	DISCOVERY	
Reserves	30.94	Amount	32.98	Peak	1999	Midpoint	1975
Discovered	59.82	Date	1998	3yr-trend	11%	RESERVES
Yet-to-Find	6.13	Years	2	Disc-Prod	24	O&GJ	18.02
Yet-to-Produce	37.07	GIANTS		DEP. RATE	Adjust
Ultimate	65.95	Amount	26.43	Current	5.8%	World Oil	31.09
Cum. Prod. 2050	65.20	%	44%	Midpoint	6.8	Factor	1.72
Reference date	12/96	Last	1993	Diff.	−1.1	Revised	03/08

Date	Production kb/d	Production Gb/y	Yet-to-Produce Gb	Dep. Rate	Giants Gb	Cum.
Pre-1930		0.01	65.94			
1930	4	0.00	65.94	0.00%		
1931	5	0.00	65.94	0.00%		
1932	6	0.00	65.93	0.00%		
1933	7	0.00	65.93	0.00%		
1934	8	0.00	65.93	0.00%		
1935	9	0.00	65.93	0.01%		
1936	10	0.00	65.92	0.01%		
1937	11	0.00	65.92	0.01%		
1938	15	0.01	65.91	0.01%		
1939	25	0.01	65.90	0.01%		
1940	27	0.01	65.89	0.01%		
1941	27	0.01	65.88	0.02%		
1942	32	0.01	65.87	0.02%		
1943	40	0.01	65.86	0.02%		
1944	26	0.01	65.84	0.02%		
1945	23	0.01	65.84	0.01%		
1946	31	0.01	65.82	0.02%		
1947	34	0.01	65.81	0.02%		
1948	44	0.02	65.80	0.02%	0.52	0.515
1949	55	0.02	65.78	0.03%		0.515
1950	77	0.03	65.75	0.04%		0.515
1951	95	0.03	65.71	0.05%		0.515
1952	114	0.04	65.67	0.06%		0.515
1953	127	0.05	65.63	0.07%		0.515
1954	154	0.06	65.57	0.09%		0.515
1955	179	0.07	65.50	0.10%		0.515
1956	189	0.07	65.43	0.11%		0.515
1957	226	0.08	65.35	0.13%		0.515
1958	243	0.09	65.26	0.14%		0.515
1959	251	0.09	65.17	0.14%		0.515
1960	284	0.10	65.07	0.16%		0.515
1961	299	0.11	64.96	0.17%		0.515
1962	313	0.11	64.85	0.18%		0.515
1963	330	0.12	64.72	0.19%		0.515
1964	360	0.13	64.59	0.20%		0.515
1965	358	0.13	64.46	0.20%		0.515
1966	355	0.13	64.33	0.20%		0.515
1967	346	0.13	64.21	0.20%		0.515
1968	327	0.12	64.09	0.19%		0.515
1969	320	0.12	63.97	0.18%	2.27	2.786
1970	309	0.11	63.86	0.18%	2.60	5.386
1971	299	0.11	63.75	0.17%	1.90	7.286
1972	306	0.11	63.64	0.18%	2.12	9.406
1973	296	0.11	63.53	0.17%	1.00	10.406
1974	272	0.10	63.43	0.16%	6.61	17.016
1975	416	0.15	63.28	0.24%	1.74	18.756
1976	542	0.20	63.08	0.31%		18.756
1977	739	0.27	62.81	0.43%		18.756
1978	1357	0.50	62.32	0.79%	1.80	20.556
1979	1729	0.63	61.68	1.01%	3.40	23.956
1980	2290	0.84	60.85	1.36%		23.956
1981	2307	0.84	60.01	1.38%		23.956
1982	2516	0.92	59.09	1.53%		23.956
1983	2934	1.07	58.02	1.81%		23.956

Date	Production kb/d	Production Gb/y	Yet-to-Produce Gb	Dep. Rate	Giants Gb	Cum.
1984	3267	1.19	56.82	2.06%	1.08	25.035
1985	3544	1.29	55.53	2.28%	0.90	25.935
1986	3713	1.36	54.18	2.44%		25.935
1987	3974	1.45	52.73	2.68%		25.935
1988	3991	1.46	51.27	2.76%		25.935
1989	4308	1.57	49.70	3.07%		25.935
1990	3800	1.39	48.31	2.79%		25.935
1991	4170	1.52	46.79	3.15%		25.935
1992	4407	1.61	45.18	3.44%		25.935
1993	4631	1.69	43.49	3.74%	0.50	26.435
1994	5542	2.02	41.47	4.65%		26.935
1995	5828	2.13	39.34	5.13%		26.435
1996	6207	2.27	37.07	5.76%		26.435
1997	6505	2.37	34.70	6.40%		
1998	6504	2.37	32.32	6.84%		
1999	6525	2.38	29.94	7.37%	Peak	
2000	6080	2.22	27.72	7.41%		
2001	5628	2.05	25.67	7.41%		
2002	5198	1.90	23.77	7.38%		
2003	4802	1.75	22.02	7.37%		
2004	4437	1.62	20.40	7.35%		
2005	4100	1.50	18.90	7.34%		
2006	3790	1.38	17.52	7.32%		
2007	3504	1.28	16.24	7.30%		
2008	3240	1.18	15.06	7.28%		
2009	2996	1.09	13.96	7.26%		
2010	2771	1.01	12.95	7.24%		
2011	2564	0.94	12.02	7.23%		
2012	2373	0.87	11.15	7.21%		
2013	2196	0.80	10.35	7.19%		
2014	2033	0.74	9.61	7.17%		
2015	1882	0.69	8.92	7.15%		
2016	1743	0.64	8.28	7.13%		
2017	1614	0.59	7.70	7.11%		
2018	1496	0.55	7.15	7.09%		
2019	1386	0.51	6.64	7.08%		
2020	1284	0.47	6.17	7.06%		
2021	1191	0.43	5.74	7.04%		
2022	1104	0.40	5.34	7.02%		
2023	1024	0.37	4.96	7.00%		
2024	949	0.35	4.62	6.98%		
2025	881	0.32	4.30	6.96%		
2026	817	0.30	4.00	6.94%		
3027	758	0.28	3.72	6.93%		
2028	704	0.26	3.46	6.91%		
2029	654	0.24	3.22	6.89%		
2030	607	0.22	3.00	6.87%		
2031	564	0.21	2.80	6.85%		
2032	524	0.19	2.61	6.83%		
2033	487	0.18	2.43	6.81%		
2034	452	0.17	2.26	6.80%		
2035	420	0.15	2.11	6.78%		
2036	391	0.14	1.97	6.76%		
2037	363	0.13	1.84	6.74%		
2038	338	0.12	1.71	6.72%		
2039	314	0.11	1.60	6.70%		
2040	293	0.11	1.49	6.69%		
2041	272	0.10	1.39	6.67%		
2042	253	0.09	1.30	6.65%		
2043	236	0.09	1.21	6.63%		
2044	220	0.08	1.13	6.61%		
2045	205	0.07	1.06	6.60%		
2046	191	0.07	0.99	6.58%		
2047	178	0.06	0.92	6.56%		
2048	165	0.06	0.86	6.54%		
2049	154	0.06	0.81	6.52%		
2050	144	0.05	0.75	6.51%		

N. AMERICA / **USA**

Cum.Prod.	173.18	MIDPOINT		PRODUCTION		DISCOVERY
Reserves	27.94	Amount	105.00	Peak	1971	Midpoint 1959
Discovered	201.12	Date	1973	3yr-trend	−3%	RESERVES
Yet-to-Find	8.88	Years	−23	Disc-Prod	12	O&GJ 22.351
Yet-to-Produce	36.82	GIANTS		DEP. RATE		Adjust
Ultimate	210.00	Amount	58.83	Current	6.0%	World Oil 22.160
Cum. Prod. 2050	208.72	%	29%	Midpoint	3.1%	Factor 1.25
Reference date	12/96	Last	1982	Diff.	2.9	Revised 03/08

	Production		Yet-to-Produce	Dep.	Giants Cum.	Notes
Date	kb/d	Gb/y	Gb	Rate	Gb	
Pre-1930		11.84	198.16			
1930	2347	0.86	197.30	0.43%	31.489	31.489
1931	2477	0.90	196.40	0.46%	0.741	32.23
						Conroe
1932	2609	0.95	195.44	0.48%	2.758	34.988
						Wilmington
1933	2743	1.00	194.44	0.51%	0.958	35.946
						Greta
1934	2878	1.05	193.39	0.54%	0.760	36.706
						Hastings
1935	3015	1.10	192.29	0.57%	0.866	37.572
						Goldsmith
1936	3152	1.15	191.14	0.60%	4.023	41.595 #1
1937	3289	1.20	189.94	0.63%	0.503	42.098
						Borregos
1938	3327	1.21	188.73	0.64%	1.012	43.11 #2
1939	3466	1.26	187.46	0.67%		43.11
1940	3697	1.35	186.11	0.72%	0.85	43.96
						Hawkins
1941	3842	1.40	184.71	0.75%		43.96
1942	3799	1.39	183.32	0.75%		43.96
1943	4125	1..51	181.82	0..82%		43.96
1944	4584	1.67	180.14	0.92%		43.96
1945	4695	1.71	178.43	0.95%		43.96
1946	4751	1.73	176.70	0.97%		43.96
1947	5088	1.86	174.84	1.05%	0.529	44.489
						San Ardo
1948	5520	2.01	172.82	1.15%	1.701	46.19
						Scurry
1949	5046	1.84	170.98	1.07%		46.19
1950	5402	1.97	169.01	1.15%		46.19
1951	6149	2.24	166.77	1.33%		46.19
1952	6268	2.29	164.48	1.37%		46.19
1953	6512	2.38	162.10	1.45%		46.19
1954	6340	2.31	159.79	1.43%		46.19
1955	6763	2.47	157.32	1.54%		46.19
1956	7150	2.61	154.71	1.66%		46.19
1957	7160	2.61	152.10	1.69%		46.19
1958	6459	2.36	149.74	1.55%		46.19
1959	7043	5.57	147.17	1.72%		46.19
1960	7019	2.56	144.61	1.74%		46.19
1961	7181	2.62	141.98	1.81%		46.19
1962	7337	2.68	139.31	1.89%		46.19
1963	7537	2.75	136.56	1.97%		46.19
1964	7640	2.79	133.77	2.04%		46.19
1965	7753	2.83	130.94	2.12%	0.55	46.74
						McArthur River
1966	8301	3.03	127.91	2.31%		46.74
1967	8801	3.21	124.69	2.51%	9.45	56.19
						Prudhoe Bay
1968	9153	3.34	121.35	2.68%		56.19
1969	9178	3.35	118.00	2.76%	1.501	57.691
						Kuparuk
1970	9507	3.47	114.53	2.94%		57.691
1971	9650	3.52	111.01	3.08%		57.691 Peak
1972	9500	3.47	107.54	3.12%		57.691
1973	9225	3.37	104.18	3.13%		57.691
1974	8945	3.26	100.91	3.13%		57.691
1975	8370	3.06	97.86	3.03%		57.691
1976	8105	2.96	94.90	3.02%		57.691
1977	8240	3.01	91.89	3.17%		57.691
1978	8660	3.16	88.73	3.44%		57.691
1979	8650	3.16	85.57	3.56%		57.691
1980	8650	3.16	82.42	3.69%		57.691

	Production		Yet-to-Produce	Dep.	Giants Cum.	Notes
Date	kb/d	Gb/y	Gb	Rate	Gb	
1984	8588	3.13	79.28	3.80%	0.547	58.238
						Pt. Arguello
1982	8655	3.16	76.12	3.98%	0.594	58.832
						Sprayberry
1983	8669	3.16	72.96	4.16%		58.832
1984	8750	3.19	69.76	4.38%		58.832
1985	8919	3.26	66.51	4.67%		58.832
1986	8790	3.21	63.30	4.82%		58.832
1987	8277	3.02	60.28	4.77%		58.832
1988	8166	2.98	57.30	4.94%		58.832
1989	7676	2.80	54.50	4.89%		58.832
1990	7355	2.68	51.81	4.93%		58.832
1991	7417	2.71	49.11	5.22%		58.832
1992	7171	2.62	46.49	5.33%		58.832
1993	6847	2.50	43.99	5.38%		58.832
1994	6617	2.42	41.57	5.49%		58.832
1995	6560	2.39	39.18	5.76%		58.832
1996	6478	2.36	36.82	6.034%		58.832
1997	6086	2.22	34.59	6.03%		
1998	5719	2.09	32.51	6.03%		
1999	5374	1.96	30.55	6.03%		
2000	5050	1.84	28.70	6.03%		
2001	4745	1.73	26.97	6.03%		
2002	4459	1.63	25.34	6.03%		
2003	4190	1.53	23.81	6.03%		
2004	3937	1.44	22.38	6.03%		
2005	3699	1.35	21.03	6.03%		
2006	3476	1.27	19.76	6.03%		
2007	3266	1.19	18.57	6.03%		
2008	3069	1.12	17.45	6.03%		
2009	2884	1.05	16.39	6.03%		
2010	2710	0.99	15.40	6.03%		
2011	2546	0.93	14.47	6.03%		
2012	2393	0.87	13.60	6.03%		
2013	2248	0.82	12.78	6.03%		
2014	2113	0.77	12.01	6.03%		
2015	1985	0.72	11.28	6.03%		
2016	1865	0.68	10.60	6.03%		
2017	1753	0.64	9.96	6.03%		
2018	1647	0.60	9.36	6.03%		
2019	1548	0.56	8.80	6.03%		
2020	1454	0.53	8.27	6.03%		
2021	1367	0.50	7.77	6.03%		
2022	1284	0.47	7.30	6.03%		
2023	1207	0.44	6.86	6.03%		
2024	1134	0.41	6.44	6.03%		
2025	1065	0.39	6.06	6.03%		
2026	1001	0.37	5.69	6.03%		
2027	941	0.34	5.35	6.03%		
2028	884	0.32	5.02	6.03%		
2029	831	0.30	4.72	6.03%		
2030	781	0.28	4.44	6.03%		
2031	733	0.27	4.17	6.03%		
2032	689	0.25	3.92	6.03%		
2033	648	0.24	3.68	6.03%		
2034	609	0.22	3.46	6.03%		
2035	572	0.21	3.25	6.03%		
2036	537	0.20	3.05	6.03%		
2037	505	0.18	2.87	6.03%		
2038	474	0.17	2.70	6.03%		
2039	446	0.16	2.53	6.03%		
2040	419	0.15	2.38	6.03%		
2041	394	0.14	2.24	6.03%		
2042	370	0.13	2.10	6.03%		
2043	348	0.13	1.98	6.03%		
2044	327	0.12	1.86	6.03%		
2045	307	0.11	1.74	6.03%		
2046	288	0.11	1.64	6.03%		
2047	271	0.10	1.54	6.03%		
2048	255	0.09	1.45	6.03%		
2049	239	0.09	1.36	6.03%		
2050	225	0.08	1.28	6.03%		

Notes
#1 Slaughter, Wasson, Slaughter; #2 Coalinga Nose, Timbalier

INDEX